Learning to Reason

Learning to Reason

An Introduction to Logic, Sets, and Relations

Nancy Rodgers

Hanover College

A Wiley-Interscience Publication
JOHN WILEY & SONS, INC.
New York • Chichester • Weinheim • Brisbane • Singapore • Toronto

This book is printed on acid-free paper. ♾

Published simultaneously in Canada.

For ordering and customer service, call 1-800-CALL WILEY.

Library of Congress Cataloging in Publication Data

Rodgers, Nancy, 1941–
Learning to reason : an introduction to logic, sets & relations / Nancy Rodgers.
p. cm.
Includes bibliographical references and index.
ISBN 0-471-37122-X (hardcover : alk. paper)
1. Logic, Symbolic and mathematical. 2. Proof theory. 3. Set theory. I. Title.

QA9.R7264 2000
511.3—dc21 00-023492

Printed in the United States of America

10 9 8 7 6 5 4 3 2 1

This book is dedicated

to the memory of

Edith and Neville Rodgers

Contents

To Students

Rarely does one hear an English major say, "I like English, but I don't like to write," yet math students often say, "I like math, but I don't like to write proofs." Some students even tremble at the sound of an approaching proof assignment. The purpose of this book is to demystify the proof process by giving you the necessary reasoning techniques and language tools for constructing well-written arguments. This skill is as essential in mathematics and computer science as in English or any other discipline.

Learning to Reason is designed for a freshman/sophomore level course with no prerequisites except a desire to improve one's reasoning skills and one's ability to read and write mathematics and symbolic languages. The book covers the process of writing proofs, a process similar to writing in other disciplines, but the topics for our themes *(theorems)* will come from three unifying concepts that run through all areas of mathematics: logic, sets, and relations.

We sometimes require prerequisites for math courses in order to ensure a certain level of mathematical maturity – a maturity where one becomes an independent thinker who can figure things out without being told what to do. One of the main goals of this book is to speed up this maturation process by focusing on how we reason with mathematical language, emphasizing those elements of the language that tend to confuse students in advanced courses. Simple-sounding concepts such as substitution are not as simple as they sound. Simple words, such as "and," "or," "not," and "implies," lose their simplicity when we combine them in a sentence. If you are not fluent in how to manipulate these basic terms from which we build our language, you will be severely handicapped when you try to do any type of mathematical reasoning.

Another goal of this book is to help you see the common thread that runs throughout the vast universe of mathematics. Without this connection, you can easily get lost in an endless maze of mathematical concepts and not be able to see the forest for the trees. Many people have the misconception that mathematics is primarily a subject in which you do computations. I must confess that I have never been a fan of doing computations. In my college days, my fellow bridge players always wanted me to keep score because I was a math major. I felt like a chef being asked to wash the dishes. A chef creates dirty dishes in the process of cooking, but the goal is not to create dirty dishes. Similarly, mathematicians often generate computations in the process of doing mathematics, but the goal is not to generate computations. The goal is to create interesting structures and relations that can be supported with logical reasoning. This is the common thread that connects all of mathematics.

Contents

In Chapter 1, we cover the basic elements of mathematical language. Mathematical language is quite simple, which may surprise those who consider mathematics to be difficult and complex. Consider the myriad ways that we can form complex sentences in everyday language. In contrast, mathematical language is constructed from only five connectives and two quantifiers. If you understand how to manipulate these seven terms and how to use substitution, then you have acquired the basic technique on which logical reasoning is based.

In Chapter 2, we examine the reasoning process and how we organize our reasoning into a well-written form that can be classified as a proof. As in any good essay, a written proof contains an introduction, a body, and a conclusion. We will study various templates for writing proofs; however, the ability to construct a proof requires a deeper level of intellectual maturity than merely following an established procedure. To construct a proof, one must explore and question, find the inner structure of the situation, analyze the various parts, and then use logical reasoning to put the different pieces together to create the proof. The sparks that leap across our synapses during this creative process strengthen our powers of reasoning, one of the major benefits of studying mathematics.

In Chapter 3, we look at how we work with sets, the building blocks of mathematical language. Since prehistoric times, when people counted with a set of sticks or stones, sets have been at the foundation of mathematics. When we count, we are counting the number of elements in a set; when we analyze the form of a figure, we are analyzing a set of points; when we look at a function, we see a relation between two sets. Sets provide the basic framework for mathematical discourse.

In Chapter 4, we examine relations, a reasoning concept common to all disciplines. There are relations among pieces of music from the same period, works of art of the same style, and books of the same genre. In no discipline, including mathematics, can we analyze an object by itself; we must compare it to other objects. Relations provide a simple way to describe mathematics: Mathematics is the study of abstract relations.

www.learningtoreason.com

Additional learning tools are available at the web site, *www.learningtoreason.com.* Please visit the site and check out the resources, which will be continually enhanced. You are invited to submit questions, comments, and suggestions.

Learning a Language

As you begin your study of the language of reason, please remember that people do not learn a language through memorizing a list of words but through hearing the words used many times in various ways. The compactness of the language of mathematics with its attendant density of meaning requires that we read mathematics at a slow but contemplative pace. More than likely, we will not grasp its full import from one reading, and even if we do grasp it, we probably will not remember it all, for human memory needs a great deal of repetition to build enough bridges for the easy retrieval of stored information. So, it is important not only to read the sections, but also to reread them and ask questions about the content until you have a deep understanding of the material in both a verbal and a visual form. Anyone who is a lover of poetry knows that each rereading of a poem can bring new insights. The same is true in mathematics.

Working Out

Anyone can develop their reasoning skills if they are willing to invest the necessary time to work out with the exercises and the concepts. To become a good athlete or a good musician requires long hours of practice, so it is not surprising that learning how to reason also requires a similar investment of time. The exercises at the end of each section are an essential component of the learning process. To develop your reasoning skills, you should work out with the exercises on a daily basis. As you work through the discussions in the text, you should also write your own questions and observations. Through this process, you will build your understanding and personally internalize the meaning of the various concepts.

Throughout this text you will find activities that introduce you to concepts in the sections following them. If you work on the activities before you read the section, you will have the opportunity to discover relationships on your own. What you discover for yourself burns an indelible image in your memory and helps you to become a creative thinker, which is one of the most important skills needed in a changing society. Problems are easy when we have examples to guide us, but the creative thinkers are those who can blaze a path and create examples for others to follow. To be a logical thinker, we must develop our ability beyond merely copying procedures from examples provided by others.

When you take the extra time to figure out a problem on your own, you are building mental bridges that you can use in the future. The long hours of work that you do in building these bridges makes a deep impression that is firmly secured in your memory bank. On the other hand, when someone shows you how to do a problem, you are learning how to run across a bridge that someone else has built, which is not the same as learning how to build a bridge on your own. Computers are very adept at running across bridges that others have built, but they lack the human creativity to build new bridges for thought processes. To develop our reasoning powers beyond the mechanistic circuits of a computer, we must learn how to be creative thinkers.

To enliven your journey into the abstract world of reasoning, you may want to get into the gamesmanship of it by considering the exercises as a highly sophisticated game of mental prowess, or, for the more physically inclined, you may want to view them as aerobic exercises for the mind. The time that you spend will be a wise investment, for whatever path you take in life, the study of the topics in this book will help you to become an independent thinker who can reason in a logical manner.

Nancy Rodgers

To Teachers

Mathematics is simpler than other disciplines – physics or history, for example – because mathematics is concerned with such a very limited aspect of reality. Why, then, does such a simple subject seem so hard to so many people? I have come to believe that it is primarily a language problem. I became painfully aware of this problem in my first abstract algebra course when I ran head-on into a brick wall of mathematical language. I remember long hours of mental labor interrupted by a recurring question: why on earth did I major in math?

The next year I had a topology teacher, Professor John Seldon, who gave us a collection of theorems to prove from *Eléments de mathématique* by Bourbaki. As I worked though Bourbaki's organization of the foundations of mathematics, I began, for the first time, to understand the beautiful simplicity of mathematical language. After that experience, my studies became much easier because I now knew how to use mathematical language to structure my thinking.

Years later, while contemplating pedagogical methods that I might use to help my students over the same hurdle, I decided to write this text. The first version was used in an Algebraic Structures class. Because of student inquiries as to why they did not have this class earlier – since it would have helped them with the proofs they struggled with in other classes – the course was moved to the freshman/sophomore level. Through their many questions over the years, I began to understand the source of the great difficulty students have in writing proofs in upper division courses. The rules of syntax that seem so obvious after we subconsciously master them through long years of study are a huge language barrier to those on the other side of the fence. Some students have a great ear for the subtleties and nuances of languages and can easily learn a foreign language; a very small percentage of students have a similar gift for learning the language of mathematics. Granted, young children learn their native tongue by listening to those around them, but as we get older, most of us can benefit greatly by understanding the basic structure and syntax of a new language we are learning.

Organization

The initial goal in developing this text was to make Bourbaki's organization of the foundations of mathematics understandable and relevant at the freshman level. In addition, the book presents a lively discussion of the reasoning process, with a primary focus on deductive reasoning, but also including inductive reasoning, visual reasoning, and translations from everyday language to pictures and symbolic representations.

Starting with the foundations of logic in Chapter 1, the text explains how to analyze and logically manipulate individual sentences. In Chapter 2, the focus is on how to structure our thinking so that we can put sentences together to form a well-reasoned proof. The text illustrates the concepts with an elementary chain of ideas concerning integers, rational numbers, and real numbers. This connected series of examples and exercises helps students learn how to structure their thinking while also developing their understanding of numbers. The techniques learned here are reinforced as we examine sets, the basic building blocks of mathematics, in Chapter 3, and relations, where the action is in mathematics, in Chapter 4. This organizational structure gives students a meaningful overview of the vast subject of mathematics, while building their reasoning skills and their understanding of the basic concepts used throughout mathematics.

Special Features

The study of logical skeletons is fleshed out in mathematical settings with overviews of the structures they support and exercises that get students actively involved in and intrigued by the intellectual game of logical reasoning. Each section is preceded with a set of activities that give students the opportunity to discover for themselves important concepts from the next section. The activities encourage independent thinking and initiative, as well as help to raise the student's curiosity and interest in the upcoming material. After each section is a finely crafted set of exercises designed to help students develop their reasoning skills as they build a personal understanding of the language and notation. The exercises focus on those areas of mathematical language that tend to confuse students in upper division courses. They have been class-tested for several years and revised to maximize their benefit. Each chapter has a review section with related definitions grouped together. The definitions are alphabetized in a comprehensive glossary at the end of the book, followed by a symbol list.

The easy-going style of the book makes it accessible to a wide range of students. The concepts are carefully developed in a conversational writing style that speaks with a gentle authority, offering students motivation and encouragement along the way. It moves along at a brisk pace with careful analyses at points most likely to cause problems. The examples are cogent and thoughtfully presented, set off by lines that clearly separate them from the discussion. There is an energy in the conciseness of the writing and layout that makes it easy for students to read and remember what they have read.

Layout

In response to the first question in the book, one of my students, Becky Cantonwine, gave the following description of the difference between mathematical language and everyday language: "Mathematical language differs from everyday language in the same way that poetry differs from prose; every word or symbol is important and necessary, and their position is important to their meanings." Albert Einstein saw the same connection in his eloquent description of pure mathematics as "the poetry of logical ideas." Like written poetry, mathematical language is enhanced through the use of poetic lineation. Gestalt holistic patterns are easier to retain in the mind's eye, so poetic lineation is used in the text to highlight featured ideas and to assist the reader in working through dense notation and the thought processes involved in the reading of a proof. Great attention has been paid to the visual tone set by the geometric form of text layout, with white space generously used to minimize the denseness of the subject matter and to feature key thoughts and signposts in the reading. The overriding issue in all layout decisions was the presentation that would make it easiest to remember. Block text with its dense wrap-around lines is not as easy to assimilate and retain as text that incorporates active white space. I have tried to make the text as simple as possible, using a minimal but sufficient amount of words in explaining the concepts.

Audience

The text is designed as a bridge course for mathematics and computer science majors at the lower or upper division level. Any student who wants to learn how to structure their thinking and develop their reasoning skills will find it easy to use as a self-study text. Teachers of upper division math courses may want to use it as a supplementary text.

Acknowledgments

I owe a great deal of gratitude to my students, who, through their many questions, have helped me refine and deepen my understanding of the foundations of mathematical language. Special thanks to:

- Austin Shadday for her long hours of detailed work in the preparation of the manuscript.
- The students who helped with the proofreading: Becky Cantonwine, Emily Shreve, Andrea Spurgeon, Jessica Kirsch, and Philipp Baeumer.
- The editors at Wiley: Stephen Quigley, Heather Haselkorn, and Andrew Prince for their faith in this project and generous extensions of deadlines.
- The reviewers for their valuable comments which helped to shape the final version: Al Cavaretta, Kent State University; John Mack, University of Kentucky; Loren Larson, St. Olaf College; Martin Erickson, Truman State University; and Larry Mand, Indiana University Southeast.
- My colleagues at Hanover College, including: Bill Markel for his support and encouragement, Ralph Seifert for his insightful comments and historical references, and especially Carol Russell for the great enthusiasm she brought to the project when she class-tested the book. When faced with the question of what would you do if you had only a year to live, Carol continued to teach up until the last possible moment because she understood the intrinsic value of teaching, both for the student and the teacher.
- The NSF for funding summer workshops that contributed to the vision for this project; with deepest thanks to David Henderson at Cornell University for enriching experiences at the Teaching Undergraduate Geometry workshops in 1996 and 1999; to Fred Roberts and Rochelle Leibowitz for providing stimulating research connections at the DIMACS Reconnect '99 Conference at Rutgers University, and to Doris Schattschneider for the inspirational model that she provided at the Symmetry and Group Theory workshop at the University of Dayton in 1995.

Finally, a very special thanks to John Ramey, Coastal Carolina University, for his inspiring use of poetic language, his valuable insights on rhetoric, composition, writing pedagogies and the imagetext, and the generous sharing of his esoteric knowledge in a wide range of fields.

◈ Chapter 1

Logical Reasoning

Logical reasoning is a form of discourse that is distinguished from other forms by its complete objectivity. In order to attain a pure state of objectivity with no room for ambiguities, the language of logic had to be developed with great precision and clearly defined rules. Personal interpretations of a story, a painting, or an historical event may vary considerably, but any two people who understand the language of logic will interpret a logical argument in essentially the same way. Unlike the tangled web of rules that we use subconsciously in our everyday discourse, the rules for logical reasoning are very exact with no exceptions to the rule.

When we reason within a logical framework, words must be manipulated according to the rules of the game. Fortunately, the rules are fairly simple because the language of logic is built from only seven basic terms: two quantifiers, *for all* and *for some*, and five operators for building compound sentences, *not*, *and*, *or*, *implies*, and *is equivalent to*. The first stage in mastering the art of logical reasoning is to learn how to manipulate these seven terms. Each of these terms is simple by itself, but the meaning can easily be misconstrued when two or more are used in the same sentence, especially since we do not always use them in a consistent way in our everyday language. Once you master the basic rules, called the laws of logic, for using these seven terms, this stage of the reasoning process will be as easy as driving a car.

The next stage is a bit more challenging, for we must learn how to 1) translate sentences phrased within the complex structure of everyday language into the simplified language of logic, 2) use the powerful tool of substitution to convert abstract knowledge into various forms, and 3) translate visual reasoning to a verbal form and vice-versa. In this chapter, we will cover the basic elements of logical reasoning, including quantifiers, logical operators, substitutions, and translations.

Activity 1.1

1. Reasoning is mentally performed within the context of a language, which provides the medium through which we organize and present our thoughts. To speak or think in any language, we must be aware of the basic structure of the language.
   a. How does mathematical language differ from everyday language?
   b. Compare the way that you learn mathematical language with the way that you learned to communicate with others in your preschool days.
   c. Compare the use of pronouns in everyday language with the use of variables in abstract languages. Do they serve the same role in the following two sentences?
      He is taller than 5 feet. $x > 5$
   d. What does "complete thought" mean to you? What elements of language are needed to express a complete thought?
   e. Make a list of nouns and a list of verb phrases that you have used in mathematics. Which have you used the most?
   f. What is a sentence? Do any of the following expressions form sentences? $1 < 2$ $1 + 2$ $1 + 2 = 3$
2. Let p and q represent sentences.
   Let $\sim p$ represent the negation of p.
   a. Does $\sim(p \textit{ and } q)$ mean the same as $(\sim p \textit{ and } \sim q)$?
      This question is very abstract.
      How should you start thinking about it?
   b. What is an abstraction? Is a number an abstraction? Is the color blue an abstraction?

≡ 1.1 Symbolic Language ≡

All written languages are based on symbols. The English language is written in terms of phonetic symbols that give pronunciation information. We can symbolically represent the addition concept with the phonetic symbol "plus" or with the ideographic symbol "+" which does not give pronunciation information. They both represent the same concept. However, in the process of logical reasoning, phonetic words can bog down our thought processes. For example, consider the following question from an algebra textbook by Al-Khowarizmi in the 9th century.

> The importance of an easily manipulated symbolism is that it enables those who are not great mathematicians in their generation to do without effort mathematics which would have baffled the greatest of their predecessors.
>
> *E. T. Bell*, 1945

> What must be the amount of a square, which, when twenty-one units are added to it, becomes equal to the equivalent of ten roots of that square?

Al-Khowarizmi's question, which would have challenged the great thinkers of the Middle Ages, can be answered by most high school students today who understand symbolic manipulations. Of course, the question would have to be posed in a symbolic form or they, too, might become entangled in the phonetic words:

> Find a solution to the equation $x^2 + 21 = 10x$.

> The function f assigns to each number in the domain the value that is the square of the number obtained by multiplying the original number by three and then adding one.

Take a moment and contemplate the adjacent sentence. How long did it take you to decipher its meaning? If you know function notation, you can comprehend the same sentence in symbolic form almost instantly: $f(x) = (3x+1)^2$

The great power of mathematical symbols is the ease with which the brain can process the information. Without the pronunciation baggage, the brain manipulates the symbols with great speed, thereby enabling us to focus on deeper questions. At the other extreme, though, too many ideographic symbols tend to shorten our attention span. A page full of nothing but symbols is not as inviting as a page where symbols are interwoven with words, so we try to find a delicate balance between the two, as illustrated in the above translation.

Unfortunately, mathematical symbols pose a language barrier to those who have not taken the time to learn their meaning, leaving many people with the impression that they are viewing a foreign language. However, it is not as difficult as it appears. All it requires is that we take the time to build a personal meaning for the various symbols.

Using Symbols

In order to use symbols in the reasoning process, we must know how the symbols can be manipulated. Even more importantly, though, we need to have a personal understanding of what the symbols represent. For example, we may be able to compute $145 \div 3$ with an algorithm, but we will not be able to use the answer in a meaningful way if we do not understand the meaning of dividing a set into subsets of equal size. If we do not build a personal meaning for symbols, we lose the base for our reasoning powers and become nothing more than a computer performing mechanical processes.

Learning a Language

When learning a foreign language, we may know the meaning of a word one week but forget it the next week. The same thing happens when we learn a symbolic language. Each symbol represents a concept, and to understand the concept, we need to think about what it represents and what it does not represent. We should work through examples for which the concept applies as well as examples for which the concept does not apply. As we use a new symbol in different examples and exercises, we will slowly build our personal understanding of it until we are comfortable using it. The more we use a concept, the deeper we implant it in our memory.

Some students pick up the symbolic language of mathematics or computer science faster than others do. Similarly, some people can sit down and play the piano by ear, while others have to struggle with years of practice. Those who learned how to play through hard work, though, often end up playing far superior to those blessed with an ear for music. It is not how fast you learn a language but how hard you work to develop a deep understanding of it.

Variables

A *variable* is a letter used to represent an arbitrary element of a given set; that set is called the *domain* of the variable.

Variables are an essential component of a symbolic language. As its name implies, a variable can vary and represent a variety of elements. Instead of talking about specific numbers, we usually talk about a generic number that is symbolized by a variable, such as x. Like pronouns in everyday language, variables serve as a place holder for substituting specific elements.

The set of elements that may be substituted for a variable is called its *domain*. In the following example, the domain for x is the set of integers:

For every integer x, $x < x + 1$.

In computer science, a variable represents a storage space in the computer's memory where a number or a string of characters can be stored. Each variable is assigned a type that represents its domain. If a variable is assigned an integer type, then only integers can be stored in that variable.

We can use any letter as a variable, but we cannot use a letter to represent two different things within the same discussion. For example, an even number is any number that can be represented in the form $2k$ where k is an integer. However, if we apply this definition to two different even numbers within the same discussion, we cannot use "k" both times, for that would imply the two numbers are equal. Instead, we use another letter:

> **Theorem:** The sum of two even numbers is even.
> **Proof:**
> Let m and n be even numbers.
> Then $m = 2k$ for some integer k.
> Also, $n = 2j$ for some integer j.
> So, $n + m = 2k + 2j = 2(k + j)$.
> Since $k + j$ is an integer, by the definition of even, $n + m$ is even.

Let m and n be even numbers.
Then $m = 2k$ for some integer k.
Also, $n = 2j$ for some integer j.

In the adjacent proof, notice how the use of variables gives us a tangible way to work with even numbers, enabling us to make logical deductions about the abstract concept of even.

Sentences

Most communications in everyday language are phrased in terms of sentences, so it is not surprising that the same is true in mathematics. To express a complete thought, we use a sentence. Conversely, sentences require complete thoughts. If we are working with incomplete thoughts, either in our head or on paper, we cannot hope to make much progress in the reasoning process.

Sentences require complete thoughts.

Our work in this chapter will focus on how we logically manipulate sentences. When we reason, the steps in our reasoning process are built from sentences, so it is essential that we know how to recognize sentences, especially those that are written in symbolic form.

✣ *Example*

Which of the following are sentences? $5 < 8$ $5 + 8$ $5 + 8 = 13$

1. "$5 < 8$" is a sentence. 5 is the subject and $<$ is the verb.
2. "$5 + 8$" is not a sentence because it does not have a verb.
3. "$5 + 8 = 13$" is a sentence. The subject is "$5 + 8$" and the verb is "=."

Relations & Operations

When we place the < symbol between two numbers, we get a sentence. These types of symbols represent *relations*. However, when we place the + symbol between two numbers, we get a number, not a sentence. The + symbol operates on two numbers and produces a new number, such as $5+8$.

Relations		Operations	
$=$	$\neq$	$+$	$-$
$\approx$	$\cong$	$\times$	$\div$
$<$	$\leqslant$	$\cup$	$\cap$
$\subset$	$\subseteq$	$\vee$	$\wedge$

A relation gives a connection between two objects, whereas a binary *operation* operates on two objects and produces a third object. Relations produce sentences, but operations produce objects, such as a number or a set. In order to write well-formed mathematical sentences, we must be able to distinguish between relations and operations. Since they are different components of mathematical language, most word processors organize their equation editor with all relations grouped under one menu and all operations grouped under another menu, as illustrated on the left.

Fragments

$$\frac{-b \pm \sqrt{b^2-4ac}}{2a}$$

$$x = \frac{-b \pm \sqrt{b^2-4ac}}{2a}$$

We may sometimes jot down fragments of sentences, such as the adjacent fragment from the famous quadratic formula, but we cannot use fragments in a proof. To complete the thought, we must add a subject and a verb. Students who do not carry along the beginning of the sentence, "$x =$," often do not know what the answer represents when they finish the computation. When we do not write in complete sentences, it is easy to get confused and lose track of what we are doing.

Subjects

A well-formed sentence must have both a subject and a verb. The most frequently used subjects in mathematical sentences are sets and numbers. We will now briefly review the different types of real numbers and examine sets later on in Chapter 3.

- Questions about "how many" elements in a finite set can be answered in terms of the *natural numbers*:

  $$1, 2, 3, 4, 5, 6, \ldots$$

The Real Numbers

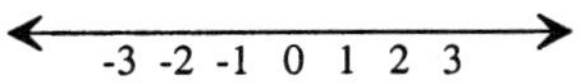

- To answer questions about "how much," such as how much length or how much area, we need a more extensive set of numbers, called the *real numbers*. We visualize the real numbers as coordinates of points on a number line, as illustrated on the left. In symbolic form, a *real number* is any number that can be represented as a decimal with a finite or infinite number of places.

- The *integers* consist of the natural numbers, their negatives, and 0:

  $$\ldots -3, -2, -1, 0, 1, 2, 3, \ldots$$

The positive integers are the natural numbers.
0 is neither positive nor negative.

- The *rational numbers* are numbers that can be represented as the quotient of two integers, such as $\frac{23}{5}$. The number .35 is a rational number because we can write it in fraction form: $\frac{35}{100}$. Using variables, we can define a rational number as follows: x is a rational number if and only if $x = \frac{a}{b}$ for some integers a and b with $b \neq 0$.

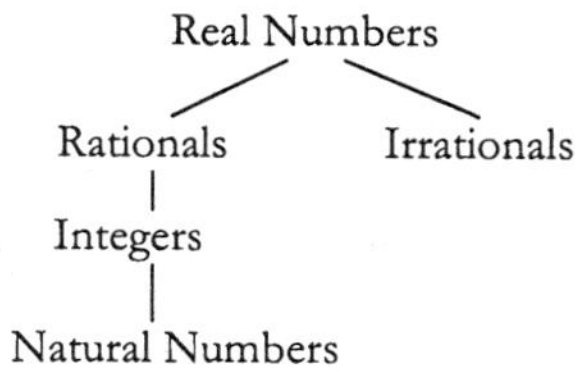

- Real numbers that are not rational, such as $\sqrt{2}$ or π, are called *irrational numbers*. Every real number is either rational or irrational.

The hierarchy of real numbers is given in the adjacent sketch. Each set is a subset of those sets that are chained above it.

Verbs

The action in everyday language comes from verbs. The same is true in mathematical language. However, most verbs in mathematics require objects, such as $x<y$ or $x=y$ or $X \subseteq Y$. In everyday language, we could have "x sings," but in mathematical language, x would have to sing to somebody, such as y. If x is a loner, we could have "x sings to x," but not just "x sings." Most mathematical verbs, such as those listed on the left, give relations between two objects.

Verbs

$\Rightarrow$
$\Leftrightarrow$
$\in$
$=$
$\approx$
$\cong$
$<$
$\leq$
$\subset$
$\subseteq$

One of the most important verbs is the implication verb, which we will examine in great detail in this chapter. This verb, which lies at the very foundation of logical reasoning, sets the structure for what we mean by a logical deduction. We use the implication to define a valid argument, which gives us the basic method for reasoning in a logical manner. We also use the implication verb to define other important verb phrases, such as "is equal to" and "is a subset of."

The most frequently used verb in mathematics is "*equals.*" In arithmetic and elementary algebra, this little verb provides the main action, with occasional help from the inequality verbs, $<$, $\leq$, $>$, $\geq$. The equals verb is used with both numbers and sets, whereas $\leq$ is used only with numbers.

The analogue of $\leq$ in set language is the subset verb, which gives a relation between two sets. *A is a subset of B*, notated as $A \subseteq B$, means that every element in A is also an element in B. This definition depends on another important verb phrase, *is an element of*, notated as $\in$.

$3 \in A$ means that 3 is an element of the set A.

Verbs that have properties similar to the equals relation, such as $\approx$, and $\cong$, are called *equivalence relations.* Verbs that impart some type of order on objects, such as $<$, $\leq$, $\subset$, and $\subseteq$ are called *order relations*. We will examine both equivalence relations and order relations in Chapter 4.

Statements

A *statement* is a sentence that is either true or false, but not both.

Some sentences, such as "7 is a lucky number," may be considered true by some people and false by others. We do not deal with this type of sentence in mathematics; instead, we restrict our discourse to sentences whose truth values are not debatable. We will use the term *statement* to denote a sentence that is either true or false, but not both. If a statement is true, then it cannot be false.

⌖ *Example*

Which of the following sentences are statements?

$$3+2=5 \qquad 3+2=6 \qquad x+2=6$$

1. "$3+2=5$" is a true sentence, so it is a statement.
2. "$3+2=6$" is a false sentence, so it is a statement.
3. "$x+2=6$" is a sentence; however, it is neither true nor false, so it is *not* a statement.

True	False
T	F
1	0
On	Off

The *truth value* of a statement is either true or false, which we will represent with T and F. In computer science, we use 1 for true and 0 for false. A computer transmits information along an electronic highway in terms of electric circuits which are either on or off. We identify the ON-state, defined as 1, with "true" and the OFF-state, defined as 0, with "false."

Statements severely limit the scope of our discourse because the truth value of many sentences is somewhere between 0 and 1. For example, the weatherman's assertion that it will be "partly cloudy" may be true only 80% of the day. These types of sentences can be analyzed with a more general type of logic known as fuzzy logic (page 60), which was developed to program artificial intelligence into computers.

Open Statements

The sentence $x+2=6$ is not a statement, but it does become a statement when we substitute an element for x.

Substitute 4 for x: $4+2=6$ (True)

Substitute 3 for x: $3+2=6$ (False)

An *open statement* is a sentence with variables that is not a statement but becomes a statement when substitutions are made for the variables.

A sentence of this type is called an *open statement*. We might be tempted to say that an open statement is any statement that has a variable. However, this is not true for we can quantify the variables by prefixing the sentence with a quantifier, as illustrated in the following example.

Example

The domain for x is the set of real numbers. Are any of the adjacent sentences open statements?

$2x + 3 = 5$

For all x, $2x + 3 = 5$.

There exists an x such that $2x + 3 = 5$.

"$2x + 3 = 5$" is an open statement. It is neither true nor false, but each time we substitute a number for x, the sentence is either true or false.

The second sentence is false, so it is not open.

The last sentence is true, so it is not open.

Even though the last two sentences in the above example have variables, they are not open statements because the variable is fixed (or bound) by the quantifier. Quantifiers are extremely important components of the reasoning process. We will examine them in detail in Section 1.2.

Solution Set

The *solution set* of an open statement in x is the set of elements from the domain of x that convert it to a true statement. To find the solution set of an equation, we solve the equation and then place the answers in a set. The solution set depends on the domain, as illustrated in the following examples.

Example

1. What is the solution set of the open statement, $x + 2 = 0$?

   Before we can answer this question, we must know the domain for x. If the domain is the set of integers, the solution set is the set whose only element is –2, which we represent with set braces as $\{-2\}$.

   If the domain is the set of natural numbers, though, the solution set is empty, which we represent with either the symbol $\{\ \}$ or ϕ.

2. What is the solution set of the open statement, $x^2 = -1$?

   Before we can answer this question, we must know the domain for x. Both i and $-i$ are solutions to the above equation: $i^2 = -1$ and $(-i)^2 = -1$. So, if the domain is the set of complex numbers (page 14), the solution set consists of i and $-i$: $\{i, -i\}$.

However, if the domain is the set of real numbers, the solution set is the empty set.

When it is not possible to list all the elements in the solution set of an open statement $p(x)$, we can represent the solution set with the following set notation:

$$\{ x \mid p(x) \}$$

We will examine set notation in more detail in Chapter 3.

Example

1. The domain for x is the set of real numbers. What is the solution set of the open statement, $x>2$?

   $\{ x \mid x>2 \}$

   Since we cannot list the elements in the solution set nor give a pattern that indicates all the members of the set, we use the adjacent set notation to express the solution set. This notation is read as "the set of all x such that $x>2$." If the reader does not know that the domain is the set of real numbers, then we should include it in the set description:

   $$\{ x \mid x>2 \text{ and } x \text{ is a real number}\}$$

   If the reader does know the domain of x, the shorter form gives a simpler image for focusing our thinking.

2. The domain for x is the set of real numbers and the domain for y is the set of real numbers. What is the solution set of the open statement, $x+3y=7$?

   $\{ (x, y) \mid x+3y=7\}$

   We cannot list all the elements in this set, so we use the adjacent set notation. Since we have two variables, the elements of the solution set are ordered pairs.

   $(1, 2)$ is a member of this set since $1+3(2)=7$.
   $(2, 1)$ is a not a member of this set since $2+3(1)\neq 7$.

Compound Sentences

When we link two sentences with a connective like *and*, we create a compound sentence. For example, we can use *and* to connect the sentence $2+3=5$ with the sentence $4+5=9$:

$$2+3=5 \text{ and } 4+5=9$$

"2 + 3 = 5" is called a *component sentence* of the compound sentence. In logic, we use only four connectives for building compound sentences: *and, or, implies, is equivalent to*. These terms are called *logical operators*.

Addition operates on 2 numbers and produces a new number.

And operates on 2 sentences and produces a new sentence.

In the adjacent box, notice the similarity between the addition operation on numbers and the *and* operation on sentences. Adding two numbers and combining two sentences are very different types of activities, but at the base level, the structure of what they do is the same. They are both binary operations, which is why we call *and* a logical operator.

Another important logical operator is the negation. Given a sentence, like $2+3=6$, we can make a new sentence by taking its negation:

It is not true that $2+3=6$.

$2+3=5$ and $4+5=9$.

$x<2$ or $x>5$.

$x<2$ implies that $x<3$.

$x<2$ is equivalent to $-x>-2$.

It is not true that $2+3=6$.

Negation is a *unary* logical operator, whereas the other four connectives are *binary* logical operators. As you probably know, "unary" means "one" and "binary" means "two." Negation forms a new sentence from a given sentence; the other four connectives form a new sentence from two given sentences, as illustrated on the left. It is rather surprising how much of our reasoning depends on these five logical operators. When we examine them in detail in Section 1.3, we will work with them in an *abstract* form, similar to abstract algebra.

Symbolic Sentences

In elementary algebra, we use letters to represent numbers and ideographic symbols to represent operations on numbers.

$$a+b=b+a$$
$$a\times(b+c)=a\times b+a\times c$$

Like an x-ray machine, this symbolic representation reveals the inner structure of arithmetic, making it easy to recognize and remember general rules for working with operations on numbers.

5 Logical Operators

$\sim p$:	not p
$p\wedge q$:	p and q
$p\vee q$:	p or q
$p\Rightarrow q$:	p implies q
$p\Leftrightarrow q$:	p is equivalent to q

To find general rules for reasoning with compound sentences, we do a similar type of abstraction. Instead of working with specific sentences, we will use the variables p and q to represent arbitrary sentences and the adjacent symbols to represent the five operations on sentences.

Using this abstract representation of compound sentences, we can formulate basic rules for manipulating the five logical operators. These rules enable us to automate our reasoning about the logical operators so that we have more time to ponder deeper questions. However, to apply the rules to specific sentences, we must be able to see the abstract structure of a compound sentence.

⊕ *Example*

What is the structure of the following compound sentence?

$$(2+3=5) \text{ and } (4+5\neq 7)$$

1. Let p and q represent the following sentences.
   p: $2+3=5$ q: $4+5=7$
   Then $p\wedge{\sim}q$: $(2+3=5)$ and $(4+5)\neq 7$

2. We could also let p: $2+3=5$ and q: $4+5\neq 7$
   Then $p\wedge q$: $(2+3=5)$ and $(4+5)\neq 7$

We can view the above compound sentence as having either the structure $p\wedge q$ or the structure $p\wedge{\sim}q$, depending on whether we want to focus on the outside structure of the sentence or look deeper into its internal structure. The different views of the structure of a sentence are similar to viewing the outside structure of the human body or taking an x-ray view of its skeletal structure.

p(x) notation

We will use the function notation $p(x)$, read as "p of x," to represent an open statement in the variable x. For example, we could let $p(x)$ represent "$x^2 + 4x - 1 = 5$." The notation $p(x)$ has two layers of variables: p is a variable that represents a sentence and x is a variable that represents a number. Whenever a new notation seems a little strange, we should work with examples and before long it will seem like a perfectly natural way to communicate. Function notation is based on the substitution principle. To translate $p(3)$, we substitute 3 for each occurrence of x.

$$p(x)\text{: } x^2 + 4x - 1 = 5$$

$$p(3)\text{: } 3^2 + 4(3) - 1 = 5$$

Formal Logic

In formal logic, a statement is called a *proposition*. Since the logical operators operate on propositions, the study of the rules for manipulating logical operators is called *propositional logic*. Open statements are called *predicates*, and the study of predicates is called *predicate logic*. Symbolic sentences are called *well-formed formulas*, sometimes abbreviated as wffs. Like the rules for grammar in everyday language, formal logic systems have syntax rules that govern how symbols can be strung together. For example, we cannot juxtapose two logical operators, such as $p \vee \wedge q$. Neither can we juxtapose two sentences pq without a logical operator.

Since this text is an informal introduction to logic, we will not use these formal terms. However, we will look informally, from a common sense viewpoint, at predicate logic in Section 1.2, propositional logic in Section 1.3, and both in Section 1.4. The latter section covers the laws of logic, which are called *tautologies* in formal logic. A law of logic is a compound statement that is always true, such as *p or ~p*.

Visual Reasoning

Since the earliest cave drawings, pictures have served an important role in communication. Aristotle, one of the deepest thinkers in the history of western thought, believed that images were essential for thoughts. His use of the word "soul" in the adjacent quote is perhaps more akin to our concept of intellect than the modern interpretation of soul. His observation has been supported by recent scientific evidence which indicates that a large portion of the human brain is dedicated solely to processing visual information. We seriously handicap ourselves if we do not use these enormous resources when we try to reason in a logical manner. Visualizations – whether formed internally in the mind or externally with a pencil or computer – provide one of the most powerful tools for the reasoning process.

> To the thinking soul images serve as if they were contents of perception. That is why the soul never thinks without an image.
>
> *Aristotle*

In the symbiotic relationship between words and pictures, words help us understand our pictures, and pictures help us understand our words. After we intuitively understand a picture, we then reason with words to provide a verbal foundation for our visual insight. Pictures can be misleading and measurements may not be exact, so we should back up our visual understanding with verbal reasoning. On the other hand, it is usually difficult to do verbal reasoning without a visual understanding of what we are thinking about. The first step in solving most problems (other than a computational algorithm) is to visualize the various components of the problem and their relation to each other.

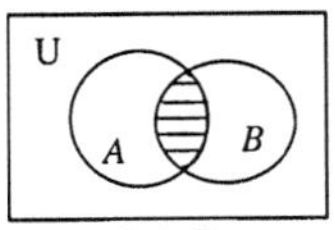

$A \cap B$

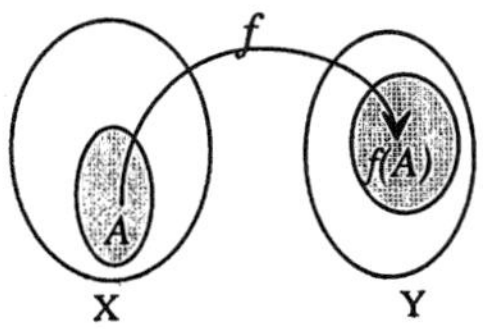

Visual reasoning thrives on not only what we see with our eyes, but also what we see in our mind, independent of our senses. When we see a mental image, we should draw it on paper so that we can carefully explore it. The sketches that we draw do not have to be detailed graphical representations. It is rather amazing how simple drawings, like the above sketch of the intersection of two sets or the adjacent sketch of the domain and range of a function, can help us focus our thinking.

To understand a mathematical concept, we need a visual picture of what it represents, for most mathematical concepts have their roots in some type of visualization. The visual concept of a ruler is essential for understanding the full meaning of a positive real number. In the adjacent ruler picture, we visualize a positive real number as the distance from the origin to the point it represents.

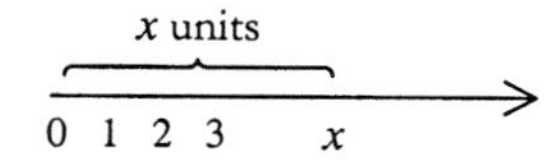

Negative Numbers

Negative numbers were not originally considered to be numbers because they could not be visualized as a length. However, these ghosts of a number frequently appeared as missing solutions to simple equations, such as $x+5 = 0$. After Fibonacci observed in the 13th century that a negative sum of money could be interpreted as a loss, various symbols were introduced to handle numerical losses. In the following centuries, these symbols were used as symbolic solutions to equations, but they were not considered numbers. It was not until the 19th century that the negative numbers were finally accepted as full-fledged numbers. Their acceptance was forced by the creation of a logical foundation which gave them a logical existence.

After the verbal conception of the negative numbers, the visual picture of the real numbers was expanded to a full number line, with the negative numbers visualized as the mirror image of the positive numbers. A real number could now be visualized as representing the *directed* distance from the origin to the point it represents. The number 3 is 3 units from the origin in the positive direction, whereas -3 is 3 units from the origin in the negative direction.

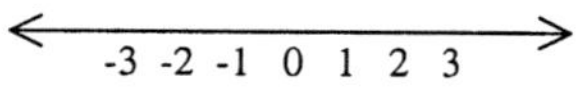

Complex Numbers

Like the negative numbers, the complex numbers also went through centuries of rejection until a simple visual picture made the existence of these ghostly numbers materialize. Like the negative numbers, complex numbers were needed for solving equations. Since there is no real number to solve the equation, $x^2 = -1$, the symbol $\sqrt{-1}$ was used as a symbolic solution, which was later labeled as i by Euler. The symbol $\sqrt{-1}$ was called an *imaginary number* because it was a complete product of the imagination. A similar type of number frequently surfaced in the quadratic formula:

$$\text{If } x^2 - x + 1 = 0, \text{ then } x = \frac{1 \pm \sqrt{-3}}{2}.$$

If we extend the standard rules of algebra to the imaginary numbers, we can factor $\sqrt{-3}$ as follows:

$$\sqrt{-3} = \sqrt{3(-1)} = \sqrt{3} \cdot \sqrt{-1}$$

We have now isolated the problem to $\sqrt{-1}$. Substituting i for $\sqrt{-1}$, we can write $x = \frac{1 \pm \sqrt{-3}}{2}$ as follows:

$$x = \tfrac{1}{2} \pm \frac{\sqrt{3}}{2} i$$

In a similar manner, all nonreal solutions produced by the quadratic formula can be represented in the form $x + iy$, where x and y are real numbers. These numbers are called *complex numbers*. Unlike the real numbers which provide lengths for real objects, no real models were apparent for complex numbers. Consequently, the complex numbers were not considered legitimate numbers, but they were needed to fill a technical void.

> We discard the symbolic sign $\sqrt{-1}$, which we repudiate completely, and which we may abandon without regret, because one does not know what this alleged sign signifies, nor what meaning one should attribute to it.
>
> *A. L. Cauchy,* 1847

Even though the imaginary numbers were not considered real numbers, they were used in blind manipulations because they often produced real results if one followed the basic rules of algebra. When one had the good fortune to multiply two imaginary numbers together, the ghosts would vanish and the computation would end up back in the land of real numbers:

$$(2i)(3i) = -6$$

$$(x - 3i)(x + 3i) = x^2 + 9$$

A major paradigm shift was started in 1673 when J. Wallis had the visual intuition to imagine the complex numbers as points in a plane. Other thinkers started to make the same visual connection, supplying more detail to the picture.

Complex Numbers

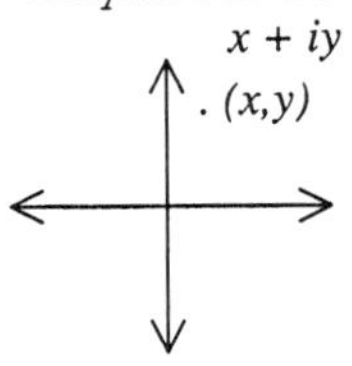

In the complex number $x + iy$, the imaginary i keeps y segregated from x, which is the same role that the ordered pair (x, y) serves. Thus, we can visualize the complex numbers as points in a plane in the same way that we visualize the real numbers as points on a line. The assignment of $x + iy$ as the number representative of the point (x, y) provides a simple visual representation of the complex numbers that not only deepens our understanding of these numbers, but also gives us powerful visual tools for working with them.

The visualization of numbers as points on a line or points in a plane gives our mind an intuitive picture which makes the concept more tangible. However, we must make a technical distinction between points and numbers. A real number is not a point on a number line. It is a coordinate of a point, or perhaps we should say that it is a name that we assign to a point. The same point can have different symbolic names, such as .5 or $\frac{1}{2}$.

With the new visibility of the complex numbers, mathematicians began to search for a verbal way to legitimatize them by building a logical foundation that supported their existence. This goal was finally accomplished towards the end of the 19th century. With their new certification as legitimate numbers, a rich new area of mathematics was created.

Complex numbers were soon put to good use by physicists, who gave them real applications in electricity and magnetism. Today, complex numbers are a standard mathematical reasoning tool, widely used by physicists, engineers, and mathematicians. The imaginary numbers are a brilliant example of the creative power that comes from the marriage of the visual imagination with logical reasoning in an abstract cathedral.

Less than

The operations of addition and multiplication were extended to the complex numbers in a way that preserved the basic properties these operations have on the set of real numbers. However, the $<$ relation does not extend to the complex numbers because the complex numbers are not lined up in a row like the real numbers.

The $<$ relation was extended to the negative numbers in accord with Fibonacci's financial debt interpretation. If your debt is \$2, you have less money than if your debt is \$1: $-2<-1$

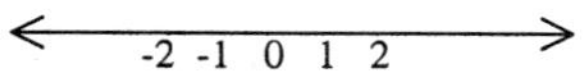

$$1 < 2$$
$$-1 > -2$$

This reversing of order between positive numbers and negative numbers gives us a mirror image picture of the negative real numbers. We can visualize the $<$ relation on the number line in terms of the relative positions of the numbers:

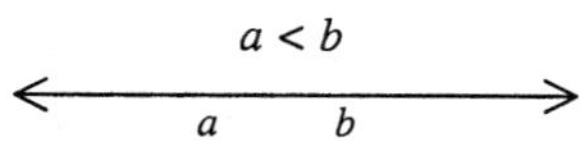

$$a<b$$

if and only if

a is to the left of b.

We will now use this picture to visually examine the effect of combining inequality sentences with *and* and *not*. Everyone knows what *and* means and what *not* means, but we sometimes use these words incorrectly, especially when we jump to conclusions without thinking about what we are saying.

⌖ *Example*

The domain for x is the set of real numbers. Write the solution set for the given open statement in interval notation.

1. Open statement: $x<2$

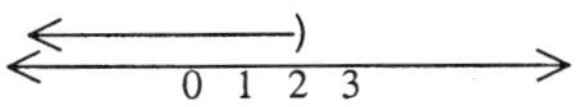

The solution set is illustrated on the left. In interval notation, we represent the solution set as $(-\infty, 2)$, which is read as "the open interval from negative infinity to 2." The open end parenthesis means 2 is not included in the set. We always use an open parentheses with the ∞ symbol because it does not represent a real number. Hence, $-\infty$ is not included in the interval $(-\infty, 2)$.

2. Open statement: It is not true that $x<2$.

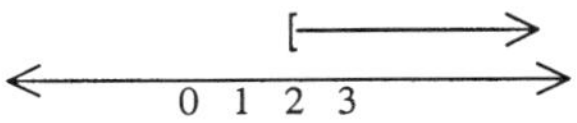

The solution set is the numbers not in the solution set of the previous example. We use the notation $[2, \infty)$ to represent this set, which is read as "the closed interval from 2 to infinity." The square bracket means that 2 is in the set.

3. Open statement: $x<2$ and $x<3$.

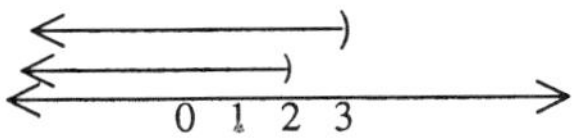

First, we visualize the solution set of each component sentence, as illustrated on the left. Because of the *and* operator, a number must be in both of these solutions sets in order to be in the requested solution set. Hence, the solution set for "$x<2$ and $x<3$" is the open interval from negative infinity to 2: $(-\infty, 2)$. To say that "$x<2$ and $x<3$" is equivalent to saying that "$x<2$." The latter form is preferable since it is simpler.

4. Open statement: It is not true that ($x<2$ and $x<3$).

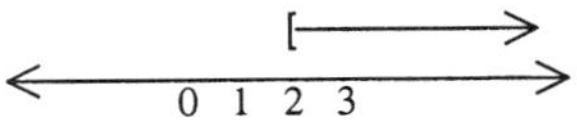

We must reason in the order indicated by the parentheses. First, we determine when ($x<2$ and $x<3$) is true, which is the interval $(-\infty, 2)$. Thus, the requested solution set is all numbers not in $(-\infty, 2)$, which is the closed interval from 2 to infinity: $[2, \infty)$.

5. Open statement: $\sim(x<2)$ and $\sim(x<3)$.

Again, we must reason in the order indicated by the parentheses. First, we find the solution set for the negation of each component sentence:

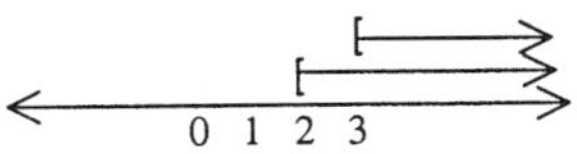

Solution set of $\sim(x<2)$: $[2, \infty)$
Solution set of $\sim(x<3)$: $[3, \infty)$

The solution set for this problem is all numbers in both of the above sets. So the solution set to the open statement is the interval $[3, \infty)$.

Reasonable Bites

As young children, we learned to cut up our food into digestible bites. Logical reasoning requires the same type of process. Given a problem, we try to cut it into simpler pieces, figure out the solution at the simpler level, and then put the pieces together, as we did in the previous example. To find the solution of $\sim p(x)$ and $\sim q(x)$, we first found the solution set of $\sim p(x)$ and the solution set for $\sim q(x)$. We then put these two pieces together to get the requested solution.

Reasoning Order

When we reason, we not only have to understand the meaning of the words used, but we also have to apply the meanings in the correct order. In (4) of the previous example, we first applied the meaning of "and" and then the meaning of "not," but in (5), we applied the definitions in the reverse order, which gave us a different answer. This happens frequently in the reasoning process; two definitions are involved, and we have to know which one to apply first. The order is usually indicated by the position of parentheses or by a comma.

Abstract Reasoning

In an *abstraction*, we merge various concrete examples under the rubric of a concept that expresses a property the examples have in common. For example, the number 3 is an abstraction of a quantitative property that various sets have in common. A variable is an abstraction of an arbitrary element in its domain.

When we prove a theorem, we try to be as abstract as possible so that our deductions have a wide range of applications. If we can prove that an abstract statement is true, we can then deduce that the statement is true for each example that satisfies the abstraction, which is why abstract reasoning is such a powerful method. With each abstract theorem that we prove, we have essentially proved a multitude of theorems, one for each example that satisfies the abstraction.

Does $\sim(p$ *and* $q)$ mean the same as $\sim p$ *and* $\sim q$?

When we contemplate an abstract question, we should think about various examples in order to understand what the abstraction represents. If we are confused by the abstract nature of the question, we should translate it in terms of examples. To contemplate the abstract question on the left, we could let p and q represent the following sentences:

p: $x<2$ $\quad$ q: $x<3$

The abstract question can now be translated as follows:

Does ~(p *and* q) mean the same as (~p *and* ~q)?

Does ~($x<2$ and $x<3$) mean the same as ~($x<2$) and ~($x<3$)?

In the examples on page 17, we saw that these two compound sentences have different solution sets. Therefore, they cannot have the same meaning. This example shows that a compound sentence of the abstract form ~(p *and* q) does not have the same meaning as ~p *and* ~q.

Sometimes we can find the answer to an abstract question with a well-chosen example, but we cannot use examples to prove that two compound statements have the same meaning. Later in this chapter, we will learn how to determine if two compound statements have the same meaning. In Chapter 2, we will study techniques for proving abstract statements.

Questions?

We conclude this section with the most important symbol in any area of logical reasoning, the question mark: ?

Why is this true?

What does this mean?

What else does it apply to?

Is there a way to generalize this result?

How do I describe what's going on here?

Is there a relation between these two things?

What is the underlying structure that makes this work?

. . .

In mathematics
the art of asking questions
is more valuable than
solving problems.

Georg Cantor, 1867

In 1867, Georg Cantor, one of the greatest mathematicians of the 19th century, published his doctoral thesis entitled, *In mathematics the art of asking questions is more valuable than solving problems.* This outlook undoubtedly contributed to his phenomenal success in creating new mathematics. By asking very simple and basic questions, Cantor developed a Theory of Sets which revolutionized the language of mathematics, providing a unifying concept for organizing the rapidly growing knowledge in mathematics. His creativity in developing a mathematical structure for working with infinite sets provided deep insights into mysterious properties of infinity (Section 3.6).

Students usually spend more time answering questions posed by others. However, we should be aware that asking good questions is an important part of the creative process of reasoning. The ultimate goal of logical reasoning is to extend our knowledge, and we cannot seek answers unless we have good questions to guide us.

We learn, discover, and create by asking questions. A good question can trigger a burst of energy in the mind that may open up slumbering portions of our brain with the imbued excitement of a possible discovery. Whenever a question pops in our mind, we should write it down, or else we may forget it and lose forever the valuable insight we might have acquired in our pursuit of its answer.

Writing as a Tool

One of the best tools to develop our reasoning skills is the process of writing. A pencil is an essential tool in a tool kit for reasoning. In addition to writing our questions, we should use this marvelous little instrument to help us find the answers. The process of writing gives us concrete visuals on which we can focus our thinking, reveals logical gaps that we need to bridge, and leaves a record where we can check and double-check our thinking. Things that seem obviously true at first glance may not hold up under a second reading.

In addition to writing homework, we should use writing to:

- Build our understanding.
- Explore our imagination.
- Illustrate our mental images.
- Record our questions.
- Search for answers.

Another advantage of writing is that it helps us remember concepts better. When we read a text, it may seem redundant to copy a definition. However, scientific evidence indicates that what we write is stored in the brain in a memory bank different from the storage area for information we have read or heard, thereby giving us double access to it. What we have written is easier to remember.

Writing also helps us deepen our understanding of a concept. Anyone can look at a definition and memorize it, but memorization has nothing to do with understanding. By writing a definition in our own words and then checking to make sure that it has the same meaning, we build a personal meaning of the concept. If our writing reveals that we have overlooked a key component of the definition, then we know on which parts we should work to build our understanding.

Writing about areas that confuse us helps to isolate the general confusion into specific questions on which we can focus. Copying examples from a text may help us see missing links that we didn't see when we read the example. When we finally see the missing links, making up similar examples and writing them on paper will help reinforce what we learned.

Although writing normally refers to the writing of words, we will expand the meaning here to include the drawing of pictures. Drawing a sketch often gives us ideas and thoughts which we can then form into words. In addition to expressing our thinking, we use writing to help us form our thoughts.

When you reach a satisfactory conclusion in the writing process (*or run head-on into a brick wall*), share your writing (*or frustrations*) with others in order to get feedback and further develop your reasoning skills. The more time you spend writing, the more progress you will make in developing your reasoning skills.

Exercise Set 1.1

1. Translate the following sentence into math symbols. Which do you prefer, the word form or the symbolic form?

   "The product of a number with the sum of two other numbers is equal to the sum of the following: the product of the first number times the second number and the product of the first number times the third number."

2. Identify the nouns, verbs, and logical operators in each sentence.

   a. $A = \{1,2,3\}$ and $A \subseteq B$ c. $x \in A$ implies that $x \notin B$.

   b. $x<2$ or $5<x$. d. $x+5 = 12$

3. How do you translate $3<x<5$ in a grammatically correct form? Identify any nouns, verbs, or logical operators in the sentence.

4. Let *x, y, a, b*, and *c* represent real numbers. Determine if the given expression is a sentence.

   a. $x = y$ b. $x+y$ c. $x<y$ d. $x \div y$ e. $-b \pm \sqrt{b^2 - 4ac}$

5. Let *A*, *B*, and *C* represent sets. Determine if the given expression is a sentence. Check Appendix *D* for any unfamiliar symbols.

   a. $A = B$ b. $A \subseteq B$ c. $A \cup B$ d. $A \cup B \subseteq C$

6. For the given domain, make a list of symbols that form a sentence when you place them between *x* and *y*.

   a. *x* and *y* represent numbers b. *x* and *y* represent sets

7. Is the expression a statement, an open statement, or neither?

   a. $2<1$ d. For all *x*, $2<x$.

   b. $2<x$ e. For some *x*, $2<x$.

   c. $2 \times (3+7)$ f. Ten is the most important number.

8. The domain of x is the set $\mathbb{N}$ of natural numbers.
   Make up an open statement $p(x)$ that has the given solution set.
   a. Its solution set is empty.
   b. Its solution set is the whole domain.
   c. Its solution set is neither empty nor the whole domain.

9. Let p: $2+4=7$, q: $3+5=8$. Translate each compound sentence and determine if it is true or false.
   a. $\sim p \wedge \sim q$ b. $\sim(p \wedge q)$ c. $\sim p \vee \sim q$ d. $\sim(p \vee q)$

10. Do you think the following sentences have the same meaning? Test your answer by making up examples for p and q using sentences from everyday life. Then make up examples using sentences from mathematics.
    a. Does $\sim(p \wedge q)$ have the same meaning as $\sim p \wedge \sim q$?
    b. Does $\sim(p \vee q)$ have the same meaning as $\sim p \vee \sim q$?

11. Translate each sentence and determine its truth value. The domain for x is the set of real numbers. $p(x)$: $x>3$ $q(x)$: $x>9$.
    a. $p(7)$ or $q(7)$
    b. $p(7)$ and $q(7)$
    c. For every x, $p(x)$ and $q(x)$.
    d. There exists an x such that $p(x)$ and $q(x)$.

12. Let $p(x)$: $x>3$ and $q(x)$: $x>9$. Illustrate the solution set for $p(x)$ and the solution set for $q(x)$ on a number line. Use your sketch to express the solution set of the given open statement in interval notation. Make sure that you reason in the order indicated by the parentheses, simplifying the task in a step by step manner.
    a. $x>3$ or $x>9$
    b. $x>3$ and $x>9$
    c. $\sim(x>3$ or $x>9)$
    d. $\sim(x>3$ and $x>9)$
    e. $\sim(x>3)$ or $\sim(x>9)$
    f. $\sim(x>3)$ and $\sim(x>9)$

13. Discuss the following questions.
    a. What is an abstraction?
    b. What is an abstraction of a sentence?
    c. What should you do when you get stuck on an abstract problem and have no idea of what to do?
    d. What is the advantage of figuring out something on your own instead of having someone else explain it to you?
    e. How do you develop your reasoning skills?

14. Why are problems in math textbooks called exercises?

Activity 1.2

Before reading the next section, see what you can figure out for yourself. The domain for x and y is the set of real numbers.

1. Let $p(x)$: $x + 1 = 4$. Translate each statement and determine its truth value. Do any of the statements have the same meaning?
   a. ~(For all x, $p(x)$).
   b. For all x, ~$p(x)$.
   c. There exists an x such that ~$p(x)$.
   d. ~(There exists an x such that $p(x)$).
2. Let $p(x)$: $x < 3$. Repeat the previous exercise.
3. Discuss what it means for the given sentence to be true. Then discuss what it means for the sentence to be false.
   a. For all x, $p(x)$.
   b. There exists an x such that $p(x)$.
4. Do any of the following statements have the same meaning?
   a. ~(For all x, $p(x)$). c. There exists an x such that ~$p(x)$.
   b. For all x, ~$p(x)$. d. ~(There exists an x such that $p(x)$).
5. Is the given sentence true or false?
   a. For every x, there exists a y such that $x + y = 0$.
   b. There exists a y such that for every x, $x + y = 0$.
   c. For every positive x, there exists a positive y such that $y < x$.
   d. There exists a positive y such that for every positive x, $y < x$.

≡ 1.2 Two Quantifiers ≡

Two of the basic terms from which we build the language of logic are the universal and existential quantifiers. They are called quantifiers because they give information on the quantity of elements in the solution set of an open statement. In everyday language, we reference the quantifiers as "every" and "some." The quantifiers are easy to work with if you master a few basic rules and learn how to recognize them when they are phrased in everyday language.

Universal Quantifier

$\forall x, p(x)$ is true if and only if every element in the domain of x converts $p(x)$ into a true statement.

The universal quantifier is a prefix which indicates that the solution set of an open statement is the whole domain.

For all x, $x+2=2+x$.

Symbolically, we represent the universal quantifier with an upside down A ($\forall$).

$$\forall x, x+2=2+x$$

The universal quantifier can be phrased as *for all x*, *for every x*, *for each x*, and in the forms given in the following example.

Example

The domain of x is the set of real numbers.
The following statements have the same meaning.

1. $\forall x, x+2=2+x$
2. *For all x,* $x+2=2+x$.
3. Let x *be a* real number. Then $x+2=2+x$.
4. If x *is a* real number, then $x+2=2+x$.
5. Let x *be an arbitrary* real number. $x+2=2+x$.
6. $x+2=2+x$ *for every x*.

In symbolic notation, we always write quantifiers as a prefix. However, in everyday language, we often write a quantifier at the end of a sentence, as illustrated in the last line of the above example.

Existential Quantifier

$\exists x, p(x)$ is true if and only if there exists at least one x in the domain of x such that $p(x)$ is true.

The existential quantifier is a prefix which indicates that the solution set of an open statement has at least one element in it.

There exists an x such that $x+3=15$.

We represent the existential quantifier with a backwards E: $\exists$

$$\exists x, x+3=15$$

The existential quantifier can be phrased as *there exists an x such that*, *for some x*, and in the wordings given in the following example.

⊕ *Example*

The domain of x is the set of real numbers.
The following statements have the same meaning.

1. $\exists x,\ x+3=2$
2. *There exists* an x such that $x+3=2$.
3. *There is* an x such that $x+3=2$.
4. *There is at least one* x such that $x+3=2$.
5. *For some* x, $x+3=2$.
6. $x+3=2$ *for some* x.

When we see "some" or "there is" in a sentence, it indicates the presence of the existential quantifier.

⊕ *Example*

Translate each statement in terms of variables and quantifiers.

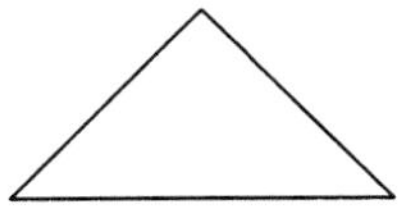

Some triangles are isosceles.

1. Some triangles are isosceles.

   "Some" represents the existential quantifier, so we introduce a variable for it.

   Let X represent an arbitrary triangle.
   $\exists X$, X is isosceles.

2. For every real number, there is a larger real number.

   "For every" represents the universal quantifier, and "there is" represents the existential quantifier. Having identified the quantifiers, we introduce a variable for each quantifier.

   Let x and y be real numbers. $\forall x\ \exists y,\ x<y$.

The truth value of a quantified statement depends on the domain for the variable. For example, consider the following statement:

$$\text{For some } x,\ x+5<1.$$

This statement is true if the domain of x is the set of integers, but it is false if the domain is the set of natural numbers.

The existential quantifier guarantees the existence of at least one element from the domain that makes the statement true; however, it does not say that there is only one or just a few. If an open statement is true for all elements in the domain, it is automatically true for at least one element in the domain, provided, of course, that the domain is not empty.

Example Let $p(x)$: $x+0=x$, where x is a real number.

$\forall x,\ p(x)$ is true. Also, $\exists x,\ p(x)$ is true.

Multiple Quantifiers

When we have more than one variable in a sentence, we need multiple quantifiers to convert it to a statement. To write the basic rules for working with multiple quantifiers, we will use the notation $p(x,y)$ to represent an open statement in the variables x and y. Unless stated otherwise, the domain for x and y is the set of real numbers.

Example Let $p(x,y)$ represent the following sentence. Find the truth value of $p(2,6)$ and $p(1,3)$.

$$p(x,y)\colon x+y=8.$$
$$p(2,6)\colon\ 2+6=8$$
$$p(1,3)\colon\ 1+3=8$$

$p(2,6)$ is true, but $p(1,3)$ is false.

Order of the Quantifiers We will now investigate whether or not the order of the quantifiers affects the meaning of the sentence. In a sentence with multiple quantifiers, the meaning of each quantifier is deciphered by applying the definitions one at a time, working from left to right. The first quantifier applies to the rest of the sentence, including any quantifiers that come after it. In the following statement, we first translate it formally, and then we interpret its meaning.

Statement A $\exists x\ \exists y,\ x+y=8$

There exists an x such that the following is true: $\exists y,\ x+y=8$.

There exists an x and there exists a y such that $x+y=8$.

Statement B

$\exists y\ \exists x,\ x+y=8$
There exists a y such that the following is true: $\exists x,\ x+y=8$
There exists a y and there exists an x such that $x+y=8$.

$\exists x\ \exists y,\ p(x,y)$
means the same as
$\exists y\ \exists x,\ p(x,y)$.

In the previous two statements, reversing the order of the two existential quantifiers does not change the meaning of the statement. If we let $p(x,y)$ represent an open statement, we can reason in a similar manner and conclude that, in general, reversing the order of two existential quantifiers does not affect the meaning of the statement. This rule is summarized in the adjacent box.

Now consider what happens when we reverse the order of two universal quantifiers.

Statement A

$\forall x\ \forall y,\ x+y=y+x$
For every x, the following is true: $\forall y,\ x+y=y+x$.
For every x and for every y, $x+y=y+x$.

Statement B

$\forall y\ \forall x,\ x+y=y+x$
For every y, the following is true: $\forall x,\ x+y=y+x$.
For every y and for every x, $x+y=y+x$.

$\forall x\ \forall y,\ p(x,y)$
means the same as
$\forall y\ \forall x,\ p(x,y)$.

In the above two statements, reversing the order of the two universal quantifiers does not change the meaning. This result can be generalized to an arbitrary open statement $p(x,y)$. When we reverse the order of two universal quantifiers, the meaning does not change. This rule is summarized in the adjacent box.

Now let's examine what happens when we have mixed quantifiers. When a statement has both a universal and an existential quantifier, we cannot translate the statement by inserting "and" between the two quantifiers as we did in the previous examples.

Statement A

$\forall x\ \exists y,\ x+y=8$
For every x, the following is true: $\exists y,\ x+y=8$

Statement B

$\exists y\ \forall x,\ x+y=8$
There exists a y such that the following is true: $\forall x,\ x+y=8$

Do the previous two statements have the same meaning? If they do, they must have the same truth value, so let's deconstruct each statement and determine its truth value.

Statement A

$\forall x \, \exists y, \, x+y=8$

We start with the quantifier on the left:

1. Let x be an arbitrary real number.

   Does there exist a real number y_o such that $x+y_o=8$? One way to prove its existence is to construct it. Working backwards from what we want to derive, $x+y_o=8$, we can figure out how to construct y_o. Since x has already been introduced, we can use x in our construction of y_o.

We use a subscript with y_o to remind us that we are looking for a specific y that works for the given x.

2. Set $y_o = 8-x$. Since x is a real number, $8-x$ is a real number. So y_o is a real number.
3. $x+y_o = x+(8-x) = 8$ (*Substitute* $8-x$ *for* y_o.)

We have demonstrated that for every real number x, there is a real number y such that $x+y=8$, so the above statement is true.

Statement B

$\exists y \, \forall x, \, x+y=8$

Is there a fixed number y_o such that for every x, $x+y_o=8$? If so, what is it? As in the previous example, we start our analysis with the first quantifier on the left; however, this time we cannot use x in our construction of y_o.

We reason with quantifiers from left to right.

1. Set $y_o =$ ________.
2. Let x be a real number.

   The difference in the order of the quantifiers affects what we can use when we construct y_o. We cannot define y_o in terms of x because x is not introduced until Step 2. There is no way that we can find a fixed y_o that will work for every x. For example:

   Set $y_o = 0$ Is $\forall x, \, x+0=8$ true? No
   Set $y_o = 5$ Is $\forall x, \, x+5=8$ true? No

   Whatever number we select for y_o, we can always find an x such that $x+y_o \neq 8$. In particular, we could let $x = 1-y_o$. Then $x+y_o = 1$, so $x+y_o \neq 8$. Therefore, the above statement is false.

The two statements in the previous example have different truth values:

$$\forall x\ \exists y,\ x+y=8 \text{ is true.}$$

$$\exists y\ \forall x,\ x+y=8 \text{ is false.}$$

$\forall x\ \exists y,\ p(x,y)$
does not have the same meaning as
$\exists y\ \forall x,\ p(x,y)$.

Therefore, these two statements do not have the same meaning. This example shows that the structures of these statements do not impart the same meaning:

$$\forall x\ \exists y,\ p(x,y)$$
$$\exists y\ \forall x,\ p(x,y)$$

Since students sometimes make the mistake of assuming these statements do have the same meaning, let's look at a geometric example where we can use visual reasoning to see why they have different meanings.

⊕ *Example*

The domain for x is all the points in a fixed plane and the domain for ℓ is all the lines in the same plane. Is the following statement true?

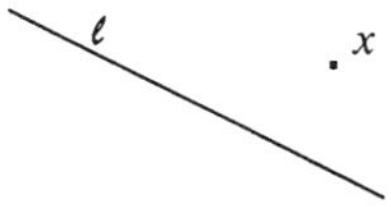

For every line ℓ, there exists a point x such that x is not on ℓ.

The above statement is true. Note that this statement has the following structure:

$$\forall \ell\ \exists x,\ x \text{ is not on } \ell.$$

What happens if we reverse the order of the quantifiers?

$$\exists x\ \forall \ell,\ x \text{ is not on } \ell.$$

There exists a point x such that for every line ℓ, x is not on ℓ.

This statement is not true because no matter which point we select for x, there will be some lines in the plane that go through x. Since these two statements have different truth values, they do not have the same meaning.

When using the existential quantifier, we often say *for some x* instead of *there exists an x such that*. However, when a sentence has mixed quantifiers, as in the above example, the meaning is easier to interpret if we use *there exists* instead of *some*.

$\forall x \, \exists y, \, p(x,y)$
does not have the same meaning as
$\exists y \, \forall x, \, p(x,y)$.

As demonstrated in the previous examples, reversing the order of the two different quantifiers changes the meaning of a sentence. When reasoning with a statement that has mixed quantifiers, we must carefully read the statement from left to right, for the order of the quantifiers makes a significant difference in the meaning of the statement. This difference is summarized below.

$\forall x \, \exists y, \, p(x,y)$

The adjacent statement guarantees that for every x, we can find a y_o that works for the given x. To prove a statement of this form, we can use x to construct y_o:

1. Let x be an element in the domain of x.
2. Set $y_o =$ ____.

$\exists y \, \forall x, \, p(x,y)$

This statement guarantees that we can find a y_o such that the remaining statement is true for all x in the domain. Since y_o cannot depend on x, we cannot use x to construct y_o. To prove a statement of this form, we should structure our reasoning as follows:

1. Set $y_o =$ ____.
2. Let x be an element in the domain of x.

Negating Quantifiers

On the surface, negations seem very simple. The negation of a true statement is false and the negation of a false statement is true. However, negations can be confusing when combined with quantifiers or connectives such as *and*, *or*, and *implies*.

To negate a sentence, we can prefix it with *"it is not true that."* With symbols, though, we usually indicate the negation by drawing a slash through the verb symbol:

It is not true that $x+1$ is equal to 3.

$$x+1 \neq 3.$$

When a sentence is quantified, it makes a difference if a negation is before the quantifier or after the quantifier:

$\sim(\forall x, \, x+1=3)$ does not mean that $\forall x, \, x+1 \neq 3$.

$\sim p(x)$ denotes the negation of the open statement $p(x)$.

$\sim \forall x, \, p(x)$ denotes the negation of the statement $\forall x, \, p(x)$.

⌖ *Example*

Do the following statements have the same meaning?

1. It is not true that for every triangle X, X is isosceles.
2. For every triangle X, it is not true that X is isosceles.

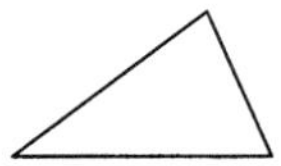

Some triangles are not isosceles.

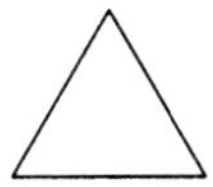

Some triangles are isosceles.

If we place our finger over the negation in the first sentence, we have a false sentence, so its negation is true. This sentence has the same meaning as "some triangles are not isosceles."

If we move the negation across the quantifier, though, we produce a false statement. Some triangles are isosceles, so the second statement is false.

Since these two statements have different truth values, they do not have the same meaning.

$\sim\forall x,\ p(x)$

has the same meaning as

$\exists x,\ \sim p(x)$.

The above example demonstrates that $\sim\forall x,\ p(x)$, which is the form of the first statement, does not have the same meaning as $\forall x,\ \sim p(x)$, which is the form of the second statement.

How then do we negate a universally quantified sentence? If it is not true that every x makes $p(x)$ true, then there must exist an x that makes $p(x)$ false. Thus, the proper way to negate a universally quantified statement is given by the adjacent rule. When we bring a negation across a universal quantifier, it must change to the existential quantifier.

$$\sim\forall x,\ x+1=3 \text{ has the same meaning as } \exists x,\ x+1\neq 3$$

$\sim\exists x,\ p(x)$

has the same meaning as

$\forall x,\ \sim p(x)$.

How do we negate an existentially quantified sentence? If it is not true that there exists an x that makes $p(x)$ true, then *every* x must make $p(x)$ false. When we bring a negation across the existential quantifier, it must change to the universal quantifier.

$$\sim\exists x,\ x+2=x \text{ has the same meaning as } \forall x,\ x+2\neq x.$$

Notice the similarity in the above two rules for negating a quantified statement. If we move a negation across either quantifier, we must change the quantifier.

When negating more than one quantifier, we apply the appropriate rules one step at a time, as illustrated in the next example.

⊕ *Example* Negate the given sentence.

1. There is an x such that $f(x) = y$.

   $\sim\exists x, f(x) = y$ has the same meaning as $\forall x, \sim[f(x) = y]$.

   So the negation can be worded as: *For all x, $f(x) \neq y$.*

2. There exists a y such that for every x, $x + y = 0$.

$$\sim\exists y\ \forall x,\ x+y=0$$
$$\forall y \sim\forall x,\ x+y=0$$
$$\forall y\ \exists x,\ \sim(x+y=0)$$
$$\forall y\ \exists x,\ x+y\neq 0$$

   For every y, there exists an x such that $x + y \neq 0$.

3. There exists a real number x such that for every integer y, $x > y$.

$$\sim\exists x\ \forall y,\ x>y$$
$$\forall x\ \exists y,\ \sim(x>y)$$
$$\forall x\ \exists y,\ x\leq y$$

   The variables x and y have different domains. However, when we move the negation across a quantifier, it does not affect the domain of the variable. So we translate the above symbolic form as follows:

   For every real number x, there exists an integer y such that $x \leq y$.

Students sometimes leave "such that" dangling in a place that is not grammatically correct. The wording of the existential quantifier is the complete phrase "there exists an x such that." When we negate an existential quantifier, we must replace the complete phrase with the universal quantifier.

In everyday language, we often use inflection to convey our meaning when we negate a quantified sentence. For example, suppose that a few students got an A on a test. In response to the question, *"Did everyone get an A?"* the teacher might respond with the appropriate inflection: *"No, everyone did not get an A."* With a different inflection, the same words would convey a different meaning. The appropriate logical response would be: *"No, some people did not get an A."* Since we do not use inflection to convey meaning in logical reasoning, we must be very careful not to fall into the loose habits of everyday speech when negating a quantified statement.

Different Letters

When using multiple variables, we must be aware that different letters do not always represent different elements. For example, the following statement is false:

For all real numbers x and y, $x<y$ or $y<x$.

This statement is false because x could be equal to y. When we want letters to represent different elements, we must inform the reader. For example, we could say:

For all distinct real numbers x and y, $x<y$ or $y<x$.

For all real numbers x and y where $x \neq y$, $x<y$ or $y<x$.

For all real numbers x and y, if $x \neq y$, then $x<y$ or $y<x$

Exercise Set 1.2

1. The domain for x and y is the set of real numbers. Determine if each sentence is true or false. If both sentences are true, determine if one is "stronger" than the other.
   a. There exists a y such that for every x, $y<x$.
      For every x, there exists a y such that $y<x$.
   b. There exists a y such that for every x, $x+y=4$.
      For every x, there exists a y such that $x+y=4$.
   c. There exists a y such that for every positive x, $y<x$.
      For every positive x, there exists a y such that $y<x$.
   d. There exists a y such that for every x, $x+y=x$.
      For every x, there exists a y such that $x+y=x$.
2. Consider the meaning of the following two statements:

   I. $\exists y\ \forall x,\ p(x,y)$

   II. $\forall x\ \exists y,\ p(x,y)$

   a. If (I) is true, does (II) have to be true?
   b. If (II) is true, does (I) have to be true?

3. Make up simple examples of sets A, B and C so that (I) is true and (II) is false.

   I. For every y in A, there exists an x in B such that xy is in C.

   II. There exists an x in B such that for every y in A, xy is in C.

4. Let $p(x,y)$: y was the mother of x. Translate each sentence in terms of everyday language. Do they have the same meaning?

   a. $\forall x\ \exists y,\ p(x,y)$ b. $\exists y\ \forall x,\ p(x,y)$

5. Make up a sentence $p(x,y)$ where x and y are real numbers. Then determine the truth value of the following:

   a. $p(1,2)$ b. $p(2,1)$ c. $\forall x\ \exists y,\ p(x,y)$ d. $\exists y\ \forall x,\ p(x,y)$

6. Let $p(x,y)$ represent an open statement with variables x and y. True or false?

   a. $\exists x\ \exists y,\ p(x,y)$ has the same meaning as $\exists y\ \exists x,\ p(x,y)$.

   b. $\forall x\ \forall y,\ p(x,y)$ has the same meaning as $\forall y\ \forall x,\ p(x,y)$.

   c. $\forall x\ \exists y,\ p(x,y)$ has the same meaning as $\exists y\ \forall x,\ p(x,y)$.

7. Determine if the sentence is true or false. Then write its negation in a form where the negation is not a prefix for a quantifier.

   a. For every real number x, $3x = 4$.

   b. For every real number x, $3x \neq 4$.

   c. There exists a real number x such that $x^2 = -1$.

   d. For every complex number x, $x^2 \neq -1$.

   e. There exists a real number y such that for every x, $x + y = 4$.

   f. There exists an integer y such that for every real number x, $y < x$.

8. The domain for all variables is the set of real numbers. Is the statement true or false? If false, write its negation so that the negation is not a prefix for a quantifier.

   a. There exists a c such that for every x, $x + c = 2$.

   b. For every x, there exists a c such that $x + c = 2$.

   c. There exists an m such that for every x, $x \leq m$.

   d. For every x, there exists an m such that in S, $x \leq m$.

9. Write the negation of each statement. Move the negation across the quantifiers in a logically correct manner.

   a. For every integer y, there is a real number x such that $g(x) = y$.

   b. For every y in B, there exists an x in A such that $f(x) = y$.

   c. For all sets A and B, there is a function f that maps A onto B.

10. The domain for x has 100 elements. If the given statement is true, how many elements are in the solution set for $p(x)$?

    a. $\forall x,\ p(x)$ b. $\sim\forall x,\ p(x)$ c. $\exists x,\ p(x)$ d. $\sim\exists x,\ p(x)$

Activity 1.3

1. Given statements p and q, the four possible cases for their truth values are listed in the adjacent table. For each case, enter the truth value of the compound sentence according to how you use the connective in everyday language. Translate $p \Rightarrow q$ as "if p, then q."

p	q	$p \wedge q$	$p \vee q$	$p \Rightarrow q$
T	T			
T	F			
F	T			
F	F			

2. Your friend gives you half the money for a lottery ticket. You promise her, *"If this ticket wins, then I will give you a million dollars."* One of the following four cases must occur.

   Case 1: Your ticket wins and you give her a million dollars.

   Case 2: Your ticket wins, but you do not give her a million.

   Case 3: Your ticket does not win, but you give her a million.

   Case 4: Your ticket does not win, so you don't give her a million.

   a. In which cases would she be justified in calling you a liar?

   b. Let p and q represent the following sentences.

      p: Your ticket wins the lottery.

      q: You give her a million dollars.

      In the adjacent table, record your answers from part (a). Do your answers agree with your implication table in the previous exercise?

p	q	$p \Rightarrow q$
T	T	
T	F	
F	T	
F	F	

3. Let x be a real number and let $p(x)$: If $x < 3$, then $x < 7$. Determine the truth value (true or false) of each sentence.

   a. $p(1)$ b. $p(4)$ c. $p(11)$ d. For all x, if $x < 3$, then $x < 7$.

4. Does the first sentence have the same meaning as either of the other two sentences?

   a. p or q $\quad$ $\sim p \Rightarrow q$ $\quad$ $\sim q \Rightarrow p$

   b. $p \Rightarrow q$ $\quad$ $\sim p \Rightarrow \sim q$ $\quad$ $\sim q \Rightarrow \sim p$

   c. $\sim(p$ and $q)$ $\quad$ $\sim p$ and $\sim q$ $\quad$ $\sim p$ or $\sim q$

   d. $\sim(p$ or $q)$ $\quad$ $\sim p$ or $\sim q$ $\quad$ $\sim p$ and $\sim q$

   e. $\sim(p \Rightarrow q)$ $\quad$ $\sim p \Rightarrow \sim q$ $\quad$ p and $\sim q$

≡ 1.3 Five Logical Operators ≡

Logical operators are used to build compound sentences from given sentences. The basic language used in logical reasoning can be built from five logical operators. Their formal names and symbolic representations are given on the left.

Negation	~	it is not true that
Conjunction	$\wedge$	and
Disjunction	$\vee$	or
Implication	$\Rightarrow$	implies
Equivalence	$\Leftrightarrow$	is equivalent to

- The *negation* symbol is a short squiggle: ~
- The *and* symbol is an A without the middle bar: $\wedge$
- The *or* symbol is the $\wedge$ symbol flipped upside down: $\vee$
- The *implies* symbol is an arrow: $\Rightarrow$
- The *equivalence* symbol is a double-headed arrow: $\Leftrightarrow$

The outside structure of most valid arguments depends on the manipulation of these five little words. They are the key to how we structure our thinking. Within this structure, we then work with the meaning of individual sentences. In order to reason in a logical manner, we must understand both the syntax and semantics of the five logical operators.

Syntax

Syntax rules tell us how we can juxtapose words or symbols to form sentences or well-formed formulas in a particular language. In a computer language, syntax rules tell us how we can string symbols together. In everyday language, we string together words and punctuation symbols according to the rules of syntax, which is an important part of grammar.

Many people feel that grammar should be judged insofar as it follows the principles of logic. Mathematics, from this viewpoint, is the ideal use of language.

David Crystal
Cambridge Encyclopedia of Language

The syntax rules for the logical operators are fairly simple. First of all, the logical operators can only be used with sentences. These terms are called operators because they operate on sentences and produce a new compound sentence. They are sometimes called logical connectives because they connect sentences. However, the negation operator does not connect sentences; it only operates on one sentence.

The negation operator serves as a prefix for a sentence.

$\sim(x \in A)$ is read as "*it is not true that x is an element of A.*"

Let A be a set.
Incorrect Syntax: $\sim A$
Correct Syntax: $\sim(x \in A)$

We sometimes read $\sim p$ as "*not p*," but we do not write it that way since it does not have correct syntax in everyday language. If p represents a sentence, we can write $\sim p$; however, if A represents a set, we cannot write $\sim A$. On the other hand, since $x \in A$ is a sentence, we can write $\sim(x \in A)$.

The other four logical operators must be placed between two sentences. For this reason, they are called binary operators. If p and q represent sentences, we can write $p \wedge q$; however, we cannot write $p \sim q$.

Let A and B be sets.

Incorrect Syntax: $x \in A \wedge B$

Correct Syntax: $x \in A \wedge x \in B$

In everyday language, we can place *and* between two nouns, but we cannot do this when we use *and* as a logical operator. If A and B represent sets, we cannot write $A \wedge B$, but we can write $x \in A \wedge x \in B$. When we apply logical rules for manipulating the word *and*, we must use *and* as a logical operator. For example, suppose that we need to negate the following sentence:

x is in both A and B.

We can translate this sentence so that *and* is a logical operator:

x is in A and x is in B.

In the above form, we can apply the rule for negating an and-sentence (see Section 1.4).

Incorrect Syntax: $x \in A \vee B$

Correct Syntax: $x \in A \vee x \in B$

We have a similar problem with *or* since we sometimes use *or* between two nouns in everyday language. So when we use *or* as a logical operator, we must carefully check our syntax. We cannot write $x \in A \vee B$, but we can write $x \in A \vee x \in B$.

Incorrect Syntax: $A \Rightarrow B$

Correct Syntax: $x \in A \Rightarrow x \in B$

The implies operator can only be used between two sentences. The sentence $p \Rightarrow q$ can be read as either "*p implies q*" or as "if p, then q." We normally use the latter form when we formulate conjectures or state theorems. If A and B represent sets, then A is not a sentence, so we cannot write "if A, then B." However, we can write "if $x \in A$, then $x \in B$."

Incorrect Syntax: $x \in A = x \in A$

Correct Syntax: $x \in A \Leftrightarrow x \in A$

The equivalence operator also goes between two sentences. If p and q represent sentences, we can write $p \Leftrightarrow q$, which is read as "*p is equivalent to q.*" Students sometimes confuse the equivalence operator with the equals relation because they are intimately related; the equals relation is defined in terms of the equivalence operator. However, there is an important syntactic difference between them. We only use the equals relation between two sets or between two numbers. We do not use it between two sentences. For example, if A represents a set, we can write $A = A$, but we do not write $x \in A = x \in A$. Instead, we write $x \in A \Leftrightarrow x \in A$.

When more than one logical operator is used in a sentence, we use parentheses to indicate the order in which the operators are performed. For example, $\sim(p \wedge q)$ means to first form the compound sentence $p \wedge q$ and then takes its negation. Without parentheses, we will interpret $\sim p \wedge q$ to mean $(\sim p) \wedge q$.

Semantics

Semantics is the study of meaning. To engage in logical reasoning, we have to know the exact meaning of the logical operators for they set the structure for most types of valid arguments. The meaning of a logical operator is determined by the truth value it produces when used in a compound statement. The conditions under which an operator produces a true statement are summarized in the adjacent table.

Let p and q represent statements.

$\sim p$	is true *means* p is false.
$p \wedge q$	is true *means* both p and q are true.
$p \vee q$	is true *means* at least one part is true.
$p \Rightarrow q$	is true *means* if p is true, then q must be true.
$p \Leftrightarrow q$	is true *means* either both parts are true or both parts are false.

As you can see, the logical operators are simple concepts by themselves. However, when more than one operator is used, especially with a negation, we can be easily mislead into a fallacious deduction.

The negation operator reverses the truth value.

If p is true, $\sim p$ is false.
If p is false, $\sim p$ is true.

We usually indicate the negation with a slash through the symbol for the verb: $\not\subseteq$

$$A \not\subseteq B \text{ means } \sim(A \subseteq B)$$

Even though we normally use the slashed symbol, it can get us into trouble when we start making logical deductions. If we need to translate the meaning of $A \not\subseteq B$, we should first write it as $\sim(A \subseteq B)$ and then substitute the definition of $A \subseteq B$. In this form, we can see the proper way to negate the sentence.

Conjunction

To form the conjunction of two sentences, we place *and* between them: $A \subseteq B$ and $B \subseteq C$. In order for an and-sentence to be true, both parts must be true.

$p \wedge q$ is true *means* both p and q are true.

When applied to two sentences, the meaning of *and* is the same as in everyday language.

Disjunction

To form the disjunction of two sentences, we place *or* between them: $(A \subseteq B)$ or $(A \subseteq C)$. An or-sentence is false only if both parts are false.

$p \vee q$ is true *means* at least one of the two is true.

When p and q are both true, $p \vee q$ is true. However, in everyday language, we sometimes label this interpretation as false. For example, if a restaurant menu states that "salad or vegetable is included," we interpret it to mean that we cannot have both for the advertised price. This usage of *or* is called the *exclusive or.*

p or *q*
means the same as
if $\sim p$, then *q*.

On the other hand, we sometimes use *or* in everyday language in an inclusive sense. When a club advertises a discount for members or senior citizens, a member who is also a senior citizen would no doubt also qualify. In logic, *or* always means the inclusive sense. To indicate an exclusive *or*, we say "p or q but not both." Some programming languages represent the exclusive-or with the symbol XOR.

We can always interpret an or-sentence as an implication. To say that "p or q is true," means that "if p is false, then q is true." We will now investigate the meaning of an implication.

Implication

The implication is perhaps the most important word in logic because it sets the structure for the meaning of a valid argument. In the implication $p \Rightarrow q$, we call p the *hypothesis* and q the *conclusion*.

The definition of $\Rightarrow$ sounds simple on the surface, but students sometimes try to read more into it, which causes logical errors.

$p \Rightarrow q$ is true *means* if p is true, then q must be true.

When we say that an implication is true, we are stating that *if* the hypothesis is true, then the conclusion must be true.

We are *not* saying that if the hypothesis is false, then the conclusion must be false.

Furthermore, we are not saying the conclusion has to be true nor are we saying the hypothesis has to be true. In fact, if the hypothesis p is false, the sentence $p \Rightarrow q$ is automatically true. This may sound strange, but it is consistent with the way we use implications in everyday language. For example, consider the following lottery example.

⊕ *Example*

Suppose that you buy a lottery ticket with a friend and you keep it in your possession with the following promise:

If this ticket wins, then I will give you a million dollars.

Either your ticket wins or it does not win, and either you give your friend a million dollars or you do not give her a million dollars. Under which cases did you lie to your friend?

Case 1: If your ticket wins and you give your friend a million dollars, then your original statement was obviously true.

Case 2: If your ticket wins and you do not give your friend a million dollars, your friend will have cause for concern because your original statement was indeed a lie.

Case 3: Now suppose that your ticket does not win, but you give your friend a million dollars. Perhaps you won on another lottery ticket. Does that make your original statement a lie? Certainly not, because you did not make any claim as to what you would do if that particular ticket did not win. In this case, your original statement is true.

Case 4: Now suppose that your ticket does not win and you do not give your friend a million dollars, which is the most likely scenario. Will your friend say that you lied? Absolutely not. In this case, your original statement is true.

In your original statement you only said what would happen if you won the lottery. You made no commitment about what you would do if you did not win the lottery. So if you don't win the lottery, your statement puts no restrictions on you. You may give your friend a million dollars [Case 3] or not give her a million dollars [Case 4]. Either way your original statement is true. The only case that makes it false is Case 2.

---

p	q	$p \Rightarrow q$
T	T	T
T	F	F
F	T	T
F	F	T

Hopefully, this discussion has convinced you of the rationale for the truth values of an implication, which are summarized in the adjacent truth table. The first time around, most students write false for Case 3 and Case 4 when they fill in the truth table on page 35. It does sound strange unless we stop and think about it. If you stop someone on the street and ask them if a false statement implies a false statement, more than likely they will say no; however, as the previous example demonstrates, the logically correct answer is yes.

The implication is used so frequently in the reasoning process that we have to understand its exact meaning; we have to understand why Case 3 and Case 4 are true. If we understand the meaning of *or*, we can reason that $\sim p$ *or* q means the same as *if* p*, then* q. Most people would not question that $\sim p$ *or* q is true in both Case 3 and Case 4; consequently $p \Rightarrow q$ must be true in these two cases.

When we assert that p *implies* q is true, we are telling the listener that Case 2 cannot occur. Any of the other three cases could happen. When the hypothesis of an implication is false, the implication is automatically true. The following implication is true because its hypothesis is false:

If $2>3$, then $4>9$.

This sentence seems a bit absurd for it imparts no useful information to the reader; however, it occurs naturally in the context of the following sentence, which does sound very reasonable:

For every real number x, if $x>3$, then $x^2>9$.

Since the above sentence is true for every x, it must be true when we substitute 2 for x:

If $2 > 3$, then $4 > 9$.

q if p

In everyday language, implications are often expressed in other forms such as "q if p." For example, a client may tell the programmer:

"The customer gets a discount *if* the customer is over 60."

q if p *means the same as* $p \Rightarrow q$.

The "if" in the above sentence is flagging the if-part, so this sentence has the same meaning as the following implication.

"If the customer is over 60, then the customer gets a discount."

In general, "q, if p" has the same meaning as "if p, then q." The if-part of "q, if p" is the if-part of the implication.

p only if q

"Only if" is sometimes confused with "if," but they have different meanings. Suppose that a client tells the programmer:

"A customer gets a discount *only if* the customer is over 60."

p only if q *means the same as* $p \Rightarrow q$.

"Only if" means that we cannot have a case where the customer gets a discount and the customer is not over 60. In other words:

If a customer gets a discount,
then the customer must be over 60.

In general, a sentence of the form "p only if q" can be translated as the implication, "if p, then q."

⊕ *Example*

Rewrite each sentence as an implication.

1. $x>3$ only if $x>1$.

   Translation: If $x>3$, then $x>1$.

2. $x>1$ if $x>3$.

   Translation: If $x>3$, then $x>1$.

3. x is in A only if x is in B.

   Translation: If x is in A, then x is in B.

4. x is in A if x is in B.

   Translation: If x is in B, then x is in A.

If and only if

p if and only if q *means the same as* $p \Rightarrow q$ and $q \Rightarrow p$.

The sentence "p if and only if q" is an abbreviated way of saying:

$$p \text{ if } q \text{ and } p \text{ only if } q.$$

We can translate the two component sentences as follows:

$$q \Rightarrow p \text{ and } p \Rightarrow q.$$

So, the meaning of "p if and only q" is a two-way implication.

Example

Translate the given sentence in terms of implications.

1. $x \in A$ if and only if $x \in B$.

   Translation: $x \in A \Rightarrow x \in B$ and $x \in B \Rightarrow x \in A$.

2. $a < b$ if and only if $-a > -b$.

   Translation: If $a < b$, then $-a > -b$, and
   if $-a > -b$, then $a < b$.

Necessary and Sufficient

p is sufficient for q *means the same as* $p \Rightarrow q$.

The words "necessary" and "sufficient" have been associated with implications since Aristotle laid the foundations for logic in the 4th century B.C.E. Since they are often used in everyday language, we need to know how to translate them in terms of an implication. Suppose that a teacher says the following:

> Getting an A on the final is *sufficient*
> to get an A in the course.

This statement can be translated as follows:

> If you get an A on the final,
> then you get an A in the course.

In general, "p is sufficient for q" means "if p, then q."

On the other hand, suppose that a teacher says the following:

> q is necessary for p
> *means the same as*
> $p \Rightarrow q$.

Getting an A on the final is *necessary*
to get an A in the course.

This statement can be translated as follows:

If you get an A in the course,
then you must get an A on the final.

In general, "q is necessary for p" means "if p, then q."

Example

Translate the implication in terms of "necessary" and "sufficient."

1. If $x>3$, then $x>1$.

   $x>3$ is sufficient for $x>1$.
   $x>1$ is necessary for $x>3$.

2. If x is in A, then x is in B.

   x being in A is sufficient for x to be in B.
   x being in B is necessary for x to be in A.

> sufficient necessary
> $p \Rightarrow q$

The first part of an implication is the sufficient part and the second part is the necessary part:

$p \Rightarrow q$
p is sufficient for q
q is necessary for p

> $p \Rightarrow q$
> *means the same as*
> If p, then q.
> q, if p.
> p only if q.
> p is sufficient for q.
> q is necessary for p.

In the adjacent box, we have a summary of the various ways to translate an implication. In the reasoning process, we usually think in terms of implications, so when we run across one of the other forms, we usually translate it in terms of an implication. When we combine "necessary" and "sufficient" in the same sentence, we get a two-way implication which has the same meaning as "if and only if."

p is necessary and sufficient for q.
p is necessary for q and p is sufficient for q.
$q \Rightarrow p$ and $p \Rightarrow q$
p if and only if q

Equivalent Sentences

p	q	$p \wedge q$	$p \vee q$	$p \Rightarrow q$	$p \Leftrightarrow q$
T	T	T	T	T	T
T	F	F	T	F	F
F	T	F	T	T	F
F	F	F	F	T	T

For the sentence *p is equivalent to q* to be true, both component sentences must have the same truth value; they must both be true or both false. Let's compare the meaning of the four binary operators in the adjacent truth table.

$p \Leftrightarrow q$ is true	means	Case 1 or 4 must occur.
p and q is true	means	Case 1 must occur.
p or q is true	means	Case 4 cannot occur.
$p \Rightarrow q$ is true	means	Case 2 cannot occur.

Meaning & Cases

The meaning of a compound sentence is determined by the cases that can occur. If we say that $p \Rightarrow q$ is true, we are telling the listener that Case 1, 3, or 4 must occur.

---

r is *equivalent* to s
if and only if
they have the same truth values.

When we reason in a logical manner, we often need to translate sentences from one form to another form that has the same meaning. Truth tables give us a systematic way to determine whether or not compound sentences that are composed of the same component sentences have the same meaning.

Let r and s represent abstract compound sentences that are composed of the same component sentences. For example, we could let r: *p and q* and s: *q and p*. The abstract sentences r and s have the same meaning if and only if they have the same truth value in each possible case. So, if they do have the same meaning and we connect them with the equivalence operator, we will have a statement that is always true. Thus, the equivalence operator tells us when two sentences have the same meaning:

---

r has the same meaning as s
if and only if
r is equivalent to s.

---

r and s have the same meaning
if and only if
$r \Leftrightarrow s$ is always true.

p	q	$p \wedge q$	$q \wedge p$	$(p \wedge q) \Leftrightarrow (q \wedge p)$
T	T	T	T	T
T	F	F	F	T
F	T	F	F	T
F	F	F	F	T

To determine if *p and q* has the same meaning as *q and p*, we can construct a truth table for $(p \wedge q) \Leftrightarrow (q \wedge p)$. In the adjacent truth table, we label a column for each of the component sentences, then record the truth values for each of the four cases. In each case, $p \wedge q$ has the same truth value as $q \wedge p$. So, we have only trues in the last column. Thus, *p and q* is equivalent to *q and p*.

Since the meaning of a compound sentence is determined by its truth values, equivalent sentences have the same meaning. Consequently, they can be used interchangeably. On the following pages, we list the most frequently used forms of equivalent sentences.

Commutative Property

If an operation has the commutative property, the order in which the operation is performed does not matter. For all real numbers x and y, $x + y = y + x$, so addition is commutative. Since $5-3 \neq 3-5$, subtraction is not commutative.

p and *q* *is equivalent to* *q* and *p*.
p or *q* *is equivalent to* *q* or *p*.
$p \Leftrightarrow q$ *is equivalent to* $q \Leftrightarrow p$.

And is commutative: p and $q \Leftrightarrow q$ and p. When we write an and-sentence, the order of the component sentences does not matter. Neither does the order matter for an or-sentence. To determine if two sentences have the same truth values, it does not matter which one we list first, so the equivalence operator is also commutative.

The logical operators, *and*, *or*, and *is equivalent to*, are each commutative because the order in which we list the component sentences does not affect the meaning of the compound sentence. However, the order does affect the meaning in an implication.

Converse

When we reverse the order in an implication, the new sentence is called the *converse* of the original implication:

$q \Rightarrow p$ is the converse of $p \Rightarrow q$.

Implication: If $x \in A$, then $x \in B$.
Converse: If $x \in B$, then $x \in A$.

Compare the truth values of $p \Rightarrow q$ with the truth values of its converse $q \Rightarrow p$ in the adjacent truth table. To compute the truth values for $q \Rightarrow p$, we must figure out when the q-column implies the p-column:

p	q	$q \Rightarrow p$	$p \Rightarrow q$
T	T	T	T
T	F	T	F
F	T	F	T
F	F	T	T

Case 1: T⇒T, which is true.
Case 2: F⇒T, which is true.
Case 3: T⇒F, which is false.
Case 4: F⇒F, which is true.

Since $q \Rightarrow p$ does not have the same truth values as $p \Rightarrow q$, these two implications are not equivalent. When we say $p \Rightarrow q$, we are saying that Case 2 cannot occur, but when we say $q \Rightarrow p$, we are saying that Case 3 cannot occur. These implications have different meanings. We cannot use them interchangeably.

$p \Rightarrow q$
is not equivalent to
$q \Rightarrow p$

$q \Rightarrow p$ is not equivalent to $p \Rightarrow q$.

The implication is not a commutative operation. We must carefully note the order of the sentences in an implication. When we reverse their order, the meaning is changed.

Rephrasing an Implication

Suppose that you make the following statement to a friend:

If I finish my homework, then I am going to the movie.

Do you mean that if you do not finish your homework, then you are not going to the movie? Let's compare the structure of these two sentences:

Does $\sim p \Rightarrow \sim q$ have the same meaning as $p \Rightarrow q$?

p	q	$\sim p$	$\sim q$	$\sim p \Rightarrow \sim q$
T	T	F	F	T
T	F	F	T	T
F	T	T	F	F
F	F	T	T	T

Now that we have a systematic method, we can quickly answer this question without the excessive thinking that was needed when we first contemplated the meaning of $p \Rightarrow q$. All we have to do is construct a truth table and compare the truth values. In the adjacent truth table, we insert a column for $\sim p$ and a column for $\sim q$ to help us calculate the truth values in the last column. If we do these steps in our head, we might possibly make a mistake and we would have no written record to check our thinking. To compute the truth values in the last column, we ask the following question:

In which cases does the $\sim p$ column imply the $\sim q$ column?

Case 1: F$\Rightarrow$F, which is true.
Case 2: F$\Rightarrow$T, which is true.
Case 3: T$\Rightarrow$F, which is false.
Case 4: T$\Rightarrow$T, which is true.

$p \Rightarrow q$
is not equivalent to
$\sim p \Rightarrow \sim q$

The truth values in Case 3 and Case 4 are not the same as the truth values for $p \Rightarrow q$. So, these two implications do not have the same meaning.

$p \Rightarrow q$ does not have the same meaning as $\sim p \Rightarrow \sim q$.

✧ *Example*

Do the following sentences have the same meaning?

If $x \in A$, then $x \in B$.
If $x \notin A$, then $x \notin B$.

To answer this question, we do not contemplate the sets A and B, nor do we contemplate the meaning of "is an element of." The structure of the first sentence is $p \Rightarrow q$ and the structure of the second sentence is $\sim p \Rightarrow \sim q$. Since these abstract sentences are not equivalent, the above sentences do not have the same meaning.

The more logical operators in a sentence, the more complex the structure and the more likely we are to misread the sentence, unless we carefully consider the logical interplay. The meaning of a compound sentence is determined by its structure. If we replace the component sentences with p's and q's as in the last example, we can see the structure and not be sidetracked by the meaning of the component sentences.

Contrapositive

As we saw in the previous discussion, we cannot rephrase an implication by reversing the order of the sentences, nor by negating the first and second parts of the implication. Consider, though, what happens when we do both, reverse the order and also negate the component sentences. This form is called the *contrapositive* of the original implication.

p	q	$\sim q$	$\sim p$	$\sim q \Rightarrow \sim p$
T	T	F	F	T
T	F	T	F	F
F	T	F	T	T

$$\sim q \Rightarrow \sim p \text{ is the contrapositive of } p \Rightarrow q.$$

In the adjacent truth table, when we compute the truth values for $\sim q \Rightarrow \sim p$, we ask the following question:

In which cases does the $\sim q$ column imply the $\sim p$ column?

Case 1: F ⇒ F, which is true.
Case 2: T ⇒ F, which is false.
Case 3: F ⇒ T, which is true.
Case 4: T ⇒ T, which is true.

$p \Rightarrow q$
is equivalent to
$\sim q \Rightarrow \sim p$

In each of the four cases, the implication $\sim q \Rightarrow \sim p$ has the same truth value as the implication $p \Rightarrow q$. Therefore, these two sentences are equivalent.

$$p \Rightarrow q \text{ has the same meaning as } \sim q \Rightarrow \sim p.$$

At last, we have found a proper way to rephrase an implication. We often use this translation when we are making logical deductions. If we are trying to derive $p \Rightarrow q$, it may be easier to derive in its contrapositive form.

⊕ *Example*

Rephrase each implication in terms of its contrapositive.

1. If $x \in A$, then $x \in B$.

   Contrapositive: If $x \notin B$, then $x \notin A$.

2. If $x^2 \not> 4$, then $x \not> 2$.

   Contrapositive: If $x > 2$, then $x^2 > 4$.

3. If a quadrilateral is a rectangle, its diagonals are congruent.

   Contrapositive: If the diagonals of a quadrilateral are not congruent, then the figure is not a rectangle.

Common Errors

The two most common errors in rephrasing an implication are replacing it with its converse or with $\sim p \Rightarrow \sim q$:

$$p \Rightarrow q \quad \text{is not equivalent to} \quad q \Rightarrow p.$$

$$p \Rightarrow q \quad \text{is not equivalent to} \quad \sim p \Rightarrow \sim q.$$

If we say: If x is in A, then x is in B.
It does not mean: If x is not in A, then x is not in B.
Nor does it mean: If x is in B, then x is in A.

⌖ Example

Do any of the following sentences have the same meaning?

1. a. If x is even, then x^2 is even.
   b. If x^2 is even, then x is even.
   c. If x is not even, then x^2 is not even.
   d. If x^2 is not even, then x is not even.

   (d) is the contrapositive of (a), so they have the meaning.
   (c) is the contrapositive of (b), so they have the meaning.

2. a. If x is in A, then x is not in B.
   b. If x is in B, then x is not in A.
   c. If x is not in B, then x is in A.
   d. If x is not in A, then x is in B.

   (a) and (b) have the same meaning.
   (c) and (d) have the same meaning.

Rephrasing Or

An or-sentence is closely related to an implication. If we think about the meaning of *or*, it is not too difficult to see that *p or q* has the same meaning as $\sim p \Rightarrow q$. If p is false, then q has to be true. Let's verify this with a truth table. In which cases does the $\sim p$ column imply the q column?

p	q	$\sim p$	$\sim p \Rightarrow q$
T	T	F	T
T	F	F	T
F	T	T	T
F	F	T	F

Case 1: F⇒T, which is true.
Case 2: F⇒F, which is true.
Case 3: T⇒T, which is true.
Case 4: T⇒F, which is false.

Since the truth values for $\sim p \Rightarrow q$ are identical to the truth values for $p \vee q$, they are equivalent sentences. Even though these two sentences look different, they impart the same information to the reader.

p or q is equivalent to $\sim p \Rightarrow q$

p or q is equivalent to $\sim q \Rightarrow p$

$$p \text{ or } q \text{ has the same meaning as } \sim p \Rightarrow q.$$

The contrapositive of $\sim p \Rightarrow q$ is $\sim q \Rightarrow p$. Since $\sim q \Rightarrow p$ has the same meaning as $\sim p \Rightarrow q$, we have another way to translate an or-sentence.

$$p \text{ or } q \text{ has the same meaning as } \sim q \Rightarrow p.$$

Using either of the above equivalences, we can rephrase an or-sentence as an implication.

⊕ *Example*

Translate each sentence in terms of an implication

1. $x<2$ or $x>5$.

   Equivalent Forms: If $x \not< 2$, then $x>5$.

   If $x \not> 5$, then $x<2$.

2. $x \in A$ or $x \in B$.

   Equivalent Forms: If $x \notin A$, then $x \in B$.

   If $x \notin B$, then $x \in A$.

3. $x \notin A$ or $x \in B$.

   Equivalent Form: If $x \in A$, then $x \in B$.

The last example shows the technique for writing an implication as an or-statement. For all sentences p and q, the following is true:

$$p \text{ or } q \Leftrightarrow \sim p \Rightarrow q$$

So, the above equivalence is true if we substitute $\sim p$ for p:

$$\sim p \text{ or } q \Leftrightarrow \sim(\sim p) \Rightarrow q$$

$$\sim p \text{ or } q \Leftrightarrow p \Rightarrow q$$

We can use the above equivalence to translate an implication as an or-sentence.

Rephrasing an Equivalence

When both an implication and its converse are true, their conjunction is true: $(p \Rightarrow q) \wedge (q \Rightarrow p)$. Let's compare this sentence with the equivalence $p \Leftrightarrow q$.

In the adjacent truth table, we insert columns for $p \Rightarrow q$ and $q \Rightarrow p$ to help us compute the truth values for $(p \Rightarrow q) \wedge (q \Rightarrow p)$:

p	q	$p \Rightarrow q$	$q \Rightarrow p$	$(p \Rightarrow q) \wedge (q \Rightarrow p)$
T	T	T	T	T
T	F	F	T	F
F	T	T	F	F
F	F	T	T	T

Case 1: T⇒T, which is true.

Case 2: F⇒T, which is false.

Case 3: T⇒ F, which is false.

Case 4: T⇒T, which is true.

The truth values for $(p \Rightarrow q) \wedge (q \Rightarrow p)$ are the same as the truth values for $p \Leftrightarrow q$, so these two sentences have the same meaning. An equivalence can always be rephrased as a double implication, which is why we use a double-headed arrow for its symbol.

◈ *Example*

Rephrase each equivalence in terms of implications.

1. $x < 4 \Leftrightarrow -x > -4$

   Equivalent Form: $(x < 4 \Rightarrow -x > -4)$ and $(-x > -4 \Rightarrow x < 4)$.

2. $x \in A \Leftrightarrow x \in B$

   Equivalent Form: $(x \in A \Rightarrow x \in B)$ and $(x \in B \Rightarrow x \in A)$.

3. A triangle is isosceles ⇔ two of its angles are congruent.

   Equivalent Form: If a triangle is isosceles, then two of its angles are congruent, and if two of its angles are congruent, then the triangle is isosceles.

$p \Leftrightarrow q$
means the same as
$p \Rightarrow q$ and $q \Rightarrow p$.

$p \Leftrightarrow q$
means the same as
p is equivalent to q
p if and only if q
p implies q and q implies p
p is necessary and sufficient for q

As we saw earlier (page 42), $p \Rightarrow q$ and $q \Rightarrow p$ can be reworded as *p if and only if q* and as *p is necessary and sufficient for q.* So both of these forms can be used to translate an equivalence. The various ways for rephrasing $p \Leftrightarrow q$ are summarized in the adjacent box.

Definitions in mathematics are always worded in terms of an equivalence. The term being defined is equivalent to its definition, which means that they can be used interchangeably. For example, a triangle is isosceles *if and only if* two of its sides are congruent.

Example

Rephrase the following equivalence in various forms:

$x \in A \iff x \in B$

1. $x \in A$ is equivalent to $x \in B$.
2. $x \in A$ if and only if $x \in B$.
3. If $x \in A$, then $x \in B$, and if $x \in B$, then $x \in A$.
4. x being in A is necessary and sufficient for x to be in B.

We can also rephrase the above equivalence in terms of contrapositives:

If $x \notin B$, then $x \notin A$, and if $x \notin A$, then $x \notin B$.

Negating an And-Sentence

Is $\sim(p \wedge q)$ equivalent to $\sim p \wedge \sim q$? Instead of constructing a truth table, let's see if we can find a case in which they have different truth values. Suppose that p is false and q is true:

$\sim(p \wedge q)$	$\sim p \wedge \sim q$
~(F and T)	~F and ~T
~F	T and F
T	F

$\sim(p$ and $q)$ *is equivalent to* $\sim p$ or $\sim q$

Since these two sentences have different truth values for this particular case, they are not equivalent. Thus, we cannot negate an and-sentence by distributing the negation across the parentheses. Let's analyze how to properly negate it.

If $\sim(p \wedge q)$ is true, then $p \wedge q$ must be false, which means that either p is false or q is false, so $\sim p \vee \sim q$ must be true. Let's compare the truth values of these two sentences in the adjacent truth table. We compute $\sim(p \wedge q)$ from the column on its left. To compute the last column, we must first mentally compute the value of $\sim p$ and the value of $\sim q$. The truth values in the last two columns are identical, so $\sim(p \wedge q)$ is equivalent to $\sim p \vee \sim q$.

p	q	$p \wedge q$	$\sim(p \wedge q)$	$\sim p \vee \sim q$
T	T	T	F	F
T	F	F	T	T
F	T	F	T	T
F	F	F	T	T

$\sim(p \wedge q)$ has the same meaning as $\sim p \vee \sim q$.

This equivalence gives us a rule for negating an and-sentence. When we take a negation inside the parentheses of an and-sentence, we must change *and* to *or*.

⊕ Example

Negate each sentence.

1. $x \in A$ and $x \in B$.

   Negation: It is not true that $x \in A$ and $x \in B$.
   $x \notin A$ or $x \notin B$.

2. $1 < x < 4$

   Negation: It is not true that $1 < x < 4$.
   $\sim(1 < x \text{ and } x < 4)$
   $1 \not< x$ or $x \not< 4$.
   $x \leq 1$ or $4 \leq x$.

   It is easy to visualize the negation of the above sentence, which is illustrated in the adjacent sketch. However, students sometimes describe this set incorrectly. An element x of the indicated set in the illustration does not have the property that "$x \leq 1$ *and* $4 \leq x$." The property must be described in terms of *or*.

Negating an Or-Sentence

> $\sim(p \text{ or } q)$
> *is equivalent to*
> $(\sim p \text{ and } \sim q)$

What does the negation of an or-sentence mean? If $\sim(p \vee q)$ is true, then $p \vee q$ must be false, which means that both p and q must be false. In other words, $\sim p \wedge \sim q$ must be true. You are asked to demonstrate that $\sim(p \vee q)$ is equivalent to $\sim p \wedge \sim q$ in the next exercise set.

$\sim(p \vee q)$ has the same meaning as $\sim p \wedge \sim q$.

When we take a negation inside the parentheses of an or-sentence, we must change *or* to *and*.

⊕ Example

Negate each sentence.

1. $x \in A$ or $x \in B$

   Negation: It is not true that $x \in A$ or $x \in B$.
   $x \notin A$ and $x \notin B$.

2. $x < 2$ or $5 < x$

   Negation: It is not true that $x < 2$ or $5 < x$.
   $x \not< 2$ and $5 \not< x$
   $x \geq 2$ and $5 \geq x$
   $2 \leq x \leq 5$

The rules for negating *and* and *or* are known as DeMorgan's Laws in honor of the English mathematician Augustus DeMorgan, who formalized these rules in the 19th century. When we take a negation inside the parentheses of an and-sentence, we must change *and* to *or*. Similarly, when we take a negation inside the parentheses of an or-sentence, we must change *or* to *and*. A common error is to hurriedly distribute the negation and not change the connective:

$\sim(p \text{ and } q)$ is not equivalent to $\sim p$ and $\sim q$.

$\sim(p \text{ or } q)$ is not equivalent to $\sim p$ or $\sim q$.

Negating an Implication

Most statements that we try to prove in mathematics are phrased in terms of implications. To prove that an implication is not true, we must prove its negation is true. Consequently, we often have to negate implications.

$\sim(p \Rightarrow q)$
is equivalent to
p and $\sim q$

What does it mean to say that $p \Rightarrow q$ is false? The only case in which $p \Rightarrow q$ is false is when p is true and q is false. Consequently, $p \wedge \sim q$ must be true. When we negate an implication, it becomes an and-sentence:

$\sim(p \Rightarrow q)$ has the same meaning as $p \wedge \sim q$.

You are asked to justify the above equivalence with a truth table in (1) of the next exercise set. Some students have difficulty remembering how to negate an implication. If you understand that the only time an implication is false is when the first statement is true and the second statement is false, then you know the rule for negating an implication.

⊕ *Example*

Negate each sentence.

1. For every x, if $x<3$, then $x<1$.

   Negation: For some x, $x<3$ and $x \not< 1$.

2. For every x, if $x \in A$, then $x \in B$.

   Negation: There exists an x such that $x \in A$ and $x \notin B$.

3. For every quadrilateral F, if its opposite sides are congruent, then F is a rectangle.

   Negation: There exists a quadrilateral F such that its opposite sides are congruent and F is not a rectangle.

4. Let f be a function. If f is continuous, then f is differentiable.

   Negation: There exists a function f such that f is continuous and f is not differentiable.

5. For every real number x, if $|x-a|<\delta$, then $|f(x)-f(a)|<\varepsilon$.

   Negation: There exists a real number x such that $|x-a|<\delta$ and $|f(x)-f(a)|\nless\varepsilon$.

Negating an Equivalence

To negate an equivalence, we can use the rule for negating *and* and the rule for negating *implies*. First, we rephrase the equivalence as a double implication.

$$\sim(p \Leftrightarrow q)$$
$$\sim(p \Rightarrow q \text{ and } q \Rightarrow p)$$

The outside structure of the above sentence is the negation of an and-sentence, $\sim(r$ and $s)$, so we first apply the rule for negating *and*:

$$\sim(p \Rightarrow q) \text{ or } \sim(q \Rightarrow p)$$

$\sim(p \Leftrightarrow q)$

is equivalent to

$\sim(p \Rightarrow q)$ or $\sim(q \Rightarrow p)$

When an equivalence is false, one of the implications must be false, which gives us the adjacent rule for negating an equivalence. If we want to negate it further, we can now apply the rule for negating an implication:

$$(p \text{ and } \sim q) \text{ or } (q \text{ and } \sim p)$$

Instead of memorizing the two forms for negating an implication, we should be able to reason as follows:

- If a double implication is not true, at least one of the implications is not true: $\sim(p \Rightarrow q)$ or $\sim(q \Rightarrow p)$
- If two sentences are not equivalent, one must be true and the other one false: $(p$ and $\sim q)$ or $(q$ and $\sim p)$

◈ Example

Negate each sentence.

1. $(x \in A \Leftrightarrow x \in B)$

   Negation: $\sim(x \in A \Rightarrow x \in B)$ or $\sim(x \in B \Rightarrow x \in A)$

   $(x \in A$ and $x \notin B)$ or $(x \in B$ and $x \notin A)$

2. $(x<2 \Leftrightarrow x^2<4)$

   Negation: $\sim(x<2 \Rightarrow x^2<4)$ or $\sim(x^2<4 \Rightarrow x<2)$

   $(x<2 \text{ and } x^2 \geq 4)$ or $(x^2<4 \text{ and } x \geq 2)$

---

3 Component Sentences

The following examples have 3 component sentences, which we represent with p, q, and r.

Example

Are the following sentences equivalent?

$(p \text{ or } q)$ and r *is not equivalent to* p or $(q$ and $r)$

1. Is $(p$ *or* $q)$ *and* r equivalent to p *or* $(q$ *and* $r)$?

   If p, q and r are each true, both compound sentences will be true, but consider the case when p is true, q is true, and r is false:

$(p$ or $q)$ and r	p or $(q$ and $r)$
(T or T) and F	T or (T and F)
T and F	T or F
F	T

   Since the two compound sentences have different truth values in this case, they are not equivalent. They do not have the same meaning. When we use both *and* and *or* in a sentence, we must carefully consider where we place the parentheses for it makes a difference.

$(p \Rightarrow q) \Rightarrow r$ *is not equivalent to* $p \Rightarrow (q \Rightarrow r)$

2. Is $(p \Rightarrow q) \Rightarrow r$ equivalent to $p \Rightarrow (q \Rightarrow r)$?

   Consider the case when p is false, q is true, and r is false:

$(p \Rightarrow q) \Rightarrow r$	$p \Rightarrow (q \Rightarrow r)$
$(F \Rightarrow T) \Rightarrow F$	$F \Rightarrow (T \Rightarrow F)$
$T \Rightarrow F$	$F \Rightarrow F$
F	T

   These two compound sentences have different truth values in the above case, so they are not equivalent. If we use two implications in a sentence, we must carefully consider where to place the parentheses because it makes a difference in the meaning.

3. Is $(p$ *or* $q)$ *or* r equivalent to p *or* $(q$ *or* $r)$?

To show that two compound sentences are not equivalent, we need to produce only one case in which they have different truth values, as in the previous two examples. However, to show that two abstract sentences are equivalent, we must check all possible cases of the truth values. For a single statement p, there are 2 possibilities for its truth value; p must be either true or false. With two statements p and q, there are 4 possible cases for their truth values. With three statements p, q and r, the cases double again:

p	q	r	$p \vee q$	$(p \vee q) \vee r$	$q \vee r$	$p \vee (q \vee r)$
T	T	T	T	T	T	T
T	F	T	T	T	T	T
F	T	T	T	T	T	T
F	F	T	F	T	T	T
T	T	F	T	T	T	T
T	F	F	T	T	F	T
F	T	F	T	T	T	T
F	F	F	F	F	F	F

r could be true with each of the 4 cases for p and q.

r could be false with each of the 4 cases for p and q.

The most efficient way to check the 8 possible cases is to construct a truth table. In the adjacent truth table, we first list the four possible cases for p and q and then we list them again. In the third column we make r true for the first 4 cases and false for the last 4 cases.

We insert a column for $p \vee q$ and a column for $q \vee r$ to help us in our computations.

$(p$ or $q)$ or r
is equivalent to
p or $(q$ or $r)$

As we work our way through the four computations for each of the 8 cases, we find that $p \vee (q \vee r)$ always has the same truth value as $(p \vee q) \vee r$. So these two compound sentences are equivalent. When we use *or* twice in a sentence, it does not matter where we place the parentheses. Both ways have the same meaning.

Associative Property

An operation has the *associative* property if the grouping does not matter when the operation is applied twice. For all real numbers x, y and z, $(x+y)+z = x+(y+z)$, so addition is associative. On the other hand, $(8-3)-1 \neq 8-(3-1)$, so subtraction is not associative. As we saw in the last example, *or* is an associative operation:

$(p$ and $q)$ and r
is equivalent to
p and $(q$ and $r)$

$$(p \text{ or } q) \text{ or } r \Leftrightarrow p \text{ or } (q \text{ or } r)$$

Is the *and* operation associative? If $(p$ *and* $q)$ *and* r is true, then each of the three component sentences must be true. The same is true for p *and* $(q$ *and* $r)$. So, *and* is also associative.

$$(p \text{ and } q) \text{ and } r \Leftrightarrow p \text{ and } (q \text{ and } r)$$

We can omit the parentheses when we have two *and's* or two *or's* in a sentence since the parentheses do not affect the meaning of the sentence.

$$x \in A \text{ and } x \in B \text{ and } x \in C.$$
$$x \in A \text{ or } x \in B \text{ or } x \in C.$$

However, we cannot omit the parentheses in the following sentence because the implication is not associative.

$$x \in A \Rightarrow (x \in B \Rightarrow x \in C)$$

Distributive Property

In elementary school, we learned that multiplication distributes over addition:

$$p \times (q + r) = (p \times q) + (p \times r)$$

p	q	r	$p \wedge q$	$p \wedge r$	$(p \wedge q) \vee (p \wedge r)$	$q \vee r$	$p \wedge (q \vee r)$
T	T	T	T	T	T	T	T
T	F	T	F	T	T	T	T
F	T	T	F	F	T	T	T
F	F	T	F	F	T	T	T
T	T	F	T	F	T	T	T
T	F	F	F	F	T	F	T
F	T	F	F	F	T	T	T
F	F	F	F	F	F	F	F

If we substitute *and* for x and *or* for +, we can ask if *and* distributes over *or*. In other words, is the following sentence true?

$$p \text{ and } (q \text{ or } r) \iff (p \text{ and } q) \text{ or } (p \text{ and } r)$$

If we work our way through the five computations for each of the 8 cases, we find that $p \wedge (q \vee r)$ always has the same truth value as $(p \wedge q) \vee (p \wedge r)$. So these two compound sentences are equivalent.

Thus, *and* does distribute over *or*.

In a similar manner, we can use a truth table to demonstrate that *or* distributes over *and*.

$$p \text{ or } (q \text{ and } r) \iff (p \text{ or } q) \text{ and } (p \text{ or } r)$$

p and (q or r)
is equivalent to
(p and q) or (p and r)

p or (q and r)
is equivalent to
(p or q) and (p or r)

We can also reason as follows. If p is true, then both of the above compound sentences are true. On the other hand, if p is false, both sides are true only if q and r are both true. So they have the same truth values.

These two distributive properties give us an important tool for rephrasing sentences. If you have trouble remembering them, write the distributive property for multiplication over addition and make the appropriate substitutions.

$$p \times (q + r) = (p \times q) + (p \times r)$$
$$p \text{ and } (q \text{ or } r) \iff (p \text{ and } q) \text{ or } (p \text{ and } r)$$
$$p \text{ or } (q \text{ and } r) \iff (p \text{ or } q) \text{ and } (p \text{ or } r)$$

⊕ *Example* Use the distributive property to rephrase each sentence.

1. $x \in A$ and ($x \in B$ or $x \in C$)

   Equivalent Form: ($x \in A$ and $x \in B$) or ($x \in A$ and $x \in C$)

2. $x \in A$ or ($x \in B$ and $x \in C$)

   Equivalent Form: ($x \in A$ or $x \in B$) and ($x \in A$ or $x \in C$)

Exercise Set 1.3

1. Do any of the sentences have the same meaning as the first sentence? Justify your answer with a truth table.

a.	$p \Rightarrow q$	$q \Rightarrow p$	$\sim p \Rightarrow \sim q$	$\sim q \Rightarrow \sim p$
b.	$\sim(p \Rightarrow q)$	$\sim p \Rightarrow \sim q$	$\sim q \Rightarrow \sim p$	p and $\sim q$
c.	p or q	$\sim p \Rightarrow q$	$\sim q \Rightarrow p$	$p \Rightarrow q$
d.	$\sim(p$ or $q)$	$\sim p$ or q	$\sim p$ or $\sim q$	$\sim p$ and $\sim q$

2. State the contrapositive of each implication.

   a. If $\sim p$, then $\sim q$. c. If $x \in A$, then $x \in B$. e. If $\sim p$, then q.

   b. If q, then p. d. If $x \notin B$, then $x \notin A$ f. If $x < 2$, then $x < 3$.

3. Do any of the sentences have the same meaning as the first sentence?

   a. If $x \in A$, then $x \in B$.
      If $x \notin A$, then $x \notin B$.
      If $x \notin B$, then $x \notin A$.
      $x \notin A$ or $x \in B$.

   b. If $x \notin A$, then $x \in B$.
      If $x \in A$, then $x \notin B$.
      If $x \notin B$, then $x \in A$.
      $x \in A$ or $x \in B$.

4. Translate each sentence as an implication.

   a. $x < 2$ only if $x < 7$. c. r only if s. e. $x \in D$ only if $x \in C$.

   b. $x < 5$ if $x < 3$. d. r if s. f. $x \in D$ if $x \in C$.

5. Let $A = \{1,2,3\}$ and $B = \{1,2,3,4\}$. Is the given sentence true for every natural number x?

   a. If x is in A, then x is in B.

   b. x is in A if x is in B.

   c. x is in B if x is in A.

   d. x is in A only if x is in B.

   e. x is in B only if x is in A.

   f. x is in A and x is in B.

6. Translate each sentence as an implication.
   a. $x<4$ is necessary for $x<3$.
   b. $x<1$ is sufficient for $x<2$.
   c. x is in A is a sufficient condition for x to be in B.
   d. x is in A is a necessary condition for x to be in B.
7. Rephrase each or-sentence as an implication.
   a. r or s
   b. $\sim r$ or s
   c. x is in C or x is in D.
   d. x is not in C or x is in D.
8. Rephrase each equivalence in terms of implications.
   a. $x \in C \Leftrightarrow x \in D$ ($\in$ *means "is an element of."*)
   b. $x \in A \cap B$ if and only if $x \in A$ and $x \in B$. ($\cap$ *means "intersection."*)
   c. $x \in B$ is a necessary and sufficient condition for x to be in C.
9. Negate each sentence. Do not leave a negation as a prefix for a compound sentence.
   a. $3<x$ and $x<7$
   b. $x<3$ or $x>7$
   c. If $x \in A$, then $x \in B$.
   d. For every x, if $x<7$, then $x<3$.
   e. For every x, if $x \in C$, then $x \in B$.
   f. For every x, if $|x-1|<4$, then $|f(x)-f(1)|<3$.
10. Given two compound sentences that are formed from the same component sentences, what does it mean to say that the two sentences have the same meaning?
11. Without using truth tables, explain in your own words why:
   a. $\sim(p \Rightarrow q)$ has the same meaning as p *and* $\sim q$.
   b. p or q has the same meaning as $\sim p \Rightarrow q$.
12. Do the compound statements have the same meaning? If not, give a case in which they have different truth values.
   a. $(p \Rightarrow q) \Rightarrow r$ $\qquad$ $p \Rightarrow (q \Rightarrow r)$
   b. p or (q or r) $\qquad$ (p or q) or r
   c. p and (q or r) $\qquad$ (p and q) or r
13. If a compound statement has n component statements, how many cases are there for possible truth values? *Hint*: Answer the question for $n = 2$, $n = 3$, $n = 4$, $n = 5$. Look for a pattern.

14. In a computer, a bit can assume the value of 0 or 1. Using 1 and 0 for T and F, complete the adjacent truth table for AND, OR, and XOR.

    XOR represents the exclusive use of the word "or."

p	q	p AND q	p OR q	p XOR q

15. A bit string is a sequence of bits, such as 100011. Given two bit strings of the same length, define "bitwise AND" by applying AND to the bits that are in the same position. As illustrated in the adjacent box:

    Bitwise AND

    100011
    111010
    100010

    100011 AND 111010 = 100010.

    Bitwise OR and bitwise XOR are defined in an analogous manner. Compute the following:

    a. 11011 AND 10010
    b. 11011 OR 10010
    c. 11011 XOR 10010
    d. (1011 AND 1010) OR 0011

16. In Fuzzy Logic, we consider varying degrees of truth that are represented by numbers from 0 to 1. A statement with a truth value of 1 is 100% true, while a statement with a truth value of .8 is 80% true and a truth value of 0 means that the statement is completely false. For example, let p and q represent the following sentences: p: It is cloudy today. q: It is raining today.

    a. Suppose that the truth values of p and q are .8 and .4. What would you use for the truth value of $\sim p$? p AND q? p OR q?

    b. Let p and q be sentences in fuzzy logic. Generalize your above answers and make up definitions for the truth values of the following: $\sim p$ p AND q p OR q

    c. Use your definitions in part (b) to complete the adjacent table. Do your values agree with your values from exercise 14? If not, try to find new definitions that will generalize the standard meaning. *Hint*: You may want to consider minimums and maximums.

p	q	p AND q	p OR q
1	1		
1	0		
0	1		
0	0		

    d. Suppose that the truth value of p is .8 and the truth value of q is .4. Using your definitions from part (c), compute the truth values of the following: $\sim(p$ AND $q)$ $\sim p$ OR $\sim q$

    For all sentences p and q in fuzzy logic, will the above two compound sentences have the same truth value?

Activity 1.4

1. A function f is continuous at $x = a$ if there is no jump or break in the graph at $x = a$. The conversion of this simple visual image into a verbal form challenged mathematical thinkers for quite a long time. You see, a break in a graph could be "infinitely small" – perhaps only one point is missing. So the challenge in verbalizing this concept is to grab hold of the very elusive concept of "infinitely small" with words that we can logically manipulate. The following epsilon-delta definition of continuity is one of the most famous definitions in mathematics. This definition was introduced in the 1870s by Karl Weierstrass as a hands-on translation of an earlier definition by Augustin Cauchy in 1821. It is rather amazing to see how easily we can capture the essence of the "infinitely small" using only one implication and three quantifiers.

   f is continuous at $x = a$ if and only if the following is true:

   For every positive ε, there exists a positive δ such that for every x, if $|x - a| < \delta$, then $|f(x) - f(a)| < \varepsilon$.

   a. State what it means for f to *not* be continuous at $x = a$. Bring the negation completely inside so that you have a good working form for demonstrating that a function is not continuous at a particular point a. If you have mastered the rules we covered earlier, this task should be quite simple.

   b. Draw a picture of a function that is not continuous at $x = a$. Now pick a positive number ε that is less than $\frac{1}{2}$ of the vertical height of the jump in the graph at $(a, f(a))$. On the y-axis, sketch all the points y such that $|y - f(a)| < \varepsilon$.

   c. Now pick a positive δ. On the x-axis, sketch all the points x such that $|x - a| < \delta$. Is the following statement true for the δ that you picked?

   For every x, if $|x - a| < \delta$, then $|f(x) - f(a)| < \varepsilon$.

   If not, can you find a δ so that it is true?

   d. Explain why your statement in part (a) is true for the example that you drew in part (b).

2. In the following questions, first give your initial impression as to whether or not any of the pairs of statements have the same meaning. Then make up various examples for $p(x)$ and $q(x)$ and test your answers. For example, you could let $p(x)$ be $x > 3$. Then try to find a $q(x)$ so that one of the statements is true and the other one is false. Do the following statements have the same meaning?

a. $\forall x, p(x)$ and $q(x)$ $\quad$ $\forall x, p(x)$ and $\forall x, q(x)$
b. $\exists x, p(x)$ and $q(x)$ $\quad$ $\exists x, p(x)$ and $\exists x, q(x)$
c. $\forall x, p(x)$ or $q(x)$ $\quad$ $\forall x, p(x)$ or $\forall x, q(x)$
d. $\exists x, p(x)$ or $q(x)$ $\quad$ $\exists x, p(x)$ or $\exists x, q(x)$

≡ 1.4 Laws of Logic ≡

A *law of logic* is an abstract compound statement that is always true.

The laws of logic give us the basic rules for manipulating the seven basic terms, which are essential verbal skills for doing logical reasoning. A *law of logic* is an abstract compound statement that is always true, regardless of the truth values of its component statements. For example, *p or ~p* is always true, so it is a law of logic. On the other hand, *p or ~q* is not always true, so it is not a law of logic.

We can verify that an abstract statement is a law of logic by constructing its truth table. For example, the following truth table shows that $(p \Rightarrow q) \Leftrightarrow (\sim q \Rightarrow \sim p)$ is always true, so it is a law of logic.

p	q	$p \Rightarrow q$	$\sim q \Rightarrow \sim p$	$(p \Rightarrow q) \Leftrightarrow (\sim q \Rightarrow \sim p)$
T	T	T	T	T
T	F	F	F	T
F	T	T	T	T
F	F	T	T	T

We can also verify that an abstract statement is a law of logic by reasoning with the definitions of the logical operators, as illustrated in the following example.

⊕ *Example* Verify that $p \Rightarrow (p \text{ or } q)$ is a law of logic.

If p is true, then by the definition of *or*, *p or q* must be true. So, by the definition of an implication, we can conclude that $p \Rightarrow (p \text{ or } q)$ is always true.

We will now review the most frequently used laws of logic, summarized in the following categories:

- Laws for Logical Operators
- Laws for Quantifiers
- Laws for Valid Arguments

Laws for Logical Operators

The following list is a summary of the basic equivalences for manipulating the five logical operators. When an equivalence is a law of logic, we can use the statements on each side of the equivalence interchangeably.

Laws of Logic – *for Logical Operators*

Commutative:	p and q	$\Leftrightarrow$	q and p
	p or q	$\Leftrightarrow$	q or p
	$(p \Leftrightarrow q)$	$\Leftrightarrow$	$(q \Leftrightarrow p)$
Associative:	(p and q) and r	$\Leftrightarrow$	p and (q and r)
	(p or q) or r	$\Leftrightarrow$	p or (q or r)
Distributive:	p and (q or r)	$\Leftrightarrow$	(p and q) or (p and r)
	p or (q and r)	$\Leftrightarrow$	(p or q) and (p or r)
Rephrasing $\Rightarrow$:	$p \Rightarrow q$	$\Leftrightarrow$	$\sim q \Rightarrow \sim p$
Rephrasing Or:	p or q	$\Leftrightarrow$	$\sim p \Rightarrow q$
Rephrasing $\Leftrightarrow$:	$(p \Leftrightarrow q)$	$\Leftrightarrow$	$(p \Rightarrow q)$ and $(q \Rightarrow p)$
Negations:	$\sim(\sim p)$	$\Leftrightarrow$	p
	$\sim$(p and q)	$\Leftrightarrow$	$\sim p$ or $\sim q$
	$\sim$(p or q)	$\Leftrightarrow$	$\sim p$ and $\sim q$
	$\sim(p \Rightarrow q)$	$\Leftrightarrow$	p and $\sim q$
	$\sim(p \Leftrightarrow q)$	$\Leftrightarrow$	$\sim(p \Rightarrow q)$ or $\sim(q \Rightarrow p)$

Commutative Laws

A commutative law states that the order does not matter for a particular operation. *And*, *or*, and *is equivalent to* are commutative. *Implies* is not commutative.

$p \Rightarrow q$ is *not* equivalent to $q \Rightarrow p$.

Associative Laws

An associative law states that the grouping does not matter when the same operation is applied twice. *And* and *or* are associative, but *implies* is not:

$$p \Rightarrow (q \Rightarrow r) \quad \text{is } \textit{not} \text{ equivalent to} \quad (p \Rightarrow q) \Rightarrow r.$$

By associativity, we can omit the parentheses in the following sentences: $p \wedge q \wedge r \qquad p \vee q \vee r$

Distributive Laws

And distributes over *or* (and *or* distributes over *and*) in the same way that multiplication distributes over addition.

$$x \in A \text{ and } (x \in B \text{ or } x \in C)$$
$$(x \in A \text{ and } x \in B) \text{ or } (x \in A \text{ and } x \in C)$$

Rephrasing Implies

We often rephrase an implication in terms of its contrapositive. For example, the following sentences are equivalent:

If $x \in A$, then $x \in B$.
If $x \notin B$, then $x \notin A$.

Rephrasing Or

An or-statement can be rephrased as an implication.

$x \in A$ or $x \in B$
If $x \notin A$, then $x \in B$.

Rephrasing an Equivalence

An equivalence can be rephrased in terms of two implications.

$x \in A$ is equivalent to $x \in B$
$(x \in A \Rightarrow x \in B)$ and $(x \in B \Rightarrow x \in A)$.

Double Negation

The more negations in a sentence, the longer it takes for the brain to process. We should always eliminate double negations:

It is *not* true that $x \neq 3$.
$x = 3$

Negating And

When we negate an and-sentence, *and* must change to *or*.

It is not true that $x \in A$ and $x \in B$.
So, $x \notin A$ or $x \notin B$.

Negating Or

When we negate an or-sentence, *or* must change to *and*.

It is not true that $x \in A$ or $x \in B$.
So, $x \notin A$ and $x \notin B$.

Negating Implies

Students often make mistakes when negating implications. To remember this frequently used rule, recall the truth values for *implies.* If $p \Rightarrow q$ is false, Case 2 must occur, which means that p is true and q is false.

It is not true that $x \in A \Rightarrow x \in B$.
So, $x \in A$ and $x \notin B$.

The above negation is not equivalent to $x \notin A \Rightarrow x \notin B$.

Negating an Equivalence

An equivalence can be rephrased as a double implication, so if an equivalence is false, one of the implications must be false.

It is not true that $x \in A \Leftrightarrow x \in B$.
$\sim(x \in A \Rightarrow x \in B)$ or $\sim(x \in B \Rightarrow x \in A)$.

Laws for Quantifiers

Each of the preceding equivalences is also valid if we replace the abstract statements p and q by open statements $p(x)$ and $q(x)$. For example, we negate an implication that has variables in the same way as we negate a statement without variables.

$$\sim(p(x) \Rightarrow q(x)) \Leftrightarrow p(x) \text{ and } \sim q(x)$$

In addition to the previous laws, we also have the following laws for quantified statements.

Laws of Logic – *for Quantifiers*

Negations:
$$\sim(\forall x, p(x)) \Leftrightarrow \exists x, \sim p(x)$$
$$\sim(\exists x, p(x)) \Leftrightarrow \forall x, \sim p(x)$$

Distributive:
$$\forall x, p(x) \text{ and } q(x) \Leftrightarrow \forall x, p(x) \text{ and } \forall x, q(x)$$
$$\exists x, p(x) \text{ or } q(x) \Leftrightarrow \exists x, p(x) \text{ or } \exists x, q(x)$$

Unlike the previous laws, the above laws cannot be verified with a truth table since $p(x)$ changes as x changes. We can, though, reason with the meaning of the quantifiers to verify that they are always true.

Negating Quantifiers

In Section 1.2, we discussed how to properly negate a quantified statement. When we bring a negation across a quantifier, wè must change the quantifier.

It is not true that for every x, $x \in A$.
There exists an x such that $x \notin A$.

It is not true that there exists an x such that $x \in A$.
For every x, $x \notin A$.

Distributing the Quantifiers

The rules for interfacing the quantifiers with the logical operators are not used as frequently as the negations, but they are often used incorrectly. The universal quantifier can be distributed across *and*, but not across *or*, whereas the existential quantifier can be distributed across *or*, but not across *and*. Let's examine why this happens.

The universal quantifier enjoys a special relationship with the and-connective; it gives us a way to express *and* in a more concise form when the component sentences have a pattern, as illustrated in the following example.

Example

Consider the statement: For all n, $n < n+1$.

1. If the domain for n is $\{1,2,3\}$, this statement means:

   $1 < 2$ and $2 < 3$ and $3 < 4$.

2. If the domain for n is $\{1,2,3, \ldots\}$, this statement means:

   $1 < 2$ and $2 < 3$ and $3 < 4$ and $4 < 5$ and $\ldots$

3. If the domain for n is the set of real numbers, the statement can only be represented with the universal quantifier.

As illustrated in the previous example, the universal quantifier is a generalization of *and*. Consequently, we can distribute a universal quantifier across an and-statement.

Example

The following sentences have the same meaning:

$\forall x$, x is in A and x is in B.

($\forall x$, x is in A) and ($\forall x$, x is in B).

To verify that the adjacent rule is always true, we will first show that the left side of the equivalence implies the right side, and then show that the right side implies the left side.

$\forall x,\ p(x)$ and $q(x)$
is equivalent to
$\forall x,\ p(x)$ and $\forall x,\ q(x)$.

$\Rightarrow$ Suppose that $\forall x,\ p(x)$ *and* $q(x)$ is true.
Then for every x, both $p(x)$ and $q(x)$ are true.
So, $\forall x,\ p(x)$ is true. Also, $\forall x,\ q(x)$ is true.
Thus, $\forall x,\ p(x)$ *and* $\forall x,\ q(x)$ is true.

$\Leftarrow$ Conversely, suppose that $\forall x,\ p(x)$ *and* $\forall x,\ q(x)$ is true. Since $p(x)$ is true for all x and $q(x)$ is true for all x, $p(x)$ *and* $q(x)$ is true for all x.
Thus, $\forall x,\ p(x)$ *and* $q(x)$ is true.

Both implications are true, so the two sides are equivalent. Since we can distribute the universal quantifier across an and-statement, we might be tempted to say that we can also distribute it across an or-statement. However, consider the following example.

⊕ *Example*

Let x be a natural number. The following statement is true:

$$\forall x,\ x \text{ is even or } x \text{ is odd.}$$

However, if we distribute the universal quantifier across the *or*, we produce a false statement:

$$(\forall x,\ x \text{ is even}) \text{ or } (\forall x,\ x \text{ is odd}).$$

$\forall x,\ p(x)$ or $q(x)$
is not equivalent to
$\forall x,\ p(x)$ or $\forall x,\ q(x)$.

The above example shows that $\forall x,\ p(x)$ *or* $q(x)$ does not have the same meaning as $\forall x,\ p(x)$ *or* $\forall x,\ q(x)$. The assumption that they are equivalent is a serious reasoning error.

The existential quantifier has the same relation to *or* as the universal quantifier does to *and*.

⊕ *Example*

Consider the statement: There exists an x such that $x \in A_n$.

1. If the domain for n is $\{1,2,3\}$, this statement means:

$$x \in A_1 \text{ or } x \in A_2 \text{ or } x \in A_3$$

2. If the domain for n is $\{1,2,3,\ldots\}$, this statement means:

$$x \in A_1 \text{ or } x \in A_2 \text{ or } x \in A_3 \text{ or } x \in A_4 \text{ or } \ldots$$

3. If the domain for n is the set of real numbers, the above statement can only be represented with the existential quantifier.

As demonstrated in the above example, the existential quantifier is a generalization of *or*. Consequently, we can always distribute an existential quantifier across an or-statement.

Example

The following sentences have the same meaning:

$\exists x$, x is in A or x is in B.

($\exists x$, x is in A) or ($\exists x$, x is in B).

$\exists x, p(x)$ or $q(x)$
is equivalent to
$\exists x, p(x)$ or $\exists x, q(x)$.

To verify the adjacent rule, we can argue as follows:

⇒ Assume $\exists x$, $p(x)$ *or* $q(x)$.
Then there exists an x_o such that $p(x_o)$ is true or $q(x_o)$ is true.
So, either $\exists x$, $p(x)$ is true or $\exists x$, $q(x)$ is true.
Thus, ($\exists x$, $p(x)$ *or* $\exists x$, $q(x)$ is true.

⇐ Conversely, assume $\exists x$, $p(x)$ *or* $\exists x$, $q(x)$.
Case 1: Suppose that $\exists x$, $p(x)$ is true.
Then there is an x_o such that $p(x_o)$ is true.
Since $p(x_o)$ is true, $p(x_o)$ *or* $q(x_o)$ is true.
So, $\exists x$, $p(x)$ *or* $q(x)$ is true.
Case 2: Suppose that $\exists x$, $q(x)$ is true.
Then there is an x_o such that $q(x_o)$ is true.
Since $q(x_o)$ is true, $p(x_o)$ *or* $q(x_o)$ is true.
So, $\exists x$, $p(x)$ *or* $q(x)$ is true.
Since one of the above two cases must occur, $\exists x$, $p(x)$ *or* $q(x)$ is true.

Since both implications are true, the two sides are equivalent. Because of this equivalence, we might be tempted to say that the existential quantifier distributes across the and-connective. However, consider the following example.

Example

Let x be a natural number. The following statement is false:

$$\exists x,\ x+1=5 \text{ and } x+2=5.$$

However, if we distribute the existential quantifier across the *and*, we produce a true statement:

$$(\exists x,\ x+1=5) \text{ and } (\exists x,\ x+2=5).$$

$\exists x, p(x)$ and $q(x)$ *is not equivalent to* $\exists x, p(x)$ and $\exists x, q(x)$.

The previous example shows that $\exists x$, $p(x)$ *and* $q(x)$ does not have the same meaning as $\exists x$, $p(x)$ *and* $\exists x$, $q(x)$. Students sometimes make the mistake of assuming that these two statements are equivalent. Using subscripts with an existential quantifier helps us avoid this type of reasoning error.

In the following example, note how we translate the two existential quantifiers in terms of subscripts, being careful to change the subscript for the second existential quantifier.

Example

Suppose that the following statement is true:

$$\exists x,\ x\in A \text{ and } \exists x,\ x\in B$$

Then there exists an x_0 such that $x_0 \in A$.
Also, there exists an x_1 such that $x_1 \in B$.

We cannot drop the subscripts and say $x\in A$ *and* $x \in B$ for that would imply that the same x is in both A and B.

We should carefully consider the meaning whenever we contemplate whether or not to distribute a quantifier:

> The existential quantifier does not distribute across the and-connective.
>
> The universal quantifier does not distributes across the or-connective.

When we apply more than one law of logic, we work from left to right and apply the laws one step at a time, as illustrated in the following example. These examples illustrate the step-by-step approach that we use when we reason in a logical manner. We may not write all these steps on paper, but we mentally execute them.

⊕ *Example*

Translate the following negations:

1. $\sim(\forall x,\ x \in A \text{ or } x \in B)$
   $\exists x,\ \sim(x \in A \text{ or } x \in B)$
   $\exists x,\ x \notin A \text{ and } x \notin B$

2. $\sim(\forall \in \exists \delta\ \forall x,\ |x-5| < \delta \Rightarrow |f(x)-f(5)| < \in)$
   $\exists \in \sim(\exists \delta\ \forall x,\ |x-5| < \delta \Rightarrow |f(x)-f(5)| < \in)$
   $\exists \in \forall \delta \sim(\forall x,\ |x-5| < \delta \Rightarrow |f(x)-f(5)| < \in)$
   $\exists \in \forall \delta\ \exists x,\ |x-5| < \delta \text{ and } |f(x)-f(5)| \not< \in$

Laws for Valid Arguments

The following laws of logic form the basis for the standard types of valid arguments. They can be verified with truth tables or by reasoning from the definitions of the logical operators.

Laws of Logic – *for Valid Arguments*

Law of the Excluded Middle:	p or $\sim p$
Law of Noncontradiction:	$\sim(p$ and $\sim p)$
Rule of Detachment:	$(p$ and $(p \Rightarrow q)) \Rightarrow q$
Transitive Law:	$((p \Rightarrow q)$ and $(q \Rightarrow r)) \Rightarrow (p \Rightarrow r)$
Expanding Or:	$p \Rightarrow (p$ or $q)$
Contracting And:	$(p$ and $q) \Rightarrow p$
Simplifications:	$(p$ and T$) \Leftrightarrow p$
	$(p$ or T$) \Leftrightarrow$ T
	$(p$ or F$) \Leftrightarrow p$

T represents a statement that is always true.
F represents a statement that is always false.

---

p or ~*p*

The Law of the Excluded Middle states that there is no middle ground for truth; a statement is either true or false, which corresponds to our definition of a statement. We sometimes use this law when we do cases in a proof. For example, to prove a statement about a real number x, either $x \geq 0$ or $\sim(x \geq 0)$. We can then split our proof into the case where $x \geq 0$ and the case where $x \ngeq 0$.

$\sim(p \text{ and } \sim p)$

A statement cannot be both true and false, which is stated in this rule. The Law of Noncontradiction forms the basis for a proof by contradiction, which we will examine in Section 2.6. Using the rule for negating an and-statement, note how this law turns into the previous law:

$$\sim(p \text{ and } \sim p) \Leftrightarrow (\sim p \text{ or } p)$$

$(p \text{ and } (p \Rightarrow q)) \Rightarrow q$

The Rule of Detachment, also called *modus ponens*, comes from the definition of the implication. If p is true and $p \Rightarrow q$ is true, then q must be true. Thus, the outside implication is always true.

$((p \Rightarrow q) \text{ and } (q \Rightarrow r)) \Rightarrow (p \Rightarrow r)$

The Transitive Law may be easier to remember if we rephrase it as follows: if the first implies the second and the second implies the third, then the first implies the third.

$$\text{If } (x \in A \Rightarrow x \in B) \text{ and } (x \in B \Rightarrow x \in C), \text{ then } (x \in A \Rightarrow x \in C).$$

We can verify that this law is always true using either a Truth Table or by reasoning from the definition of an implication (page 124).

$p \Rightarrow (p \text{ or } q)$

Given any true statement, we can expand it into an or-statement, attaching whatever we wish to it. If p is true, then p or q must be true, so the adjacent implication is always true.

$$(x \in A) \Rightarrow (x \in A \text{ or } x \in B)$$

$(p \text{ and } q) \Rightarrow p$

Given any true and-statement, we can always contract it to either of its component statements. If p *and* q is true, then p must be true, so the adjacent implication is always true.

$$(x \in A \text{ and } x \in B) \Rightarrow (x \in A)$$

$(p \text{ and T}) \Leftrightarrow p$

$(p \text{ or T}) \Leftrightarrow \text{T}$

$(p \text{ or F}) \Leftrightarrow p$

The simplification laws allow us to simplify compound statements when we know one of the component statements is always true or always false. Let T represent a compound statement that is always true and F represent a compound statement that is always false.

The truth value of p *and* T is the same as the truth value of p.

The truth value of p *or* T is always true.

The truth value of p *or* F is the same as the truth value of p.

For example:

$$p \text{ and } (q \text{ or } \sim q) \Leftrightarrow p$$
$$p \text{ or } (q \text{ or } \sim q) \Leftrightarrow \mathsf{T}$$
$$p \text{ or } (q \text{ and } \sim q) \Leftrightarrow p$$

Abstract Structure

In order to apply the laws of logic to a specific compound sentence, we must be able to recognize its abstract structure. Students often make errors in translating $x \notin A \cup B$ because they forget that the slash is a negation prefix and do not see the outside structure of the sentence. If we rewrite this sentence in an equivalent form without the slash, as illustrated in the following example, we can easily see which law of logic we need to apply.

⌖ *Example*

Does $x \notin A \cup B$ have the same meaning as $x \notin A$ *or* $x \notin B$?

We translate the original statement, one step at a time.

$x \notin A \cup B$

$\sim(x \in A \cup B)$

$\sim(x \in A \text{ or } x \in B)$ .... *Definition of union*

$\sim(x \in A) \text{ and } \sim(x \in B)$ ...... *Law of logic*

$x \notin A \text{ and } x \notin B$

So, $x \notin A \cup B$ does not have the same meaning as $x \notin A$ *or* $x \notin B$.

When we need to apply two laws of logic, the outside structure of the sentence determines which rule to apply first, as illustrated in the following examples.

⌖ *Example*

Translate the given sentence.

1. ~(For all x, $x \in A$ or $x \in B$.)

   Since the outside structure is the negation of a quantified sentence, we first apply the rule for negating a quantifier:

   There exists an x such that $\sim(x \in A$ or $x \in B)$.

   Next, we apply the rule for negating an or-sentence:

   There exists an x such that $x \notin A$ and $x \notin B$.

2. ~(For all x, $x \in A$ or for all x, $x \in B$.)

   Since the outside structure is the negation of an or-statement, we first apply the rule for negating *or*:

   ~(For all x, $x \in A$) and ~(for all x, $x \in B$).

   Next, we apply the rule for negating "for all."

   There exists an x such that $x \notin A$ and there exists an x such that $x \notin B$.

⊕ *Example*

Are the following sentences equivalent?

If x is not in A, then x is in B.
If x is not in B, then x is in A.

We can view the first sentence as having a structure of $p \Rightarrow q$.

p: x is not in A *q: x is in B*

Then the second sentence has the form $\sim q \Rightarrow \sim p$. Since the second sentence is the contrapositive of the first, these two sentences are equivalent.

To see the outside structure of a sentence, we may sometimes want to hide a negation, as illustrated in the above example.

Exercise Set 1.4

1. Demonstrate that the following statements are laws of logic.
   a. $(p \Rightarrow q) \Leftrightarrow (\sim q \Rightarrow \sim p)$
   b. $\sim(p \Rightarrow q) \Leftrightarrow p$ and $\sim q$
   c. $((p \Rightarrow q)$ and $(q \Rightarrow r)) \Rightarrow (p \Rightarrow r)$
2. Do any of the following sentences have the same meaning? (*Use p and q to compare their structures.*)
   a. If x is in C, then x is in B.
   b. If x is in B, then x is in C.
   c. If x is not in C, then x is not in B.
   d. If x is not in B, then x is not in C.

3. Rephrase each implication in terms of its contrapositive.
   a. If x is in C, then x is in A.
   b. If x is not in B, then x is in C.
4. Rephrase each or-sentence as an implication.
   a. x is in A or x is in B.
   b. $x \geq 0$ or $x < 0$.
   c. y is not in $f(A)$ or y is in $f(B)$.
5. Take the negation inside the parentheses and simplify.
   a. $\sim(x \in A \Rightarrow x \in B)$
   b. $\sim(x \in C \Rightarrow x \notin A)$
   c. $\sim(z \notin A$ and $z \notin B)$
   d. $\sim(x \notin C$ or $x \in D)$
   e. $\sim$(If $|x-2| < \delta$, then $|f(x)-f(2)| < \varepsilon$.)
   f. $\sim(x \in D$ if and only if $x \in B)$
6. Negate each sentence. Do not leave a negation as a prefix.
   a. There exists an x such that for every y, $x+y = 1$.
   b. For every y, there exists an x such that $f(x) = y$.
   c. For every x, if x is in B, then x is in C.
   d. For every x, if x is not in B, then x is not in C.
   e. For every $\in$, there exists a δ such that for every x, if $|x-2| < \delta$, then $|f(x)-f(2)| < \in$.
   f. There exists a c such that for every x, if x is in S, then $x \leq$ c.
   g. For every x, if x is in C, then x is in A or x is in B.
7. Write the negation of each sentence.
   a. $z \in X$, and, $z \in Y$ or $z \in Z$
   b. For all x, $x \in C$ or $x \notin D$.
   c. For all x, $x \in C$ or for all x, $x \notin D$.
   d. There exists an x such that x is in A and x is in B.
   e. There exists an x such that x is in A and there exists an x such that x is in B.
8. Translate each implication in terms of its contrapositive. Do not leave a negation as a prefix.
   a. If x is rational and y is irrational, then $x+y$ is irrational.
   b. If x is an integer, then x is even or x is odd.
   c. If $c > 0$, then $a < b \Rightarrow ca < cb$.

9. Is the given implication a law of logic? If not, give a case in which the statement is false.

   a. $(p \text{ and } q) \Rightarrow p$
   b. $(p \text{ or } q) \Rightarrow p$
   c. $p \Rightarrow (p \text{ or } q)$
   d. $p \Rightarrow (p \text{ and } q)$
   e. $(p \text{ and } q) \Rightarrow (p \text{ or } q)$
   f. $(p \text{ or } q) \Rightarrow (p \text{ and } q)$

10. Use the distributive laws to rewrite each sentence in an equivalent form.

    a. $x \in C$ and $(x \in A \text{ or } x \in B)$.
    b. $x \in B$ or $(x \in C \text{ and } x \in A)$.

11. Let $p(x)$: $x+3=5$, and $q(x)$: $7x=28$, where x is a real number. Do the following statements have the same truth value?

    $\exists x, p(x) \text{ and } q(x)$ $\quad$ $\exists x, p(x)$ and $\exists x, q(x)$

12. Let $p(x)$: $x>0$, $q(x)$: $x \leq 0$, where x is a real number. Do the given statements have the same truth value?

    a. $\exists x, p(x) \text{ and } q(x)$ $\quad$ $\exists x, p(x)$ and $\exists x.\ q(x)$
    b. $\forall x, p(x) \text{ or } q(x)$ $\quad$ $\forall x, p(x)$ or $\forall x, q(x)$

13. The domain for x is the set of real numbers. Do the given statements have the same meaning? If not, give examples of sets A and B so that the two statements have different truth values.

    a. There exists an x such that x is in A and x is in B.
       There exists an x such that x is in A and there exists an x such that x is in B.
    b. There exists an x such that x is in A or x is in B.
       There exists an x such that x is in A or there exists an x such that x is in B.
    c. For all x, x is in A and x is in B.
       For all x, x is in A and for all x, x is in B.
    d. For all x, x is in A or x is in B.
       For all x, x is in A or for all x, x is in B.

14. Do you think the given implication is a law of logic? If so, explain your reasoning. If not, explain why not.

    a. $\exists x, p(x)$ and $\exists x, q(x) \Rightarrow \exists x, p(x) \text{ and } q(x)$
       $\exists x, p(x) \text{ and } q(x) \Rightarrow \exists x, p(x)$ and $\exists x, q(x)$
    d. $\forall x, p(x)$ or $\forall x, q(x) \Rightarrow \forall x, p(x) \text{ or } q(x)$
       $\forall x, p(x) \text{ or } q(x) \Rightarrow \forall x, p(x)$ or $\forall x, q(x)$

15. Which of the equivalences in this section are the hardest for you to remember? Explain why they are true.

Activity 1.5

Given a logical expression built from p, q, and r, we can construct its truth table. Let's do the reverse process now. For each of the following truth tables, construct a logical expression that has the truth values in the last column.

1.

p	q	?
T	T	F
T	F	T
F	T	T
F	F	T

2.

p	q	?
T	T	F
T	F	F
F	T	F
F	F	T

3.

p	q	?
T	T	T
T	F	T
F	T	F
F	F	T

4.

p	q	?
T	T	F
T	F	F
F	T	T
F	F	F

5.

p	q	?
T	T	T
T	F	T
F	T	F
F	F	F

6.

p	q	r	?
T	T	T	T
T	F	T	T
F	T	T	F
F	F	T	F
T	T	F	T
T	F	F	F
F	T	F	F
F	F	F	F

≡ 1.5 Logic Circuits ≡

The rapid processing that goes on in a computer is handled through a complex network of electronic circuits where the flow and manipulation of information is controlled by electronic devices. In 1938, while working on a problem of designing electronic circuits to meet given specifications, Claude Shannon, a student at M.I.T., noticed an underlying relation between certain types of electrical circuits and the logical operators. By merging the two, he created a powerful tool for designing electronic circuits.

AND-Gates

In the adjacent sketch of an electric circuit, we have two switches labeled p and q arranged in series on the top, a battery on the left side of the circuit and a light bulb on the right side. The four possible cases for the positions of the two switches are summarized in the following table. C represents closed and O represents open. When both switches are closed, the current flows, so the light will be on. In the other three cases, the current will not flow through the circuit, so the light will be off. Notice the similarity of the structure of this table with the truth table for *and*. For this reason, this type of series switch was named an AND-gate.

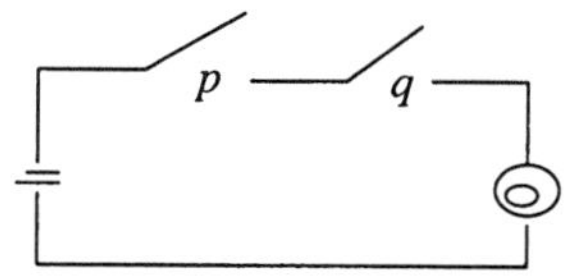

p	q	Light
C	C	On
C	O	Off
O	C	Off
O	O	Off

p	q	p and q
1	1	1
1	0	0
0	1	0
0	0	0

In the AND-table, the first two columns represent the input values for p and q, and the last column represents the output values of the circuit. 1 represents closed (or true) and 0 represents open (or false). If 1 and 0 are input, the AND-gate outputs 0.

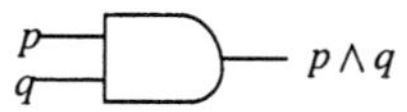

The symbol for an AND-gate is given on the left. The letters on the left side of the gate represent the input. We view the current as flowing from left to right with an output of $p \wedge q$. In circuit theory, $p \wedge q$ is written as pq; $p \vee q$ is written as $p + q$. However, to help reinforce the laws of logic studied in this section, we will continue to use $\vee$ and $\wedge$. In modern technology, series switches have been replaced by electronic devices which have the same net result for input and output.

OR-Gates

In the adjacent circuit, we have two switches in parallel. With this type of circuit, we need only one of the switches closed in order for current to flow through the light. Because of the similarity between the following table for a parallel circuit and the truth table for *or*, this type of circuit is called an OR-gate.

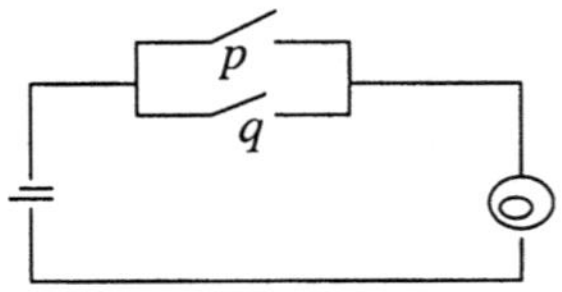

p	*q*	Light
C	C	On
C	O	On
O	C	On
O	O	Off

p	*q*	*p* or *q*
1	1	1
1	0	1
0	1	1
0	0	0

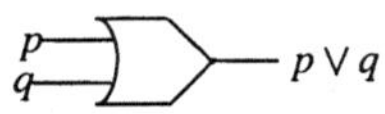

The symbol for an OR-gate is given on the left. The input values are output as $p \vee q$.

NOT-Gates

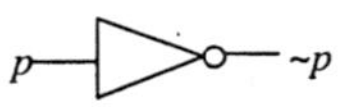

Unlike the previous two gates, a NOT-gate has only one input value; it changes an input of 1 to 0 and an input of 0 to 1, which is analogous to the impact of negation on a statement. We represent a NOT-gate symbolically as illustrated on the left. A NOT-gate is also called an *inverter*.

Combinatorial Circuits

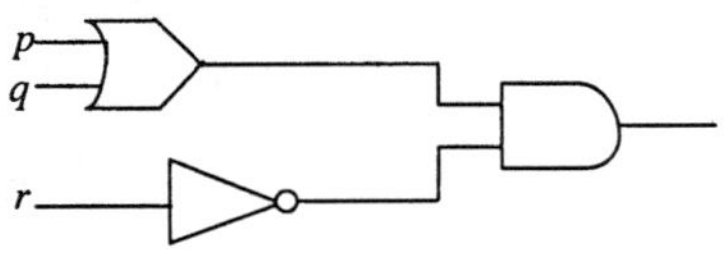

A combinatorial circuit consists of gates combined in various ways, such as the circuit illustrated on the left. When we input values for *p*, *q*, and *r*, the circuit outputs a value on the far right. To compute the output value of the total circuit, we trace through the circuit from left to right, computing the output that each gate has from its input.

For example, suppose that we input 1 for *p*, 0 for *q*, and 0 for *r*. The OR-gate turns "1 or 0" into 1; the NOT-gate turns 0 into 1. When we reach the AND-gate, the input is "1 and 1", which is output as 1. Thus, for the given values of *p*, *q*, and *r*, the circuit outputs 1.

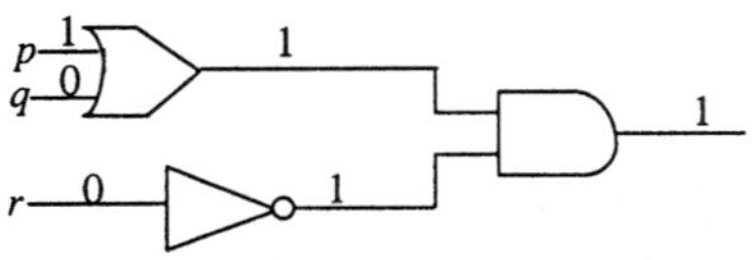

We can automate our work in computing the output of a circuit by representing it with a logical expression. First, we start with the original input and write the logical expression determined by each gate, as illustrated below. The OR-gate outputs *p or q*, the NOT-gate outputs ~*r*, and, the AND-gate takes its input and outputs (*p* or *q*) *and* ~*r*.

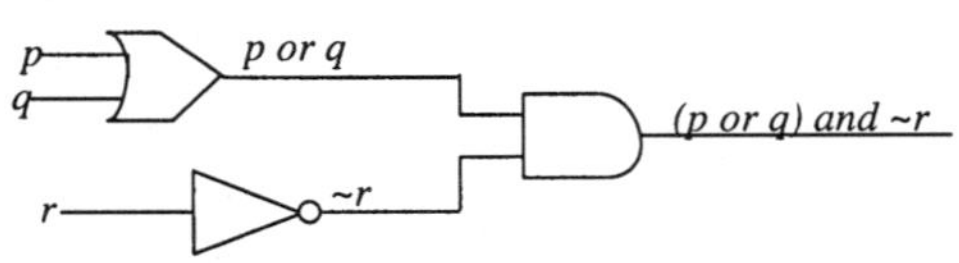

p	q	r	$p \vee q$	$(p \vee q) \wedge \sim r$
1	1	1	1	0
1	0	1	1	0
0	1	1	1	0
0	0	1	0	0
1	1	0	1	1
1	0	0	1	1
0	1	0	1	1
0	0	0	0	0

A truth table provides a systematic way to record the complete information on the behavior of a combinatorial circuit. The adjacent truth table for (p *or* q) *and* $\sim r$ lists the circuit's output for each possible input for p, q, and r. Our previous computation where we traced an input of 1, 0 and 0 through the circuit corresponds to Case 6 in the table. To get the output for an input of 1 for p, 1 for q and 1 for r, we look at Case 1 in the table, which gives an output of 0.

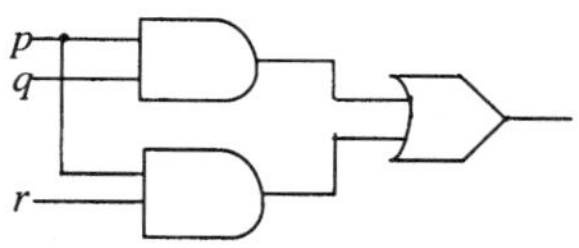

In a combinatorial circuit, we sometimes split a line to feed into more than one gate. In the adjacent sketch, the input for p feeds into two AND-gates. The dot on the p-line indicates that the branch line is connected to the p-line; this line crosses over the q-line, but the absence of a dot means that it is not connected to the q-line. In our drawings, we can branch a line into two gates. However, we cannot combine two different lines unless it passes through a gate, for we must indicate how we are combining the two lines. The output for the adjacent circuit is $(p \wedge q) \vee (p \wedge r)$.

Now let's reverse the process and construct a circuit for a given logical expression. When we construct a circuit, we work from the inside out, constructing the gates in the order that the computations are performed.

⊕ *Example*

1. Construct a circuit for $\sim p \vee \sim q$.

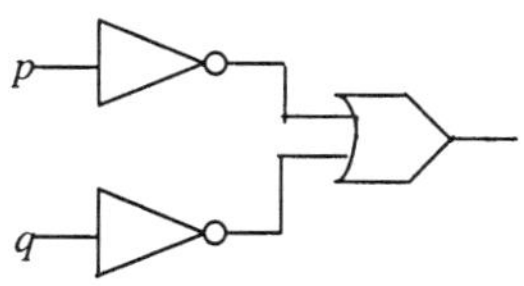

To evaluate $\sim p \vee \sim q$, we first compute $\sim p$ and then compute $\sim q$. We duplicate this process with the gates. First, we send the input for p through a NOT-gate and the input for q through a NOT-gate. Next we take their output and send it through an OR-gate, as illustrated on the left.

2. Construct a circuit for $\sim(p \wedge q)$.

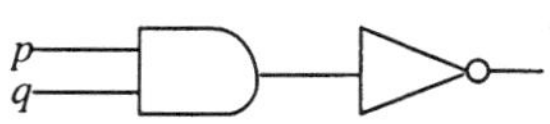

To evaluate $\sim(p \wedge q)$, we first compute $p \wedge q$. So in our circuit, we first send the input from p and the input from q through an AND-gate. The output of the AND-gate is then sent through a NOT-gate, as illustrated on the left.

Simplifying Circuits

The sentences in the above two examples are logically equivalent. For any input, the output of the circuit $\sim p \vee \sim q$ will be identical to the output of the circuit $\sim(p \wedge q)$. However, in terms of the cost and efficiency, we have a major difference. The first expression, $\sim p \vee \sim q$, requires a total of three gates, whereas

$\sim(p \wedge q)$ requires only two gates. Thus, the second circuit is a simpler way to get the same output.

In any business, cost and efficiency are a major concern, but in computer electronics, efficiency is of special concern. The difference of one little nanosecond, which is one billionth of a second, makes a huge difference when large batches of data are being processed. A programmer may use a clever algorithm in writing a program, but there may not be enough time in the universe to compute it because of the efficiency of the circuits. The inclusion of one extra gate in a circuit can make a significant different in the processing time.

$p \wedge \mathrm{T} \Leftrightarrow p$

$p \vee \mathrm{T} \Leftrightarrow \mathrm{T}$

$p \vee F \Leftrightarrow p$

The laws of logic give us a powerful tool for simplifying circuits and making them more efficient. We will now examine how we can use the negation laws, the distributive laws and the three adjacent equivalences to reduce the number of gates. The strategy for using these equivalences is to try to manipulate the expression into a form where a statement that is always true or always false appears. For example, $p \wedge \sim p$ is always false and $p \vee \sim p$ is always true. So, we look for ways to produce these terms, as illustrated in the following examples.

Example

1. In the rush of the day at a high-tech plant, you are asked to construct a circuit for $(p \wedge q) \vee (p \wedge \sim q)$. Can you simplify it?

   First, we use the distributive law to factor out p.

   $(p \wedge q) \vee (p \wedge \sim q)$

   $p \wedge (q \vee \sim q)$ .... *Distributive Law for* $\wedge$ *over* $\vee$

   $p \wedge \sim p \Leftrightarrow \mathrm{F}$

   $p \vee \sim p \Leftrightarrow \mathrm{T}$

   Since $q \vee \sim q$ is always true, we can simplify further.

   $p \wedge \mathrm{T}$ .... *Simplification*

   p .... *Simplification*

   It would be foolish to implement the circuit $(p \wedge q) \vee (p \wedge \sim q)$ because its output is identical to p.

2. Is it possible to simplify the adjacent circuit?

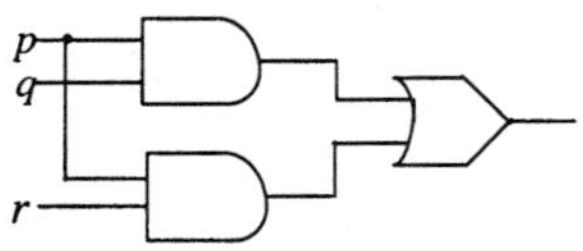

   We can represent this circuit as $(p \wedge q) \vee (p \wedge r)$. Using the distributive law, we can rewrite it as $p \wedge (q \vee r)$. Since the latter expression uses only two gates, it is simpler.

3. Simplify the following expression so that its circuit has fewer gates:

   $$\sim(p \wedge \sim q) \wedge r \wedge (\sim p \vee \sim q)$$

Since r is in only one term, we use the commutative and associative properties of *and* to move r over to the side.

$$[\sim(p \wedge \sim q) \wedge (\sim p \vee \sim q)] \wedge r$$

Bring the first negation inside the parentheses:

$$[(\sim p \vee q) \wedge (\sim p \vee \sim q)] \wedge r$$

$$[\sim p \vee (q \wedge \sim q)] \wedge r \quad .. \textit{ Distributive Law}$$

$$(\sim p \vee \text{F}) \wedge r \quad .... \textit{ Simplification}$$

$$\sim p \wedge r \quad .. \textit{ Simplification}$$

We have reduced the number of gates from 8 to 2.

Designing a Circuit

From specifications for a circuit where given input values must be transformed to specified output values, how does one design the circuit? For example, how could we design a circuit to satisfy the specifications in the adjacent truth table? Of course, efficiency will be a concern, but let's initially focus on doing it however we can. Once we have a base model that does the job, we can then focus on streamlining it to maximize the efficiency.

p	q	r	Output
1	1	1	1
1	0	1	0
0	1	1	0
0	0	1	0
1	1	0	1
1	0	0	0
0	1	0	1
0	0	0	0

Finding a logical expression that fits a given truth table can be quite challenging, unless we know a few tricks. With the following algorithm, we can always find an logical expression that will give the desired output.

1. List each case (or row) where the output is 1. For each of these cases, write an and-statement that produces 1 for that case and 0 everywhere else. In Case 5, r is false, so we use $\sim r$ in the and-statement: $p \wedge q \wedge \sim r$

2. Connect each and-statement from the previous step with *or*. This logical expression will have the desired output values, producing 1 in only the listed cases.

This technique is illustrated in the following example.

✧ *Example*

Find a logical expression for the above truth table.

1. We have an output of 1 in Case 1, Case 5, and Case 7.

   Case 1: $p \wedge q \wedge r$ outputs 1 in Case 1 and 0 everywhere else.

   Case 5: $p \wedge q \wedge \sim r$ outputs 1 in Case 5 and 0 elsewhere.

   Case 7: $\sim p \wedge q \wedge \sim r$ outputs 1 in Case 7 and 0 elsewhere.

2. Connect these 3 expressions with *or*:

$$(p \wedge q \wedge r) \vee (p \wedge q \wedge \sim r) \vee (\sim p \wedge q \wedge \sim r)$$

The output of this logical expression is the given truth table.

Karnaugh Maps

We can use the above technique to find a logical expression for any given truth table. However, the expression will usually require many more gates than are necessary. The next step is to simplify the expression. To streamline this procedure, we can use a visual device called a *Karnaugh map*.

When we apply the previous algorithm to 3 variables, there are 8 possible and-statements that we might have in Step 1, one for each case. A Karnaugh map for 3 variables is a rectangular array with a cell for each of these 8 possible and-statements, as illustrated on the left.

	$q \wedge r$	$q \wedge \sim r$	$\sim q \wedge \sim r$	$\sim q \wedge r$
p	$p \wedge q \wedge r$	$p \wedge q \wedge \sim r$	$p \wedge \sim q \wedge \sim r$	$p \wedge \sim q \wedge r$
$\sim p$	$\sim p \wedge q \wedge r$	$\sim p \wedge q \wedge \sim r$	$\sim p \wedge \sim q \wedge \sim r$	$\sim p \wedge \sim q \wedge r$

The cells are positioned so that adjacent cells differ by only one factor.

The terms in the same row of the first and last columns differ by only one factor, so we will also consider these cells as adjacent.

To simplify an expression, we shade the cells in the Karnaugh map that correspond to the AND-terms in our expression. Adjacent shaded blocks indicate terms that have a common factor. We then take adjacent cells and start factoring. The beauty of the Karnaugh map is that when we factor adjacent terms, we are always left with a factor of the form $r \vee \sim r$, which we can then simplify as T. This technique is illustrated in the following examples:

Example

Simplify the logical expression from the last example:

$$(p \wedge q \wedge r) \vee (p \wedge q \wedge \sim r) \vee (\sim p \wedge q \wedge \sim r)$$

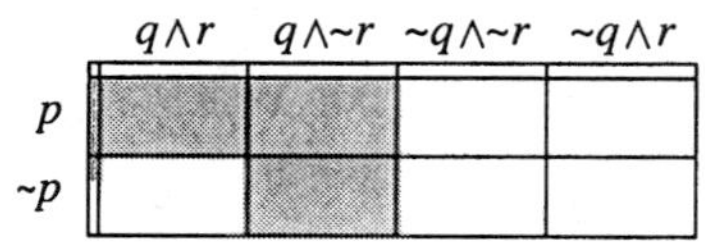

First we shade the 3 cells that correspond to the AND-terms:

$$p \wedge q \wedge r \qquad p \wedge q \wedge \sim r \qquad \sim p \wedge q \wedge \sim r$$

We can start with either the adjacent cells in the first row or the adjacent cells in the second column.

The adjacent terms in the first row can be reduced to $p \wedge q$:

$$(p \wedge q \wedge r) \vee (p \wedge q \wedge \sim r)$$
$$(p \wedge q) \wedge (r \vee \sim r) \text{ .... } \textit{Distributive}$$
$$(p \wedge q) \wedge \mathsf{T} \text{ .... } \textit{Simplification}$$
$$(p \wedge q) \text{ .... } \textit{Simplification}$$

Substitute this simplification in the original expression:

$$(p \wedge q \wedge r) \vee (p \wedge q \wedge \sim r) \vee (\sim p \wedge q \wedge \sim r)$$
$$(p \wedge q) \vee (\sim p \wedge q \wedge \sim r) \text{ .... } \textit{Substitution}$$
$$(q \wedge p) \vee (q \wedge \sim p \wedge \sim r) \text{ .... } \textit{Commutative}$$
$$q \wedge (p \vee (\sim p \wedge \sim r)) \text{ .... } \textit{Distributive}$$

Now we use the Distributive Law to expand the latter term:

$$q \wedge [(p \vee \sim p) \wedge (p \vee \sim r)] \text{ ... } \textit{Distributive}$$
$$q \wedge [\mathsf{T} \wedge (p \vee \sim r)] \text{ .... } \textit{Simplification}$$
$$q \wedge (p \vee \sim r) \text{ .... } \textit{Simplification}$$

We have reduced the number of gates from 11 to only 3.

If we use the Karnaugh maps a lot, we will learn to recognize time-saving shortcuts. For example, a 2×2 block in a Karnaugh map can always be reduced to a single variable. Watch how the terms reduce in the next two examples.

⌖ *Example*

Simplify the given logical expression.

1. Simplify: $(p \wedge q \wedge r) \vee (p \wedge q \wedge \sim r) \vee (\sim p \wedge q \wedge r) \vee (\sim p \wedge q \wedge \sim r)$

   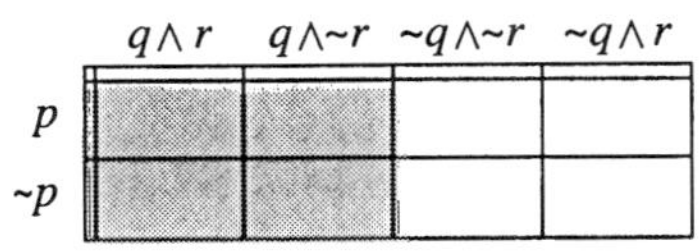

   First we shade the four cells that correspond to the above AND-terms: $p \wedge q \wedge r$, $p \wedge q \wedge \sim r$, $\sim p \wedge q \wedge r$, $\sim p \wedge q \wedge \sim r$. The shaded cells form a 2×2 block. As demonstrated in the last example, the first two terms in the first row of the block reduce to $p \wedge q$. In a similar manner, the first two terms in the second row, reduce to $\sim p \wedge q$. We can then apply the distributive law to these two terms:

$$(p \wedge q \wedge r) \vee (p \wedge q \wedge \sim r) \vee (\sim p \wedge q \wedge r) \vee (\sim p \wedge q \wedge \sim r)$$
$$(p \wedge q) \vee (\sim p \wedge q)$$
$$(p \vee \sim p) \wedge q \text{ ..... } \textit{Distributive}$$
$$\mathsf{T} \wedge q \text{ .... } \textit{Simplification}$$
$$q \text{ .... } \textit{Simplification}$$

Thus, the original expression, which was a square block in the Karnaugh map, reduces to the single variable q. Note that q is the only variable whose negation was not used in any of the original four terms.

2. Simplify: $(p\wedge q\wedge r)\vee(p\wedge \sim q\wedge r)\vee(\sim p\wedge q\wedge r)\vee(\sim p\wedge \sim q\wedge r)$

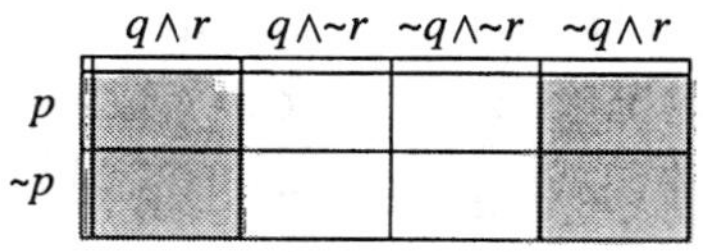

When we shade the four cells that correspond to the AND-terms, we obtain the adjacent map. The outside vertical edges are adjacent since their cells differ by only one factor. We may want to visualize the outside edges as glued together in order to see their adjacent relation. With the gluing, we see a 2×2 block. Let's work through the details and see if the original expression reduces to a single term as it did in the last example. First, we factor the common terms from each row and then simplify.

$$(p\wedge q\wedge r)\vee(p\wedge \sim q\wedge r) \;\vee\; (\sim p\wedge q\wedge r)\vee(\sim p\wedge \sim q\wedge r)$$
$$(p\wedge r)\wedge(q\vee \sim q) \;\vee\; (\sim p\wedge r)\wedge(q\vee \sim q)$$
$$(p\wedge r)\wedge \mathsf{T} \;\vee\; (\sim p\wedge r)\wedge \mathsf{T}$$
$$(p\wedge r) \;\vee\; (\sim p\wedge r)$$
$$(p\vee \sim p)\wedge r$$
$$\mathsf{T}\wedge r$$
$$r$$

Note the similarity with the last example. The original expression, which was a square block in the Karnaugh map, reduces to the single variable r. Furthermore, r is the only variable whose negation was not used in any of the original four terms.

3. Simplify: $(p\wedge q\wedge r)\vee(p\wedge q\wedge \sim r)\vee(p\wedge \sim q\wedge \sim r)\vee(p\wedge \sim q\wedge r)$

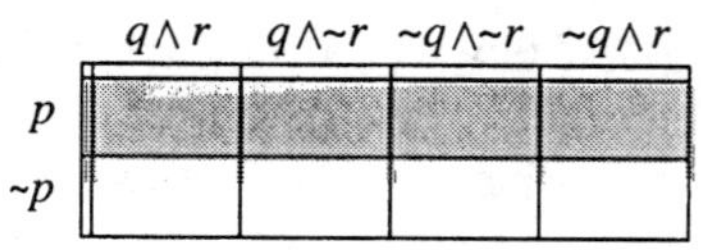

When we shade the four cells that correspond to the above AND-terms, we obtain a 4×1 block. The first two terms in the first row, can be reduced to $p\wedge q$. Similarly, the last two terms in the first row can be reduced to $p\wedge \sim q$. We can then apply the distributive law to these two terms:

$$(p\wedge q\wedge r)\vee(p\wedge q\wedge \sim r)\vee(p\wedge \sim q\wedge \sim r)\vee(p\wedge \sim q\wedge r)$$
$$(p\wedge q)\vee(p\wedge \sim q)$$
$$p\wedge(q\vee \sim q) \;\ldots\ldots\; \textit{Distributive}$$
$$p\wedge \mathsf{T} \;\ldots\ldots\ldots\ldots\; \textit{Simplification}$$
$$p \;\ldots\ldots\ldots\ldots\; \textit{Simplification}$$

The original expression, which was a 4×1 block, reduces to the single variable p. Note that p is the only variable whose negation was not used in any of the original four terms.

Anytime we have a 2×2 block or a 4×1 block shaded in a Karnaugh map, we can reduce those four terms to a single variable, namely the variable whose negation was not used in any of the original four terms.

4 Inputs

A circuit with 3 inputs has 8 possible cases for input values. If we add a fourth input, it can have a value of 1 with each of the 8 previous inputs or a value of 0. Thus, the new circuit has 16 possible cases. So, the design specificiations for a circuit with 4 inputs can be entered in a truth table that has 16 cases (rows). We can then use the algorithm on page 81 to design a circuit that has the desired outputs:

> First we construct an and-statement to produce each output of 1. Then we take the disjunction (connect with or) of these statements.

To simplify the resulting logical expression with a Karnaugh map, we need 16 cells since there are 16 cases which give 16 possibilities for the and-statements. As before, we arrange the terms so that terms in adjacent cells differ by only one factor.

	$r \wedge s$	$r \wedge \sim s$	$\sim r \wedge \sim s$	$\sim r \wedge s$
$p \wedge q$	$p \wedge q \wedge r \wedge s$	$p \wedge q \wedge r \wedge \sim s$	$p \wedge q \wedge \sim r \wedge \sim s$	$p \wedge q \wedge \sim r \wedge s$
$\sim p \wedge q$	$\sim p \wedge q \wedge r \wedge s$	$\sim p \wedge q \wedge r \wedge \sim s$	$\sim p \wedge q \wedge \sim r \wedge \sim s$	$\sim p \wedge q \wedge \sim r \wedge s$
$\sim p \wedge \sim q$	$\sim p \wedge \sim q \wedge r \wedge s$	$\sim p \wedge \sim q \wedge r \wedge \sim s$	$\sim p \wedge \sim q \wedge \sim r \wedge \sim s$	$\sim p \wedge \sim q \wedge \sim r \wedge s$
$p \wedge \sim q$	$p \wedge \sim q \wedge r \wedge s$	$p \wedge \sim q \wedge r \wedge \sim s$	$p \wedge \sim q \wedge \sim r \wedge \sim s$	$p \wedge \sim q \wedge \sim r \wedge s$

We consider the first column to be adjacent to the last column since those terms differ by only one factor. Similarly, we consider the top row to be adjacent to the last row.

Pick any two adjacent cells in the above map and you will see a variable that you can eliminate. Pick any 2×2 or 4×1 block in the above map and you will see two variables that you can eliminate. If we always start with the largest blocks, we will find an efficient method for reducing the number of gates. Instead of wasting our time trying to figure out which terms to factor first, we can go on autopilot and start with the biggest block using the visuals from a Karnaugh map.

Civilization advances by extending the number of important operations which we can perform without thinking about them.

Alfred North Whitehead

Exercise Set 1.5

1. Draw a logic circuit for each expression.

   a. $\sim p \wedge q$ b. $\sim(p \wedge q)$ c. $\sim p \vee \sim q$ d. $(p \vee q) \wedge \sim(p \vee r)$

2. Write a logical expression that represents the following circuits. Compute the output if the input signal is $p = 1$, $q = 1$ and $r = 0$.

   a.

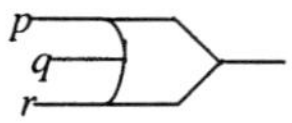

   b.

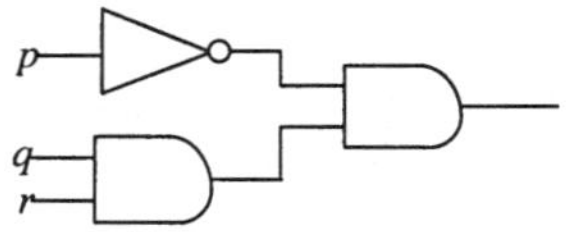

   c.

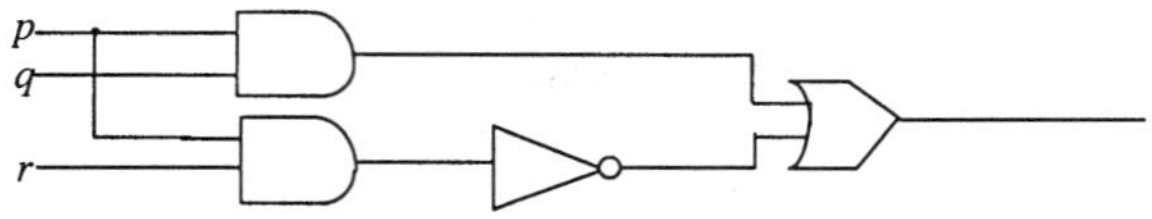

3. Simplify each expression so that its circuit requires fewer gates.

   a. $(p \wedge q) \vee (p \wedge \sim q)$

   b. $(p \wedge \sim q) \vee (\sim p \wedge \sim q)$

   c. $(p \wedge q \wedge r) \vee (\sim p \wedge q \wedge r)$

   d. $(p \wedge \sim q) \vee (\sim p \wedge q) \vee (\sim p \wedge \sim q)$

4. Construct a logical expression that has the given output table. Then use a Karnaugh map to simplify the expression.

   a.

p	q	r	Output
1	1	1	0
1	0	1	1
0	1	1	0
0	0	1	0
1	1	0	1
1	0	0	1
0	1	0	1
0	0	0	1

   b.

p	q	r	Output
1	1	1	1
1	0	1	0
0	1	1	1
0	0	1	0
1	1	0	0
1	0	0	1
0	1	0	0
0	0	0	0

   c.

p	q	r	s	Output
1	1	1	1	0
1	0	1	1	1
0	1	1	1	1
0	0	1	1	0
1	1	0	1	1
1	0	0	1	1
0	1	0	1	0
0	0	0	1	0
1	1	1	0	0
1	0	1	0	0
0	1	1	0	0
0	0	1	0	1
1	1	0	0	1
1	0	0	0	1
0	1	0	0	1
0	0	0	0	0

   d. The table in (6) on page 76.

Activity 1.6

The set [1,4] has a greatest element, whereas the set [1,4) does not have a greatest element. Let S be a set of real numbers.

1. Make up definitions for the following. Flag any quantifiers that you use in your definition.
   a. c is the greatest element in S.
   b. S has a greatest element.
   c. S has a least element.
2. Use your definitions to translate the following:
   a. c is not the greatest element in S.
   b. S does not have a greatest element.
   c. S does not have a least element.
3. Multiplication is distributive over addition:

   For all real numbers a, b, and c, $a \times (b+c) = a \times b + a \times c$.

   Use substitution to generalize the above property.

   * is distributive over # if and only if for all a, b, and c, ____
4. Use your generalization in (3) to answer the following:
   a. Is $\times$ distributive over $-$?
   b. Is $+$ distributive over $\times$?
   c. Is *and* distributive over *or*?

≡ 1.6 Translations ≡

Mathematical reasoning involves a continual translation, back and forth, from everyday language to pictures and symbolic representations. To comprehend a symbolic representation, we must be able to translate it in terms of visual pictures and the richer vocabulary of everyday language, using examples to build our personal understanding of it.

The reverse task is often more difficult, for we need to be able to translate concepts from everyday language into a precise format using variables, quantifiers, and logical operators. For example, how do we translate the concept of the "smallest" element in a set of real numbers? First, let's translate the superlative ending. A set S of real numbers has a smallest element if the following is true:

> *There is* an element in S that is smaller
> than *every* other element in the set.

The italicized words are flagging the quantifiers imbedded in our description. For each quantifier, we introduce a variable. Let's use the variable "y" to go with *"there is"* and the variable "x" to go with *"every."* We can now give a precise definition of "smallest" using variables and quantifiers.

> Let S be a set of real numbers.
> S has a *smallest* element if and only if
> there exists a y in S such that for all x in S, $y \leq x$.

Definitions

The ability to recognize a concept and create a definition for it is an important reasoning skill. When we define a word such as "smallest," the word is a shorthand notation for its definition. Consequently, we always use the "if and only if" connective in a definition. Definitions provide a quick bypass through a dense jungle of words, allowing us to speed up our reasoning process by focusing on a single word instead of its equivalent definition.

To prove that an object does not satisfy a definition, we negate its definition, as illustrated in the following example:

⊕ *Example*

What do we have to demonstrate to prove that a set S does not have a smallest element,?

~(S has a smallest element)
if and only if
~(there exists a y in S such that for every x in S, $y \leq x$).

We apply the rules for negating quantifiers:

$$\sim(\exists y \text{ in } S, \forall x \text{ in } S,\ y \leq x)$$
$$\forall y \text{ in } S,\ \exists x \text{ in } S,\ y \not\leq x$$
$$\forall y \text{ in } S,\ \exists x \text{ in } S,\ x < y$$

For every y in S, there exists an x in S such that $x < y$.

To prove that there is no smallest positive real number, we must demonstrate the above statement.

Symbols vs. Words

In the previous example, the symbolic notation is more concise. However, the version fleshed out with phonetic words emanates a more comfortable feeling. Perhaps we hear the words more when we see the phonetic spelling. Whatever the cause, when we read or write mathematics, it is helpful to balance the symbols with words. For this reason, in math textbooks we normally write out the quantifiers in words rather than represent them with the symbols $\forall$ and $\exists$. When writing by hand, the symbolic form may be more convenient.

Whenever

"Whenever" is often used in mathematical discourse in sentences such as the following:

$$|f(x)-f(1)|<\varepsilon \text{ whenever } |x-1|<\delta.$$

Negating the above sentence can be a little tricky unless we first translate the sentence in terms of the logical operators and quantifiers. After that, we can go on autopilot and apply the laws of logic.

We can translate "whenever" as "if," but "ever" contains an implicit notion of the universal quantifier, so we must also include the universal quantifier:

$$\text{For every } x,\ |f(x)-f(1)|<\varepsilon \text{ if } |x-1|<\delta.$$

Next we write the sentence in standard implication form:

$$\text{For every } x, \text{ if } |x-1|<\delta, \text{ then } |f(x)-f(1)|<\varepsilon .$$

In the above form, we can apply the rule for negating an implication (page 65).

$$\text{There exists an } x \text{ such that } |x-1|<\delta \text{ and } |f(x)-f(1)|\geq\varepsilon.$$

⊕ *Example*

Translate each sentence.

1. $x\in A$ whenever $x\in B$.

   Translation: For every x, $x\in A$ if $x\in B$.
   For every x, if $x\in B$, then $x\in A$.

2. $\sim(x\in A$ whenever $x\in B)$.

   Translation: $\sim$(For every x, if $x\in B$, then $x\in A$).
   There exists an x such that $x\in B$ and $x\notin A$.

Eliminating an Implication

We can sometimes eliminate an implication by changing the domain in the quantifiers. For example, in the following implication we move the information in the hypothesis over to the domain for x.

For every x, if $x \in B$, then $x \in A$.

For every x in B, x is also in A.

To negate the above two statements, we use different rules, but the end results have the same meaning.

~(For every x, if $x \in B$, then $x \in A$).

There exists an x such that $x \in B$ and $x \notin A$.

~(For every x in B, x is also in A.)

There exists an x in B such that $x \notin A$.

Negating Quantifiers

When we negate the universal quantifier in the last statement, we might be tempted to say "there exists an x not in B." However, we do not negate the domain when we negate a sentence. To guard against this type of error, we can translate the domain for x in a separate sentence:

For every x in B, x is also in A.

Translate: Let x be in B. $\forall x,\ x \in A$

Negation: Let x be in B. $\exists x,\ x \notin A$

So, there exists an x in B such that $x \notin A$.

Translating with Variables

When a definition is not phrased in terms of variables, we may need to translate it in order to have something tangible to manipulate. For example, to form the union of two sets, we combine their elements. On the other hand, the intersection of two sets is where they overlap. How do we translate "combine" and "overlap" in terms of variables?

At the first stage, we introduce variables A and B to represent the sets and symbols to represent union and intersection:

Let A and B be sets.

$A \cup B$ represents the union of the sets A and B.

$A \cap B$ represents the intersection of the sets A and B.

At the next stage, we state a property an element must have to be a member of set $A \cup B$, so we need to introduce a variable x to represent an arbitrary element in the new set. When we combine the elements in two sets, the property that an element x must have to be a member of the new set can be phrased as follows:

$$x \in A \text{ or } x \in B$$

> Let A and B be sets.
> $x \in A \cup B$
> if and only if
> $x \in A$ or $x \in B$.

Using the language tool of variables, we can translate the notion of "combining elements" into a very simple definition based on the word *or*. Given an element, we can determine whether or not it is a member of the set $A \cup B$ by applying the adjacent definition.

In a similar manner, we can translate the "overlap of two sets" in terms of *and*. To be in the overlap of two sets, an element x must have the following property:

> Let A and B be sets.
> $x \in A \cap B$
> if and only if
> $x \in A$ and $x \in B$.

$$x \in A \text{ and } x \in B$$

The definitions of the set operations of union and intersection are based completely on the meaning of *or* and *and*.

Even though the definitions of union and intersection are simple, students sometimes get confused on how to substitute in these definitions, especially when the order of the substitution is involved. Now we will examine one of the most powerful tools in the reasoning process, the substitution principle.

The Substitution Principle

Variables endow the language of mathematics with a sleek form, but the real power comes from the substitution principle, which, like a magician's wand, enables us to magically transform a single sentence into myriad forms. For example, we can transform the following statement of the distributive property by substituting $x+1$ for *a*, x for *b*, and 3 for *c*:

> A legitimate expression may be *substituted* for a variable as long as all occurrences of the variable are replaced by the same substitution.

$$a \times (b+c) = a \times b + a \times c$$
$$(x+1) \times (x+3) = (x+1) \times x + (x+1) \times 3$$

Substitution is a simple concept, but its simplicity can be deceiving. Most of the difficulties that students have with abstract mathematics stem from a lack of understanding of how to use the substitution principle to apply definitions and theorems to specific examples. Problems are easy when we have examples to guide us, but the creative thinkers are those who can blaze a path and create examples for others to follow. To be a logical thinker, we must develop our ability beyond merely copying procedures from examples provided by others. This requires that we learn how to use the substitution principle and freely apply it to definitions and theorems.

When we need to apply several definitions to translate a sentence, we must analyze the *outside* structure of the sentence to determine which definition to use first. We usually construct a sentence from the inside out, deciding how we want to connect the interior components. However, when we deconstruct a compound sentence, we must work from the outside in, as illustrated in the following example:

◈ *Example*

Translate the following sentence: $x \in (E \cap F) \cup C$

To construct the set $(E \cap F) \cup C$, the parentheses tell us to first form the intersection of the sets E and F, and then take this set and union it with C. However, when we apply the definitions, we must unravel it in the *reverse* direction. The outside structure of the set $(E \cap F) \cup C$ is the union of two sets, so to deconstruct the above sentence, we first apply the definition of union:

Suppose that $x \in (E \cap F) \cup C$.
Then $x \in E \cap F$ or $x \in C$. *Definition of union*

Next we replace $x \in E \cap F$ with its definition:

($x \in E$ and $x \in F$) or $x \in C$. . . *Definition of intersection*

When we substitute another letter or expression for a variable, we must be careful not to conflict with other variables. If a variable is universally quantified, we have a free hand in our choice of letters to substitute. We can even substitute a letter used in another variable. In the following example, the variables a and b are universally quantified, so we can substitute $a + b$ for a by replacing each occurrence of a with $a + b$. This does not mean that $a + b = a$.

◈ *Example*

1. For all real numbers a and b, $a + b = b + a$.

   Substitute $a + b$ for a: $(a + b) + b = b + (a + b)$

2. For all x, y, and z, $x(y + z) = xy + xz$.

   Substitute y for x: For all y and z, $y(y + z) = yy + yz$.

If a sentence has an existential quantifier, we do not have a free hand in our choice of letters to substitute. We cannot substitute y for x in the following example:

For every y, there exists an x such that $f(x) = y$.

However, we can substitute y for x if we first replace the original y with another letter, such as z:

For every y, there exists a z such that $f(z) = y$.
For every x, there exists a z such that $f(z) = x$.

Existential Quantifiers

Students generally have more difficulty with definitions that involve the existential quantifier. Perhaps it is the subtleties of substituting in the presence of an existential quantifier, or perhaps it is the existence issue, which also provides difficulties for philosophers and theologians. For whatever the reason, the ability to use the existential quantifier with ease and confidence seems to require a longer incubation period than does the universal quantifier. To build your skill in this area, let's look at an important definition that contains an existential quantifier.

Image of a Set

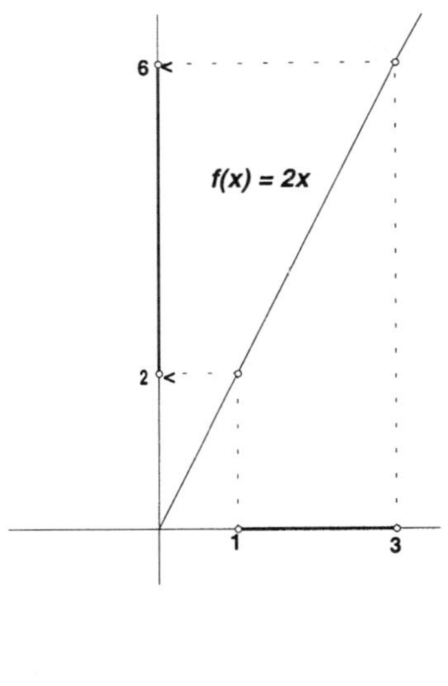

Let f be a function from X into Y and let x be an element of X. The notation $f(x)$ denotes the image of x under f. For example, let $f(x) = 2x$, where x is a real number. Since $f(3) = 2 \times 3$, the image of 3 is 6. To find the image of 3 in the adjacent graph, we move vertically until we hit the graph and then move horizontally until we hit the y-axis.

Let $A = [1,3]$. The image of the set A under f is denoted as $f(A)$. $f(A)$ is the set that consists of the images of all elements in A. In the graph, we can see that $f(A) = [2,6]$.

If $x \in A$, then $f(x) \in f(A)$. To express what it means for an arbitrary element y to be in $f(A)$, we need to use the existential quantifier:

$y \in f(A)$
if and only if
there exists an x in A such that $f(x) = y$.

We will cover this definition in more detail when we examine functions in Chapter 4 (page 364). For the time being, we will use it to practice substitutions.

Example

1. Translate: $z \in g(B)$.

   In the definition, substitute: z for y, g for f, B for A.

   $z \in g(B)$ if and only if
   there exists an x in B such that $g(x) = z$.

   We can also substitute b for x:

   $z \in g(B)$ if and only if
   there exists a b in B such that $g(b) = z$.

$y \in f(A)$
if and only if
there exists an x in A
such that $f(x) = y$.

2. Translate: $x \in f(A)$

   Before we substitute x for y in the definition,
   we substitute another letter for the original x:

   $y \in f(A)$ if and only if
   there exists an a in A such that $f(a) = y$.

   Now we substitute x for y in the above sentence:

   $x \in f(A)$ if and only if
   there exists an a in A such that $f(a) = x$.

To translate an expression that involves more than one definition, we work from the outside to the inside as we deconstruct the sentence. In the following example, we must carefully read the parentheses for they give us the outside structure, which, in turn, tells us which definition to use first.

Example

Do the following sentences have the same meaning?

$$y \in f(A) \cap f(B) \qquad y \in f(A \cap B)$$

To compare their meaning, we need to translate each sentence in terms of the definitions. The outside structure of the first sentence has the form: $y \in Z \cap W$. So, we first substitute in the definition of intersection:

1. $y \in f(A) \cap f(B)$

   $y \in f(A)$ and $y \in f(B)$ .... *Definition of intersection*

   There exists an x_0 in A such that $f(x_0) = y$, and
   there exists an x_1 in B such that $f(x_1) = y$.
   .... *Definition of image of a set*

The outside structure of the second sentence has the form: $y \in f(Z)$. So, we first substitute in the definition of $f(Z)$.

2. $y \in f(A \cap B)$

   There exists an x in $A \cap B$ such that $f(x) = y$.

   .... *Definition of image of a set*

   There exists an x such that $x \in A$ and $x \in B$ and $f(x) = y$.

   .... *Definition of intersection*

The above two translations are not equivalent because we can not factor $\exists x$ across an and-statement (page 69). So, the two sentences have different meanings. This example is discussed further on page 365.

Generalizations

One of the goals of the reasoning process is to generalize as much as possible. If we see that something is true, we try to generalize and find the broadest range of objects for which it is true. For example, if we note that $2 + 3 = 3 + 2$, $5 + 7 = 7 + 5$, etc., we may observe the similarity in the structure of these examples and wonder if we can generalize it. To generalize these examples, we substitute variables for the specific numbers:

For all numbers a and b, $a + b = b + a$.

Commutative Property of Addition

For all numbers a and b,

$a + b = b + a$.

The above statement is true for both real numbers and complex numbers, so we have generalized the examples. We have also identified an important property of the addition operation, which is called the *commutative property of addition*.

As we examine the commutative property of addition, we may wonder if we can generalize further. What is there left to generalize? We have pushed the numbers as far as we can, but what about the operations? Addition is only one of many types of operations. Does multiplication have a similar property? As we learned in elementary school, multiplication does indeed have a similar property.

Commutative Property of Multiplication

For all numbers a and b,

$a \times b = b \times a$.

Now compare the adjacent two properties and notice their similarity. How do we generalize these two statements? The only difference is the operation, so let's generalize the operation by representing it with the symbol$*$:

For all numbers a and b, $a * b = b * a$.

In order to generalize addition and multiplication with a symbol such as $*$, we need to recognize the common features that they share. They both operate on two numbers and produce a new number:

When addition operates on 4 and 7, it produces 11.

When multiplication operates on 4 and 7, it produces 28.

$*$ is a *binary operation* on a set S
if and only if
for all a and b in S,
$a*b$ is defined and $a*b \in S$.

The union operation has a similar property; it operates on two sets and produces a new set. Similarly, the intersection operation operates on two sets and produces a new set. Now, we will do a broad generalization that includes not only addition and multiplication, but also union and intersection. This general concept is called a *binary operation* on a set S. A binary operation operates on each pair of elements in S and produces a new element that is also in S. Addition, multiplication, and subtraction on the set of real numbers each satisfy the adjacent definition of a binary operation. However, division does not satisfy the definition because division by 0 is not defined. Division does, though, satisfy the definition on the set of positive rational numbers.

Let $*$ be a binary operation on S.
$*$ is *commutative*
if and only if
for all a and b in S, $a*b = b*a$.

We can now build on the definition of a binary operaiton and define the generalized commutative property as stated on the left. To determine if a particular operation is commutative, we reverse the generalization process and substitute in the definition, as illustrated in the following examples.

⊕ *Example*

1. Is multiplication commutative on the set of real numbers?

   In the previous definition, substitute $\times$ for $*$:

   $\times$ is commutative if and only if
   for all real numbers a and b, $a \times b = b \times a$.

   Since the above statement is true,
   multiplication is commutative.

2. Is subtraction commutative on the real numbers?

   $3 - 4 \neq 4 - 3$, so subtraction is not commutative.

The number 0 has a special property with respect to the addition operation.

For every real number a, $a + 0 = a$ and $0 + a = a$.

We use the word *identity* to describe this property: 0 is the identity for addition. To generalize the identity property to an arbitrary binary operation, we introduce the letter i to represent the identity. In the identity property for 0, we substitute i for 0 and $*$ for $+$, which gives the following definition:

Let $*$ be a binary operation on S.
$*$ has an identity
if and only if
there exists an i in S such that
for every a in S,
$a * i = a$ and $i * a = a$.

Let $*$ be a binary operation on a set S and let i be in S.
i is the *identity* for $*$
if and only if
for every a in S, $a * i = a$ and $i * a = a$

A binary operation on S has an identity if there exists an i in S that satisfies the above definition. In the adjacent translation, note the order of the mixed quantifiers. As we saw in Section 1.2, the order of mixed quantifiers changes the meaning of the sentence. If we reverse the order of the quantifiers, we do not have the definition of an identity.

Example

Let S be the set of real numbers.

1. Does multiplication have an identity on S?

   Substitute $\times$ for $*$ in the definition:

   > Does there exist a real number i such that for every real number a, $a \times i = a$ and $i \times a = a$?

   Set $i = 1$. For every real number a, $a \times 1 = a$ and $1 \times a = a$. So 1 is the identity for addition.

2. Does subtraction have an identity on S?

   Substitute $-$ for $*$ in the definition:

   > Does there exist a real number i such that for every real number a, $a - i = a$ and $i - a = a$?

   Set $i = 0$. $a - 0 = a$, but $0 - a \neq a$. So 0 is not the identity for subtraction. If $a - i = a$, then i must be 0, but 0 does not satisfy the second equation. Therefore, subtraction does not have an identity.

In high school algebra, we study specific binary operations, such as addition and multiplication. In higher mathematics, we generalize to abstract binary operations and investigate the type of structures that they induce on a set.

One of the goals of logical reasoning is to find patterns and relationships between concepts that may seem very different on the surface. For example, our generalization of addition to the concept of a binary operation shows a relation between the structure of the addition operation and the structure of the union operation. Even though addition and union operate on very different types of objects, the abstract concept of a binary operation brings them under the same umbrella. They even share some of the same properties since they are both commutative and associative.

We use translations to both generalize definitions and to apply definitions to specific examples. If we know how to use the substitution principle, translating a mathematical definition for specific examples is fairly straightforward. On the other hand, translating everyday language into a more precise logical form can be very challenging for everyday language is quite complex. The more you work on developing your reasoning skills, though, the easier it becomes.

Exercise Set 1.6

1. Translate in terms of variables, quantifiers and logical operators.
   a. Some elements in the set A are in the set B.
   b. Every element in the set A is in the set B.
   c. Some x are not in both A and B.
   d. x is in A but not in B.
   e. x is in A whenever x is in B.
   f. x is in A only if x is in B.
2. Translate in terms of variables, quantifiers and logical operators.
   a. These two sets have some elements in common.
   b. These two sets have no elements in common.
   c. There is one and only one x that is in both A and B.
   d. There is a unique x that is in both A and B.
3. *Definition*: x is even if and only if there exists an integer k such that $x = 2k$.

   Use the above definition to translate each sentence.
   a. mn is even.
   b. $m+n$ is even.
   c. m^2 is even.

4. Is the following definition correct? If not, why not?
   Definition: x is even if and only if $x = 2k$ for every integer k.

5. *Definition*: f maps X onto Y if and only if for every y in Y, there exists an x in X such that $f(x) = y$.
   Use the above definition to translate each sentence.
   a. g maps X onto Z.
   b. f maps Y onto X.

6. *Definition*: $y \in f(A)$ if and only if there exists an x in A such that $f(x) = y$.
   Definition: $z \in C \cup D$ if and only if $z \in C$ or $z \in D$.
   Definition: $z \in C \cap D$ if and only if $z \in C$ and $z \in D$.
   Translate each sentence by substituting in the above definitions one step at a time. Do the two sentences have the same meaning?
   a. $y \in f(A \cup B)$ $\quad$ $y \in f(A) \cup f(B)$
   b. $y \in f(A \cap B)$ $\quad$ $y \in f(A) \cap f(B)$

7. Make up definitions for the following using variables and quantifiers. The domain of all variables is the set of integers.
   a. x is an *odd* number.
   b. a is a *factor* of b. *Hint*: Why is 3 a factor of 12?
   c. b is a *multiple* of a. *Hint*: Why is 12 a multiple of 3?
   d. a *divides* b. *Hint*: Why does 3 divide 12?

8. Translate each sentence by substituting in your definitions in (7).
   a. mn is an odd number.
   b. $m+n$ is an odd number.
   c. a is a factor of $b+c$.
   d. a is a multiple of $b+c$.
   e. 4 divides $3-b$.
   f. n divides $a-b$.

9. Translate in terms of variables, quantifiers and logical operators.
   a. Between every two real numbers there is another real number.
   b. The sum of every two even numbers is even.
   c. $|f(x)-f(3)| < \in$ whenever $|x-3| < \delta$.
   d. The function f is an increasing function.
      (The domain and range of f is the set of real numbers.)
   e. The set S has a greatest element. *(S is a set of real numbers.)*
   f. The set S has a least element. *(S is a set of real numbers.)*

10. Let S be a set of real numbers. u is an *upper bound* for S if and only if u is greater than or equal to every element in S.
    a. Is 6 an upper bound of the set $[1,5)$?
    b Is 5 an upper bound of the set $[1,5)$?
    c. How many upper bounds does $[1,5)$ have?
    d. Does $[1,5)$ have a least upper bound? If so, what is it?
    e. Does the set of integers have any upper bounds?

11. Let S be a set of real numbers. Translate the following in terms of variables, writing all quantifiers at the beginning of the sentence.
    a. The set S has an upper bound.
    b. The set S does not have an upper bound.
    c. m is the least upper bound of the set S.

12. The definition of the absolute value of a real number is given on the right. Use substitution to write the definition of the following:
    a. $|a|$ b. $|x+1|$ c. $|x-y|$

$$|x| = x, \quad \text{if } x \geq 0.$$
$$|x| = -x, \quad \text{if } x < 0.$$

13. Generalize each example to a statement that gives a basic property of addition on the set of integers. Use variables to represent the numbers. Use the appropriate quantifier for each variable.
    a. $(-2)+[8+(-3)] = [(-2)+8]+(-3)$
    b. $(-3)+0 = -3$ and $0+(-3) = -3$
    c. There is an integer b such that $(-7)+b = 0$ and $b+(-7) = 0$.

14. Let $*$ be an arbitrary binary operation on a set S. Generalize each of your generalizations in (13) for $*$.

15. Let R represent an arbitrary relation on a set S. The notation $x \mathrm{R} y$ means x is related to y under the relation R. "Equals" gives a relation between two numbers. Generalize the following 3 properties of equality for the relation R. Let x, y, and z be real numbers.
    a. $x = x$ (*Reflexive Property*)
    b. If $x = y$, then $y = x$. (*Symmetric Property*)
    c. If $x = y$ and $y = z$, then $x = z$. (*Transitive Property*)

16. Let S be a set of people. Determine if the given relation R has any of the 3 properties that you generalized in (15).
    a. $a \mathrm{R} b$ if and only if a has the same birthday as b.
    b. $a \mathrm{R} b$ if and only if a lives within 1 mile of b.
    c. $a \mathrm{R} b$ if and only if a is shorter than b.

Review

Symbol A letter or figure used to represent something. Phonetic symbols, such as "plus," give pronunciation information. Ideographic symbols like + give a more concise representation that is easier to manipulate.

Sentence A string or sequence of words that satisfy the language rules for being a sentence. A well-formed sentence must have both a subject and a verb.

Statement A sentence that is either true or false, but not both. In formal logic, a statement is called a proposition.

Compound statement A sentence composed of statements connected with logical operators.

Abstract compound statement A compound statement where the component statements are represented by variables such as p and q.

Abstraction The merging of concrete examples under the rubric of a concept that expresses a property the examples have in common. An abstraction exists as an idea with no material existence. For example, the abstract number 3 describes a property that various physical sets have in common, but the number 3 has no physical existence.

Truth value Either true or false. Truth value is only used with sentences.

Variable A letter used to represent an arbitrary element of a given set, which is called the *domain* of the variable.

Domain The set of elements that can be substituted for a variable.

Open statement A sentence with variables that is not a statement but becomes a statement whenever substitutions are made for the variables. An open statement can be converted to a statement by substituting for each variable or by binding each variable with a quantifier, such as $\forall x\ \exists y, p(x,y)$.

Universal quantifier Asserts that each substitution of an element from the domain of the variable converts an open statement into a true statement. $\forall x, p(x)$ is true if and only if every element in the domain of x converts $p(x)$ into a true statement.

Existential quantifier Asserts that at least one substitution of an element from the domain of the variable converts an open statement into a true statement. $\exists x, p(x)$ is true if and only if there exists at least one x in the domain of x such that $p(x)$ is true.

Logical operators Connectives used to form a compound sentence from given component sentences: *and*, *or*, *implies*, *is equivalent to*, and *negation*.

Negation A logical operator that reverses the truth value of a statement. The negation of p is true if and only if p is false.

Conjunction A compound statement of the form: p *and* q. For an and-statement to be true, both parts must be true.

Disjunction A compound statement of the form: p *or* q. For an or-statement to be true, at least one part must be true, but both could be true.

Exclusive or A logical operator that joins two statements with the *exclusive or*: p *XOR* q. This compound statement is true only when one statement is true and the other one false.

Implication A compound statement of the form: p *implies* q. p is called the hypothesis or premise and q is called the conclusion. The only case in which an implication is false is when the hypothesis is true and the conclusion is false. To say that $p \Rightarrow q$ is true means that if p is true, then q must be true.

Contrapositive The contrapositive of $p \Rightarrow q$ is $\sim q \Rightarrow \sim p$. The contrapositive has the same meaning as the original implication.

Converse The converse of $p \Rightarrow q$ is $q \Rightarrow p$.

Equivalence A compound statement of the form: p *is equivalent to* q. For an equivalence to be true, either both parts are true or both parts are false. If two abstract compound statements composed of the same component statements are equivalent, they have the same meaning and can be used interchangeably.

Translation The process of converting words, thoughts or ideas from one form, language, or medium to another. Mathematical reasoning involves a continual translation, back and forth, from everyday language to pictures and symbolic representations.

Substitution principle In a sentence with a variable, another letter or legitimate expression may be substituted for a universally quantified variable as long as all occurrences of the variable are replaced by the same substitution. Similar substitutions can be made for an existentially quantified variable if none of the substituted letters are used with other variables.

Binary Refers to two. A binary operation, such as + or $\cup$, operates on two elements in a set and produces a new element in the set. A binary relation, such as $\leq$ or $\subseteq$, gives a relation between two elements. A binary decimal system has a base of two.

Binary operation	$*$ is a *binary operation* on a set S if and only if for all a and b in S, $a*b$ is defined and $a*b\in S$.

Natural numbers	$1, 2, 3, 4, 5, 6, \ldots$
Integers	$\ldots -3, -2, -1, 0, 1, 2, 3, \ldots$
Rational number	A number that can be represented in the form $\frac{p}{q}$, where p and q are integers with $q \neq 0$.
Real number	A number that can be represented as a decimal with a finite or infinite number of places. The visual picture of the real numbers is the points on a number line.
Irrational number	A real number that is not rational.
Complex number	A number that can be represented in the form $x+yi$ where x and y are real numbers and $\mathrm{i} = \sqrt{-1}$. The visual picture of the complex numbers is the points in a plane, where $x+yi$ is identified with the point (x, y).

Logic	A formal study of the art of reasoning and the principles for making valid deductions.
Law of logic	An abstract compound statement that is always true, regardless of the truth values of its component statements. A law of logic is also called a tautology.

Frequently used equivalences

$$p \text{ and } q \iff q \text{ and } p \quad \ldots\ldots\ldots\ldots \textit{Commutative}$$
$$p \text{ or } q \iff q \text{ or } p$$
$$(p \Leftrightarrow q) \iff (q \Leftrightarrow p)$$
$$(p \text{ and } q) \text{ and } r \iff p \text{ and } (q \text{ and } r) \quad \ldots\ldots \textit{Associative}$$
$$(p \text{ or } q) \text{ or } r \iff p \text{ or } (q \text{ or } r)$$
$$p \text{ and } (q \text{ or } r) \iff (p \text{ and } q) \text{ or } (p \text{ and } r) \quad . \textit{Distributive}$$
$$p \text{ or } (q \text{ and } r) \iff (p \text{ or } q) \text{ and } (p \text{ or } r)$$
$$p \Rightarrow q \iff \sim q \Rightarrow \sim p \quad \ldots\ldots\ldots\ldots \textit{Contrapositive}$$
$$p \text{ or } q \iff \sim p \Rightarrow q \quad \ldots\ldots\ldots\ldots \textit{Rephrasing Or}$$
$$(p \Leftrightarrow q) \iff (p \Rightarrow q) \text{ and } (q \Rightarrow p) \quad . \textit{Rephrasing} \Leftrightarrow$$
$$\sim(p \text{ and } q) \iff \sim p \text{ or } \sim q \quad \ldots\ldots\ldots\ldots \textit{Negations}$$
$$\sim(p \text{ or } q) \iff \sim p \text{ and } \sim q$$
$$\sim(p \Rightarrow q) \iff p \text{ and } \sim q$$
$$\sim(\forall x, p(x)) \iff \exists x, \sim p(x) \quad \ldots\ldots\ldots\ldots \textit{Quantifiers}$$
$$\sim(\exists x, p(x)) \iff \forall x, \sim p(x)$$
$$\forall x, p(x) \text{ and } q(x) \iff \forall x, p(x) \text{ and } \forall x, q(x)$$
$$\exists x, p(x) \text{ or } q(x) \iff \exists x, p(x) \text{ or } \exists x, q(x)$$

Chapter Review

1. Why is symbolic notation often used in logical reasoning?
   Why use "+" instead of "plus?" What is a symbol?
2. What is a *compound* sentence?
   What is an *abstract* compound sentence?
3. What is the difference between a statement and an open statement?
   Is every sentence that has a variable an open statement?
   If not, why not?
4. You should be able to do the following:
   a. Determine if two compound sentences composed of the same component sentences are logically equivalent.
   b. Write the negation of a given sentence, simplifying as much as possible.
   c. Identify different forms in which implications and equivalences are written in everyday language.
   d. Use substitution to translate a definition.
5. Explain the following so that your friends would understand it:
   a. How to negate a quantified sentence.
   b. How to negate a compound sentence.
   c. Why the truth values for an implication are defined that way.
6. Given two different compound sentences that are formed from the same component sentences, what does it mean to say that they have the same meaning? What do cases have to do with the meaning? Why do equivalent statements have the same meaning?
7. Why does the contrapositive of an implication have the same meaning as the implication, but the converse does not?
8. What is a law of logic? Give examples of laws of logic that are equivalences and examples that are not equivalences.
9. You should know how to rewrite an or-sentence as an implication.
   a. For every integer x, x is even or x is odd.
   b. x is in D or x is in E.
10. You should be able to write the contrapositive of an implication.
    a. If $x \in A$, then $x \in B$.
    b. If $\sum_{i=1}^{\infty} a_n$ converges and $\sum_{i=1}^{\infty} b_n$ converges,
    then $\sum_{i=1}^{\infty} (a_n + b_n)$ converges.

11. If p is equivalent to q, is $\sim p$ equivalent to $\sim q$?

12. Negate each sentence. Do not leave a negation as a prefix for a compound sentence.
   a. If $x \in A$, then $x \in B$.
   b. If $x \notin B$, then $x \notin A$.
   c. $x \notin A$ or $x \in B$.
   d. There exists an x such that $x \in A$ and $x \in B$.
   e. For all real numbers c and d, if $c < d$, then $f(c) < f(d)$.
   f. For every ε, there exists a δ such that for every x, $|x-1| < \delta$ implies that $|f(x) - f(1)| < \varepsilon$.
   g. For every function f defined on the interval $[0,1]$ and for every number c between $f(0)$ and $f(1)$, there exists an x in the interval $(0,1)$ such that $f(x) = c$.
   h. For every function f and all real numbers a and b, there exists a real number c between a and b such that $f'(c) = \frac{f(b)-f(a)}{b-a}$.

13. For each definition, write what it means to not have the given property. For example, in part (a), translate the following:
   x is not even if and only if ____.
   a. *Definition*: x is *even* if and only if there exists an integer n such that $x = 2n$.
   b. *Definition*: x is *rational* if and only if there exist integers p and q with $q \neq 0$ such that $x = \frac{p}{q}$.
   c. *Definition*: Let a and b be integers. *a divides b* if and only if there exists an integer k such that $b = ka$.
   d. *Definition*: $x \in A \cup B$ if and only if $x \in A$ or $x \in B$.
   e. *Definition*: $x \in A \cap B$ if and only if $x \in A$ and $x \in B$.
   f. *Definition*: $A \subseteq B$ if and only for every x, if $x \in A$, then $x \in B$.
   g. *Definition*: $A = B$ if and only for every x, if $x \in A$, then $x \in B$, and if $x \in B$, then $x \in A$.
   h. *Definition*: $A \subset B$ if and only $A \subseteq B$ and $A \neq B$.
   i. *Definition*: The set S has a *largest* element if and only if there exists an m in S such that for every x in S, $x \leq m$.
   j. *Definition*: f is a *function* if and only if for every a and b in the domain of f, if $a = b$, then $f(a) = f(b)$.
   k. *Definition*: f is a *one-to-one* function if and only if for all a and b in the domain of f, if $a \neq b$, then $f(a) \neq f(b)$.

l. *Definition*: f maps X *into* Y if and only if for every x in X, $f(x)$ is in Y.

m. *Definition*: Let f map X into Y. f maps X *onto* Y if and only if for every y in Y, there exists an x in X such that $f(x) = y$.

n. *Definition*: $y \in f(A)$ if and only if there exists an x in A such that $f(x) = y$.

o. *Definition*: The function f is *increasing* on $[a, b]$ if and only if for every c and d in $[a, b]$, if $c < d$, then $f(c) < f(d)$.

p. *Definition*: The function f is continuous at a if and only if for every positive ε, there exists a positive δ such that for every x, if $|x - a| < \delta$, then $|f(x) - f(a)| < \varepsilon$.

14. Translate each sentence by substituting in definitions from the previous exercise, one step at a time.
    a. $x \in A \cup (B \cap C)$
    b. $x \in (A \cup B) \cap C$
    c. $y \in f(C \cap \mathrm{D})$
    d. $y \in f(C) \cap f(\mathrm{D})$
    e. $y \in f(C \cup \mathrm{D})$
    f. $y \in f(C) \cup f(\mathrm{D})$

15. Translate each sentence using definitions from (13).
    a. x^2 is even.
    b. $a + b$ is even.
    c. h maps Y onto X.
    d. $g \circ f$ maps X onto Z.
    e. $x \in g(C)$

16. Do the given pair of sentences have the same meaning? If not, give an example of sets A and B where they have different truth values.
    a. For every x, $x \in A$ and $x \in B$.
       For every x, $x \in A$ and for every x, $x \in B$.
    b. There exists an x such that $x \in A$ and $x \in B$.
       There exists an x such that $x \in A$ and there exists an x such that $x \in B$.
    c. For every x, $x \in A$ or $x \in B$.
       For every x, $x \in A$ or for every x, $x \in B$.
    d. There exists an x such that $x \in A$ or $x \in B$.
       There exists an x such that $x \in A$ or there exist an x such that $x \in B$.

17. You should be able to do the following:
    a. Draw a logic circuit of a given logical expression.
    b. Find a logical expression for a given logic circuit.
    c. Given the output values of a combinatorial circuit (or truth table), find a logical expression that has the same output values.
    d. Use laws of logic to simplify a given circuit (or logical expression) so that fewer gates are needed.
18. You should be able to translate sentences, such as the following, in terms of variables, quantifiers and logical operators.
    a. Every element in A is an element in B.
    b. These two sets have some elements in common.
    c. These two sets have no elements in common.
    d. $x \in A$ whenever $x \in B$.
    e. There is a unique x such that $x \in A \cap B$.
    f. There is a unique x such that $p(x)$ is true.
19. Using variables and quantifiers, verbalize a precise definition for familiar concepts such as the following.
    a. n is even.
    b. n is odd.
    c. n is a *factor* of m.
    d. m is a *multiple* of n.
    e. n *divides* m.
    f. S has a largest element.
20. Why do we try to generalize statements?

Activity 1.7

List 3 questions of a mathematical nature that interest or intrigue you. The questions could be about something that you've never understood, something that you've always wondered about, or something that you would like to know more about.

◈ Chapter 2

Writing Our Reasoning

A proof is the culminating stage of the reasoning process in which we logically organize our reasoning into a written form that can be followed by others. As in any writing process, the first stage is to find something interesting to write about. A good essay depends on having a good thesis statement; similarly, a good proof depends on having a good theorem. Finding an interesting theorem to prove that no one else has thought of is a challenging and exhilarating part of the reasoning process that requires a lot of thoughtful prospecting. Since intellectuals have been pursuing this creative quest for over 2500 years, one might be tempted to think that everything there is to prove has already been proven. However, in any human endeavor that involves the creative spirit, the topics are never exhausted. In fact, more new theorems have been discovered in the past thirty years than in all previous history.

As with prospectors for gold, the theorem-seekers usually try to find a fertile vein of contemplation that has not yet been heavily mined. A new vein may be discovered by making up definitions to generalize properties of specific examples. By analyzing and comparing various examples, one may see clues for a possible theorem. When we find a possible theorem, we may intuitively know that it is true, but at this stage we can only call it a conjecture. Before we can label it a theorem, we must prove it using deductive reasoning.

Techniques for writing proofs can be learned by using templates and studying proofs that others have written, but the ability to construct a proof requires a deeper level of intellectual maturity than merely following an established procedure. To construct a proof, one must explore and question, find the inner structure of the situation, analyze the various parts, and then use logical reasoning to put the different pieces together and create a proof. The sparks that flash during this creative process strengthen our powers of reasoning. Building a bridge is not the same as walking across a bridge that someone else has built; similarly, reading proofs that others have constructed does not have the same developmental effect as creating a proof yourself. This chapter covers the basic techniques for constructing a proof and provides exercises to help you develop your skill in using deductive reasoning and writing your reasoning in a well-formed argument.

Activity 2.1

Is the given argument valid? If so, explain why. If not, draw a sketch of circular sets A and B that shows the argument is not valid.

1. x is in A.
   Therefore, x is in A or x is in B.
2. If x is in A, then x is in B.
   x is in A.
   Therefore, x is in B.
3. If x is in A, then x is in B.
   x is in B.
   Therefore, x is in A.
4. If x is in A, then x is in B.
   If x is in B, then x is in C.
   So, if x is in A, then x is in C.
5. x is in A or x is in B.
   Therefore, x is in A.
6. If x is in A, then x is in B.
   x is not in A.
   Therefore, x is not in B.
7. If x is in A, then x is in B.
   x is not in B.
   Therefore, x is not in A.
8. x is in A or x is in B.
   x is in not in A.
   Therefore, x is in B.

≡ 2.1 Proofs & Arguments ≡

Most sentences that we accept as true come from one of the following sources:

- We are told by someone that we believe.
- We are convinced by our own feelings.
- We are convinced by a valid argument.

> Thus, logic and intuition have each there necessary role . . . Logic, which alone can give us certainty, is the instrument of demonstration; intuition is the instrument of invention.
>
> *H. Poincaré*
> 1854–1912

Experience and intuition, the primary sources for building individual beliefs, give us guidance for what we choose to believe from the first and second sources. The first source is being used when we accept a mathematical formula as true because a teacher told us or perhaps we saw it printed in a textbook. Most students believe the formulas given them and are happy to substitute in them without ever questioning where they came from because they have a naive faith in the source. Mathematical discoveries are often made from the second source; after long hours of contemplation, a mathematician may suddenly get a "flash" that something is true. However, it is not labeled true until it can be verified from the third source with a valid argument.

In the opening statement of their highly acclaimed series on mathematics, the famous group of French mathematicians known as Bourbaki state the great importance of proofs. Mathematics is based on proofs. Unfortunately, though, proofs often have a bad reputation with students. Some students even go so far as to express the sentiment that they like mathematics but they don't like to do proofs. This type of attitude indicates that the student may not yet have had the opportunity to experience the challenge and excitement inherent in the intellectual process of constructing proofs. Indeed, this intellectual pursuit is more intriguing to some people than the most riveting courtroom drama on the big silver screen.

> Since the early Greeks, to speak of mathematics has been to speak of 'proof.'
>
> *Bourbaki*

Let's compare a lawyer's proof with a mathematical proof. Both are arguments that convince the target audience, explaining *why* the claim is true. A lawyer has proved his case if he convinces the jury; however, in legal proofs, subjectivity is often involved. Upon appeal, the next jury may have a different opinion.

In contrast, the rules for a mathematical proof are constructed so that any jury of rational thinkers will always give the same verdict when asked to pass judgment on a

Pythagorean Theorem
500 B.C.E.

If a, b and c represent the lengths of the sides of a right triangle with c representing the hypotenuse, then

$$c^2 = a^2 + b^2.$$

proposed mathematical proof. Two rational thinkers may not agree on a theory in psychology, art, history, sociology, or physics, but they always agree on a theorem in mathematics because it has a proof to back it up. The great power of mathematics is that theorems have a 100% guarantee within the framework of the system in which they were proved. We still use mathematical theorems proved by the ancient Greeks over two thousand years ago, whereas their theories of science have long since been discarded. As you can well imagine, this type of staying power requires very careful attention to the discourse used in a proof.

Sentences

When addressing a jury, a lawyer speaks in complete sentences because complete sentences are necessary to communicate complete thoughts. For the same reason, we only use complete sentences in a mathematical proof. Since proofs often contain a multitude of symbols, we must keep in mind that the symbols represent words which must form sentences. For example, the expression $A \subseteq B$ is a sentence, but $A \cup B$ is not.

Proof

A *proof* is a list of sentences where each sentence comes from one of the following three categories:

- Sentences that we assume are true.
- Sentences that we already know are true.
- Sentences that we derive from previous lines.

Proof:
Sentence
Sentence
...
Sentence

Usually we write a proof in an informal style, including reasons to help the reader see how we derive a statement from previous lines in the proof. Sometimes we supply extra discourse when we remind the reader of what we already know. Since the goal of a proof is to convince other people that a certain statement is true, we should write our proofs in a style that is easy for others to follow.

Theorems

A *theorem* is a statement that has been proved. The formal definition of a theorem is the last line of a proof, which means that it follows from the previous lines. Sometimes, though, authors do not restate the complete theorem at the end since it is usually stated at the beginning as the heading of the proof.

A statement is never labeled as a theorem unless someone has constructed a legitimate proof for it. If we think that a statement is true, but no one has yet proved it, we call the statement a *conjecture*.

Playing the Game

The first person to construct a proof for an important theorem achieves a lasting fame by forever having their name associated with it, unlike the fleeting fame of an athlete who recedes into the background when his world record is surpassed. Even though Michael Jordan enjoys world recognition today, it is fairly safe to predict that his fame will not have the longevity of Pythagoras who lived over two thousand years ago in a small Greek village and whose name is still known to most every student of high school age. In the next exercise, when we try to prove the Pythagorean Theorem, we may wonder if we too might not have been able to be the originator of such a clever proof. When that happens, we will understand the excitement of doing mathematics and why some consider it as the most sophisticated game in town.

As we examine strategies and rules for being a player in this game, please keep in mind that knowing the rules is essential, but as in basketball, knowing the rules is not enough to make one a good player. We must practice with as much interest and intensity as athletes do in their attempt to become excellent athletes. Most mathematicians, as do most athletes, play for the love of the game. A few become superstars who ask questions and find answers which have a major impact on the course of future knowledge, and their names will be remembered for centuries.

The challenge in playing the game of mathematics involves both of the following.

- Finding possible theorems.
- Constructing proofs for theorems.

Finding Theorems to Prove

Before we can write a proof, we must have something to prove. Finding interesting statements to prove requires a great deal of detective work. The most fertile grounds are in areas that interest us, areas that we are curious about, and especially areas that we have a burning desire to know about. That desire will motivate us to investigate, analyze and sift through a wide range of examples looking for clues to patterns that may be hidden beneath the surface. When we find a pattern that works for our examples, the next task is to try to prove that the pattern holds for a more general class of examples.

The mathematician at work makes vague guesses, visualizes broad generalizations, and jumps to unwarranted conclusions. He arranges and rearranges his ideas, and he becomes convinced of their truth long before he can write down a logical proof.

Paul Halmos

Through experimentation, the ancient Babylonians discovered an amazing pattern in the relation between the sides of a right triangle: $c^2 = a^2+b^2$. Pythagoras may have spent countless hours contemplating how to prove this result that the Babylonians had discovered 15 centuries before; he may finally have seen the solution in a brilliant flash of insight as he watched the

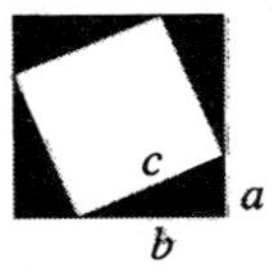

pillars in a magnificent temple cast its shadow across a square tiled floor (*see adjacent illustration*). But regardless of how the proof was actually discovered, we can be assured that the discoverer had an epiphany of great magnitude, for it is rather spectacular to see how the human mind can explain why we have this beautiful relationship between quantitative numbers and geometric triangles.

We find possible theorems to prove through either our own creative observations and experimentation, or we latch onto a discovery that someone else has made, as did Pythagoras, and we then try to prove it.

Inductive Reasoning

We frequently draw conclusions based on our experiences. Seeing that the sun rises every morning, we infer that it will always rise in the morning. This type of reasoning, when we discover a general relation from specific examples or experiences, is called *inductive reasoning*. Through inductive reasoning, the Babylonians discovered the Pythagorean Theorem. Inductive reasoning is the basis for the scientific method. The scientist looks at experimental data and tries to make a generalization which will fit the evidence supplied by the data. The validity of the generalization is then based solely on its accuracy in making predictions.

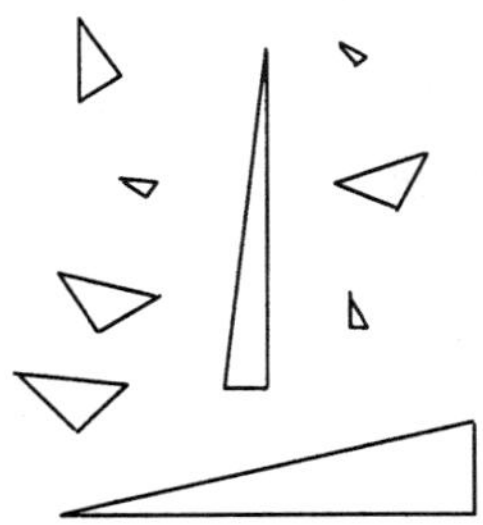

Inductive reasoning is an extremely important part of the reasoning process, but we must be aware of its limitations. First of all, most experimental data depends on some type of measurement, and any measurement such as length, weight, or speed must by necessity be an approximation. We cannot distinguish between 5 and 5.00001 on a yardstick, and even with hi-tech measuring devices that capture microscopic detail, some error is always present. If we measure the sides of a right triangle to be 3, 4, and 5 centimeters, the observed evidence shows that $c^2 = a^2 + b^2$. But if the hypotenuse measured 5.00001 instead of 5, then our formula changes to $c^2 \approx a^2 + b^2$. Granted, we may think these two numbers are very close, but closeness is relative to how much we have zoomed in on the situation. From a human perspective, there is quite a bit of difference between 5 light years and 5.00001 light years.

The second cause for concern is that the inductive method carries no guarantees. Just because Santa Claus has visited us every year for 18 years is no guarantee that he will visit us next year. Similarly, just because we measure 18 million right triangles and obtain a certain relation between the measurements is no guarantee that the same relation will appear with the next triangle that we measure.

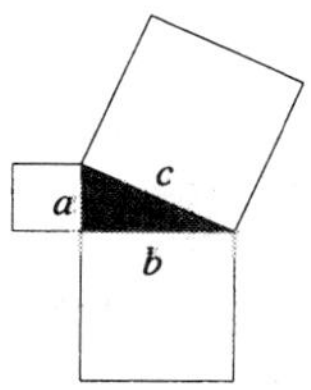

Perhaps the greatest shortcoming of the inductive process, is that it does not shed any light on why something is true. The human need to understand *why* things happen has fueled the quest for knowledge. Anyone can see that the sun rises every morning, but *why* does the sun rise? Is there something else happening that makes the sun rise? Similarly, we can measure the sides of a right triangle and see that $c^2 = a^2 + b^2$, but why does this happen? In the adjacent illustration, why is it that we can cut up the two squares along sides a and b, which represent a^2 and b^2, and fill up the square along the c-side? Inductive reasoning cannot produce answers to this type of question.

Deductive Reasoning

The search for answers to why-questions is an exciting mind sport that, unlike athletic sports, gives us deep insights that help us understand the universe in which we live. The tool that we use in this quest is called deductive reasoning. *Deductive reasoning* is the type of reasoning we use when we derive a conclusion from other sentences that we accept as true. With deductive reasoning, we show why one thing follows from another through the use of valid arguments. When the ancient Greeks developed the method of deductive reasoning in the 6th century B.C.E., they forever changed the course of human knowledge, elevating it to a higher plane where we can use the power of our mind to figure out why things happen.

We use inductive reasoning to find patterns, and then we use deductive reasoning to explain why those patterns happen. Using inductive reasoning, we look at examples with an x-ray vision, searching for patterns in the structure, patterns that may not be apparent on the surface. When we finally see a potential theorem, we then use deductive reasoning to prove that the theorem is true, or at least that it follows from sentences that we consider to be true.

The persuasiveness of a lawyer's argument may depend on its oral delivery, whereas the validity of a proof depends solely on its written form, which does not tarnish with age. Two thousand years in the future there may be people still admiring some of the proofs written in this century, as we still admire those of Pythagoras, Euclid, and other great thinkers in ancient Greece. When one is working in a pure art form like mathematics, the thought products do not get as easily dated as in other disciplines.

Structure of a Proof

The structure of a proof is similar to that of a good essay; it must have an introduction, a body, and a conclusion, which is the theorem that we want to prove. To begin the process of constructing a proof, we can structure our work by starting at

Proof
Introduction

Body

Conclusion

the bottom, leaving a large blank area in the middle for the steps that we will fill in later:

1. First, we write the conclusion, which is the theorem that we want to prove, at the bottom of the page. By writing the conclusion first, we are setting our goal, which has a distinct psychological advantage. By placing our goal at the bottom of the page, we have structured our writing space in the proper direction; we must work down to that conclusion. Furthermore, since the introduction depends on what we have in our conclusion, we must be focused on the conclusion when we write the introduction.
2. Next we write the introduction at the top of our work space.
3. Finally, we start work on constructing the body of our proof.

This writing format keeps us focused on what we need to do in the body of our proof to build logical steps from the introduction down to the conclusion.

The Introduction

In the introduction of a proof, we introduce the reader to the variables that we will use and state any assumptions that we need to make. We must always introduce variables before we use them. If the theorem contains a universal quantifier for x, we usually start by saying, "let x be a ____," as illustrated in the adjacent example. The first line can be phrased in other ways, such as "let x be an arbitrary real number" or "assume that x is a real number."

The assumptions that we make in the beginning of a proof set up the outside structure for the type of proof that we will present. If we decide later to try a different method of proof, we will need to change our assumptions. The assumption in the second line of the adjacent proof sets up the structure for a direct proof of the implication. Sometimes, though, we make a different assumption and construct an indirect proof, which we will discuss on page 135. Since there are different assumptions we can make to prove an implication, we must always clearly state our assumptions so that the reader will know which method we are using. Instead of the word "assume," we can use other words like "suppose" to indicate an assumption, but we must indicate that it is an assumption. We cannot just write "$x > 1$" in the adjacent proof.

> **Theorem:** For every real number x, if $x > 1$, then $x^2 > 1$.
>
> ---
>
> **Proof:** Let x be a real number.
> Assume that $x > 1$.
>
> . . .
>
> Hence, $x^2 > 1$.
> Therefore, if $x > 1$, then $x^2 > 1$.

Whenever we make an assumption in a proof, we should check our grammar to make sure that we have a legitimate sentence. For example, students sometimes write the following incorrect assumption, which has no meaning.

Assume $A \cup B$.

When we make an assumption, we assume that a sentence is true. We cannot assume that the set $A \cup B$ is true for $A \cup B$ is not a sentence. We could, though, make either of the following assumptions:

Assume $x \in A \cup B$.

Assume $A \cup B = C$.

The Body

The body of a proof forms a bridge that connects the beginning with the end. Even though we are building a one-way bridge from the beginning to the end, we usually work backwards from the end as far as possible so that we can see how to structure the beginning of the proof. We will discuss this process in more detail in Section 2.3. As we construct the body of the proof, we can build on what we already know, using previously proved theorems and definitions. Since a proof involves the meaning of words, we usually need to invoke definitions. To prove a statement about $A \cup B$, we will have to either use the definition of the union of two sets or cite a previous theorem about the union of two sets.

The Conclusion

The conclusion is analogous to a thesis sentence, which is usually introduced in the beginning of a good essay and then stated again at the end. Similarly, we state the theorem that we are going to prove, make the preliminary introductions, give the evidence that supports our claim, and then state the conclusion again at the end.

We usually preface a conclusion with a transition word, such as "therefore," "hence," "thus," or "so," as a signpost to let the reader know that it follows from previous sentences. "So" may sound a littler weaker than the others, but it does suggest the metaphor of sewing thoughts together. Some textbooks signal the end of a proof with Q.E.D., an abbreviation for *quod erat demonstrandum*, a Latin phrase that means "which was to be demonstrated."

The introduction and conclusion set up the outer structure of a proof. The body of a proof sets up an inner structure where we derive the conclusion through valid arguments. We will now examine how to argue in a genteel, logical manner.

Valid Arguments

In personal relationships, arguments have a rather negative connotation, but in mathematics a good argument is something to be desired. For the creator of the argument, it is a sign of great mental prowess. For the reader of the argument, it provides a stimulating exercise for the mind which the reader may be able to use in some other type of situation.

An argument is *valid* if the conclusion follows from the hypotheses.

An argument is a list of sentences called *hypotheses* followed by a sentence called the *conclusion,* which we usually flag with a transition word such as "therefore," "thus," "hence," or "so." These transition words indicate that the appended sentence *follows from* the previous sentences, which is the definition of a valid argument. In a valid argument, we logically deduce the conclusion from the sentences that precede it. If each hypothesis were true, then the conclusion would have to be true.

Valid arguments are sometimes called deductive arguments, for deductive reasoning is based on drawing conclusions from valid arguments (page 115). When we use inductive reasoning, we base our conclusion on experiments or experiences, in which case we are betting on the odds that similar things will continue to happen. With deductive reasoning, we isolate where the probability lies – with our hypotheses, not with our conclusion. If our hypotheses are true in a valid argument, we can be 100% assured that our conclusion is true, at least within the framework of the system in which we are making our deductions.

⊕ *Example*

Is the adjacent argument valid?

ARGUMENT
p or q $\sim p$ Therefore, q.

To see if an argument is valid, we first assume that both hypotheses are true.

Assume that p *or* q is true.
Assume that $\sim p$ is true.

Do these assumptions force the conclusion to be true?

Since $\sim p$ is true, p is false.
Since p *or* q is true and p is false, q must be true.

If the hypotheses are true, the conclusion must also be true.
So this argument is valid.

To show that the previous argument is valid, we demonstrated that the following implication is true:

$$[(p \text{ or } q) \text{ and } \sim p] \Rightarrow q$$

ARGUMENT
h_1 h_2 Therefore, c.

If we let h_1 and h_2 represent the hypotheses and c the conclusion, we can represent this argument as illustrated in the adjacent template. The above implication can be translated as:

$$(h_1 \text{ and } h_2) \Rightarrow c$$

An argument of the adjacent form is valid if and only if the above implication is true.

If we have more than two hypotheses from which we wish to draw a conclusion, we can generalize further by representing the hypotheses with a sequence of letters:

$$h_1, h_2, h_3, \ldots, h_n.$$

Let $h_1, h_2, \ldots, h_n$ represent the hypotheses of an argument and c represent the conclusion. The argument is *valid* if and only if the following implication is a law of logic:

$(h_1 \wedge h_2 \wedge h_3 \wedge \ldots \wedge h_n) \Rightarrow c$

This new argument, which has n different hypotheses, will be valid if and only if the following implication is true:

$$(h_1 \text{ and } h_2 \text{ and } h_3 \text{ and } \ldots \text{ and } h_n) \Rightarrow c$$

When we say a conclusion "follows from" the hypotheses, we mean that the above implication is true. In other words, if each hypothesis is true, the conclusion must also be true. The formal definition of a valid argument is given on the left.

Structure vs. Content

The validity of an argument is determined completely by the structure of the argument rather than the content of the component sentences. To determine if the following argument is valid, we do not consider whether "x is in A" or "x is in B;" we only consider the structure of the argument.

⊕ *Example*

Is the following argument valid?

Argument: x is in A or x is in B.
x is not in A.
Therefore, x is in B.

ARGUMENT
p or q $\sim p$ Therefore, q.

To see the structure of the argument, let p represent "x is in A" and q represent "x is in B," as illustrated on the left. Since its structure is the same as in the previous example on the facing page, this argument is valid.

Truth vs. Validity

The validity of an argument does not guarantee that the conclusion is true; it only guarantees that the conclusion *follows* from the hypotheses. To deduce that the conclusion is true, we must also know that the hypotheses are true.

✦ *Example*

Determine whether or not the given argument is valid.

ARGUMENT
p or q $\sim p$ Therefore, q.

1. *Argument*: $1+1=3$ or $2+2=5$.

   $1+1 \neq 3$

   Therefore, $2+2=5$.

   Since the argument has the adjacent structure, it is a valid argument.

   This argument is an example of a valid argument with a false conclusion. Note, however, that the first hypothesis of the argument is false. If both hypotheses were true, the conclusion would have to be true.

ARGUMENT
p or q p Therefore, q.

2. *Argument*: $1+1=2$ or $2+2=4$.

   $1+1=2$

   Therefore, $2+2=4$.

   The structure of this argument is given on the left. If we assume that both hypotheses are true, it does not force the conclusion to be true. So this argument is not valid.

   This argument is an example of an invalid argument with a true conclusion. If an argument is not valid, it does not mean that the conclusion is false; it only means that the conclusion does not follow from the hypotheses.

Using Logical Operators

When we make deductions, the outside structure of our reasoning process is based on the logical operators and quantifiers. We must be very comfortable with the meaning of these terms in order to have the fluency of language that is necessary to work through the steps of a proof. In addition to having a personal understanding of their definitions, we also need to have their equivalent formulations in the top drawer of our memory file. If we're still bothered by the fact that $\sim q \Rightarrow \sim p$ has the same meaning as $p \Rightarrow q$, we should go back to Chapter 1 and work on understanding why this is true. As in any sport, we need to master the rules before we can play the game.

If you are present at a game of chess, it will not suffice, for the understanding of the game, to know the rules for moving the pieces. . . . To understand the game is a wholly different matter; it is to know why the player moves this piece rather than the other.

H. Poincaré
1854–1912

Chapter 1 contains all the rules that we need to know to play the ultimate mind-sport. Of course, as Poincaré reminds us in the adjacent quote, the game involves far more than the rules.

In football, you have to know how to throw the ball, how to catch the ball, and how to run with it. Fortunately, we will not have a field of moving obstacles trying to tackle us, but we will have to find a path to get us where we want to go. In the remainder of this section, we will work on throwing the ball and catching the ball. In the succeeding sections, we will practice running with the ball. Like throwing and catching, we need basic techniques for two types of deductions:

- How to *derive* a compound sentence
- How to make a deduction *from* a compound sentence.

We will first look at deductions that involve implications.

Deductions from an Implication

EQUIVALENCES
$p \Rightarrow q$ Therefore, $\sim q \Rightarrow \sim p$. Therefore, $\sim p$ or q.

LAW OF DETACHMENT
$p \Rightarrow q$ p Therefore, q.

When we have an implication in a proof, either in the beginning or in the middle, what can we deduce? From a single implication, we can deduce its equivalent formulations. If we know $p \Rightarrow q$ is true, we can deduce that $\sim q \Rightarrow \sim p$ is true and we can also deduce that $\sim p$ or q is true. When we have other information, we may be able to make further deductions.

If we know that $p \Rightarrow q$ is true and we also know that p is true, we can deduce that q is true. This deduction comes straight from the definition of implies. If an implication is true and its hypothesis is true, then its conclusion must be true. Logicians call this form of argument *modus ponens*, a Latin term for "method of assertion." It is also called the Law of Detachment because we detach the hypothesis from the implication.

The Law of Detachment is very straightforward and easy to remember. Most students use this rule correctly; it's the next rule that may cause some confusion.

A Negative Perspective

CONTRAPOSITION
$p \Rightarrow q$ $\sim q$ Therefore, $\sim p$.

Negations always add a layer of complexity to a reasoning task. The brain needs more processing time to interpret the same sentence phrased in a negative perspective, so we may need to spend a little more time in thinking through the validity of the following argument.

The only case in which $p \Rightarrow q$ is false is when p is true and q is false. If we know that $p \Rightarrow q$ is true and we also know that q is false, then p has to be false. Logicians call this form of argument *modus tollens*, which is Latin for "method of denial." Being in a state of denial is not very fashionable in our

CONTRAPOSITION
$\sim q \Rightarrow \sim p$
$\sim q$
Therefore, $\sim p$.

positive-thinking society, so we will call it the Law of Contraposition. We can view the structure of this argument as the Law of Detachment applied to the contrapositive. If we replace the implication $p \Rightarrow q$ with its contrapositive $\sim q \Rightarrow \sim p$, as illustrated in the adjacent box, the argument has the form of the Law of Detachment. If an implication is true and its hypothesis is true, then its conclusion must be true.

⊕ *Example*

Determine if the given argument is valid.

1. *Argument*: If x is not in B, then x is not in A.
   x is not in B.
   Therefore, x is not in A.

$p \Rightarrow q$
p
Therefore, q.

   When we analyze the structure of an argument, we may sometimes want to hide a negation inside a component sentence, as we do in the following substitutions:

   p: x is not in B. q: x is not in A.

   These substitutions reveal the adjacent structure, which is the Law of Detachment. So, the argument is valid.

2. *Argument*: If x is in A, then x is in B.
   x is not in B.
   Therefore, x is not in A.

$p \Rightarrow q$
$\sim q$
Therefore, $\sim p$.

   Let p and q represent the following sentence:

   p: x is in A. q: x is in B.

   With these substitutions, the argument has the same structure as the Law of Contraposition, so it is a valid argument. If we want to avoid the Law of Contraposition, we can replace the implication with its contrapositive. We then have the same argument as in the first example.

Deriving an Implication

In the previous discussion, we examined how we can proceed when we have an implication in the beginning or in the middle of an argument, which is analogous to throwing the ball. Now we will learn techniques for catching the ball. How do we structure an argument that ends with an implication? From this perspective, we know what we want to derive, but how do we set it up?

DIRECT PROOF
Assume p.
. . .
So, q.
Therefore, $p \Rightarrow q$.

One way to set up the proof of an implication is to assume that the hypothesis is true and then derive that the conclusion must be true. We can then assert that the implication is true. This type of proof is called a *direct proof* because it is a very direct method. However, we must be careful to track the assumptions and derivations, distinguishing between those that are stand-alone derivations and those that depend on the assumption.

The assumption in a direct proof must be clearly marked with a word like "assume" or "let." When we write a sentence without the "assume" preface, we mean that it stands on its own as a true sentence. If we know that p is true and we use p to derive that q is true, we would summarize our results by saying the much stronger statement, p *and* q. Notice the difference in the conclusions of the adjacent two arguments. Whether we say "assume $x \in A$" or just "$x \in A$" has no effect on how we would derive that $x \in B$ in either Argument 1 or Argument 2, but it does affect our final conclusion. The conclusion in Argument 2 is an and-statement, not an implication. All assumptions in a proof must be clearly flagged as assumptions because they affect our final deduction.

Argument 1

Assume that $x \in A$.

. . .

So, $x \in B$.

Hence, if $x \in A$, then $x \in B$.

Argument 2

$x \in A$

. . .

So, $x \in B$.

Hence, $x \in A$ and $x \in B$.

In Argument 1, $x \in B$ is not a stand-alone conclusion. After we make an assumption, we may want to adjust our writing style to help the reader navigate these subtleties. We may want to indent the dependent lines to indicate that the indented deduction is hanging on the coattail of that assumption, or we may want to reserve the bombastic "therefore" for our big stand-alone conclusion and use the milder "so" or "hence" to flag deductions that are dependent on an assumption. We should at least start a new paragraph after we have made our final deduction from an assumption.

Another Method

When we use the method of a direct proof to derive an implication, we sometimes run into a brick wall as we try to connect the beginning with the end. As we will see in the following sections, some implications are much easier to prove when they are stated in terms of the contrapositive, which gives us another method for proving an implication. This method can be considered as either a *proof by contraposition* or an *indirect proof*. They both have the same structure.

PROOF BY CONTRAPOSITION
Assume $\sim q$.
. . .
Hence, $\sim p$.
So, $\sim q \Rightarrow \sim p$.
Therefore, $p \Rightarrow q$.

In a proof by contraposition, we do a direct proof of the contrapositive, as illustrated in the adjacent template. We do the same thing in an *indirect proof*, except that we do not state the contrapositive at the end. We may find, though, that focusing on the contrapositive helps us keep this method straight.

INDIRECT PROOF
Assume $\sim q$.
...
So, $\sim p$.
Therefore, $p \Rightarrow q$.

From the definition of an implication, we know that the only case where $p \Rightarrow q$ is false is when p is true and q is false. Both the direct proof and the indirect proof show that this case cannot occur.

- In a direct proof, we assume p is true and derive that q must be true.
- In an indirect proof, we assume q is false and derive that p is false.

⊕ Example

Set up the outside structure of a valid argument whose conclusion is "If x is in A, then x is in B."

We can set up the outside structure of the argument with either a direct proof or an indirect proof.

Direct Proof: Assume that x is in A.

...

So, x is in B.
Therefore, if x is in A, then x is in B

Indirect Proof: Assume that x is not in B.

...

So, x is not in A.
Therefore, if x is in A, then x is in B

The Transitive Law

The *Transitive Law* gives us a way to leave out the middle term when we have two implications of the following form.

The first implies the second.

The second implies the third.

TRANSITIVE LAW
$p \Rightarrow q$
$q \Rightarrow r$
Therefore, $p \Rightarrow r$.

Using the Transitive Law, we can then deduce the following.

The first implies the third.

The validity of the Transitive Law follows from the definition of an implication. In the adjacent template, assume that both of the hypotheses are true. Now focus on the conclusion. If p is false, the conclusion is automatically true. If p is true, from the first implication, we can deduce that q is true. Since we now have that q is true, we can deduce from the second implication that r is true. Thus, $p \Rightarrow r$ is true. Consequently, the Transitive Law is a valid argument.

Example

Determine if the given argument is valid.

1. *Argument*: If x is in A, then x is in B.
   If x is in B, then x is in C.
   Therefore, if x is in A, then x is in C.

   This argument has the structure of the Transitive Law. So, it is a valid argument.

2. *Argument*: If x is not in A, then x is not in B.
   If x is not in B, then x is in C.
   Therefore, if x is not in A, then x is in C.

   Because of the negations, this argument may seem more complex than the previous argument. However, with the following substitutions, this argument has the same structure as the Transitive Law.

   p: x is not in A. q: x is not in B r: x is in C.

   By the Transitive Law, this argument is valid.

3. *Argument*: If x is not in A, then x is not in B.
   If x is not in B, then x is not in C.
   Therefore, if x is in C, then x is in A.

   We can view the first two sentences as $p \Rightarrow q$ and $q \Rightarrow r$. Using the Transitive Law, we can deduce $p \Rightarrow r$:

   If x is not in A, then x is not in C.

   Translating this statement in terms of its contrapositive gives us the above conclusion. So, this argument is valid.

Deductions from Or-Sentences

DERIVING FROM OR
p or q $\sim p$ Therefore, q.

What can we deduce from an or-sentence? If we know that one of the parts of the or-sentence is false, we can then deduce that the other part is true. The structure of this type of derivation is illustrated in the adjacent template. If we translate *p or q* in its equivalent form as $\sim p \Rightarrow q$, this argument has the same structure as the Law of Detachment.

In a similar manner, if we know that *p or q* is true and we also know that q is false, we can deduce that p is true, as illustrated in the following example:

Example

Valid Argument: x is in A or x is in B.
x is not in B.
Therefore, x is in A.

Cases

CASES
p or q
$p \Rightarrow r$
$q \Rightarrow s$
Therefore, r or s.

CASES
p or q. *Case* 1. Assume p. . . . Therefore r. *Case* 2. Assume q. . . . Therefore s. Therefore, r or s.

When we run into an or-sentence in a proof and have no information on the truth values of the component sentences, we usually set up the structure for a case by case argument. With cases, we take each component of the or-sentence and try to find something that it implies. Suppose that we know *p or q* is true. If we can derive $p \Rightarrow r$ and also derive $q \Rightarrow s$, then since one of the two hypotheses must be true, we can deduce that either r or s must be true. So the adjacent argument is valid.

A case argument has several layers, so we must be careful to maintain the appropriate structure when we write this type of argument. We will need a separate subproof to derive $p \Rightarrow r$ and also a separate subproof to derive $q \Rightarrow s$. The adjacent template illustrates the structure of a proof by cases. Each part of the or-sentence determines a case which we delineate with "Case 1" and "Case 2." If we know that *p or q* is true, then for Case 1, we assume that p is true and try to see what we can derive. Note that Case 1 in the adjacent box is simply the format for proving $p \Rightarrow r$. For Case 2, we assume that q is true and try to see what we can derive. Since one of the two cases must occur, one of the conclusions must be true. Thus, we can make a general stand-alone conclusion that either the conclusion from Case 1 or the conclusion from Case 2 is true. We will discuss cases in more detail in Section 2.5.

Example

The following argument is valid.

Argument: x is in A or x is in B.

Case 1. Assume that x is in A.
. . .
So, $x < 7$.

Case 2. Assume that x is in B.
. . .
So $x > 9$.

Therefore, $x < 7$ or $x > 9$.

Deriving an Or-Sentence

In the preceding discussion, we examined how to make a deduction *from* an or-sentence. Now we will go in the reverse procedure and figure out how to derive an or-sentence at the end of our proof (or sub-proof).

DERIVING OR
Assume $\sim p$. . . . So, q. Therefore, p *or* q.

Or-sentences seem to cause more confusion than implications, possibly because of the vagueness of *or*. However, the structure of how we derive an or-sentence is quite simple. To derive p *or* q, we first note that p is either true or false.

If p is true, then p *or* q is true.

Hence, to derive p *or* q, we only need to consider the case when p is false, as illustrated in the adjacent template. With this structure, we have also derived $\sim p \Rightarrow q$, which is equivalent to p *or* q. The basic technique for deriving an or-sentence is to assume that one of the parts is false and then derive that the other part is true.

EXPANDING OR
p is true. Therefore, p or q is true.

Another method for deriving an or-sentence is to prove that one of the components is true, although in most instances this will not be possible. On the other hand, if we know a sentence is true, we can expand it into a true or-sentence, regardless of the truth value of the second sentence. This seems fairly obvious; however, we may sometimes be a little timid about adding on a sentence whose truth value we do not know. The function of the word *or*, though, is to allow this type of expansion. If p is true, then p *or* q is true regardless of the truth value of q. So the adjacent argument is valid. Note that the first sentence is a stand-alone true sentence; it is not an assumption.

You may wonder why on earth we would ever want to take a sentence p that we know is true and write it as an or-sentence where we cannot be sure which part is true. Sometimes, though, we do have a real need for this type of argument. For example, to prove that for all sets A and B, $A \subseteq A \cup B$, we can argue as follows:

⊕ *Example*

The following argument is valid.

Argument: Assume that x is in A.
Then x is in A or x is in B. . . . *Expanding Or*
So, x is in $A \cup B$.

Deductions with And-Sentences

CONTRACTING AND
p and q Therefore, p.

DERIVING AND
. . . So p is true. . . . So q is true. Therefore, p and q is true.

If we have an and-sentence in a proof, we can break it down and work with the sentences individually. If *p and q* is true, then p must be true. Thus, the adjacent argument is valid. We can always contract an and-sentence into either of its component sentences. We cannot go in the other direction, though, and expand a true sentence into a true and-sentence. If we know that p is true, we cannot conclude that *p and q* is true.

And is the strongest of the logical operators. To say that "*p and q* is true" is stronger than saying "*p or q* is true;" it is also stronger than "$p \Rightarrow q$ is true." By a stronger sentence, we mean one that gives us more information. When we say "*p and q*" is true, we know that each component sentence is true. None of the other logical operators give us such specific information. Since *and* is stronger than the other operators, we have more work to do when we derive an and-sentence. We must prove that each part stands on its own without making any assumptions in the outside structure. In the adjacent template, please note that there are no assumptions. We must derive that p is true as a stand-alone deduction. If we assume p is true and derive q, we cannot deduce *p and q* as a stand-alone statement.

⊕ *Example*

Determine if the given argument is valid.

1. *Argument*: x is not in A.
   Therefore, x is not in A or x is in B.

   This argument is an *Expanding Or* argument, so it is valid.

2. *Argument*: x is in in A and x is not in B.
   Therefore, x is not in B.

   This argument is valid.

3. *Argument*: x is in A or x is in B.
   Therefore, x is in A.

   If we assume the hypothesis is true, it does not guarantee that the conclusion is true. So, this argument is not valid.

4. *Argument*: x is in A.
   Therefore, x is in A and x is in B.

   If we assume the hypothesis is true, it does not guarantee that the conclusion is true. So, this argument is not valid.

Analyzing Structure

In the reasoning process, we often have to analyze the structure of a situation or the structure of an argument. When we analyze an argument, we can see different structures, depending on the substitution filter that we use. We may want to hide a negation inside an abstract component sentence, or we may want to feature the negation as a key part of the structure. We may want to introduce more negations via the contrapositive in order to see the structure in a different light.

Outside to Inside

When we set up the structure to prove a given statement, we work from the outside to the inside. For example, let's examine how to write the outside structure of a proof of the following statement:

If p, then r *and* s.

The outside structure of this sentence is an implication, so we first set up the framework for proving an implication.

Assume that p is true.

. . .

Hence r *and* s is true.

Therefore, if p, then r *and* s.

Assume that p is true.
  . . .
  → So r is true.
  . . .
  → So s is true.
→ Hence r and s is true.
Therefore, if p, then r *and* s.

At the next level, we set up the structure to derive r *and* s, which is given on the left. Inside this layer, we have two subproofs to complete. We must derive r so that it stands on its own within this layer. When we work on this subproof, we can use the prior assumption that p is true.

When we structure the outside argument, we leave a space in the middle to complete our work. Within the outside structure, we usually need other arguments. Frequently, we have layers of valid arguments in a proof with one valid argument a subproof of another valid argument. As we interweave the arguments, we must keep track of the results that we derive from an assumption so that each derivation can be summarized in a stand-alone conclusion. It is not as complicated as it sounds. Actually, it is what makes the process interesting, giving it both texture and depth. As long as we keep track of the structure of our layers, it will seem simple. In this section, we worked with the outside structure of basic types of arguments. In the following sections, we will work on how to bridge the gap, creating layers of arguments and weaving them together to create proofs.

What Is a Proof?

Now that we have an understanding of valid arguments, we can add more detail to our earlier description of a proof (page 112). We build a proof by constructing valid arguments that we sew together with logical reasoning:

I mean the word proof not in the sense of the lawyers, who set two half proofs equal to a whole one, but in the sense of the mathematician, where

$$\tfrac{1}{2}\ \text{proof} = 0$$

and it is demanded for proof that every doubt becomes impossible.

Carl Friedrich Gauss
1777–1855

A *proof* is a linearly ordered structure of interwoven valid arguments where each sentence is one of the following:

- An assumption used in a valid argument
- An axiom, previous theorem, or definition
- A sentence that can be derived from previous sentences by a valid argument

The final stand-alone conclusion is the theorem that has been proved.

A proof must adhere to the high standards described by Carl Gauss in the adjacent quote. If we have 999 steps in a proof that are logically correct and one little step that is not backed up with a valid argument, the whole proof collapses.

Different Proofs

A proof of a theorem is not unique. Starting at a given point, there may be many different routes that we can travel to reach the same conclusion. Most students have a tendency to seek a straight path from the beginning to the end, but sometimes the straight path is not the shortest. When we connect the beginning of a proof with the end, a straight path may lead us up over the top of a rugged mountain and down the other side, when there at the bottom of the mountain may have been a simple path around the base that led to the same conclusion.

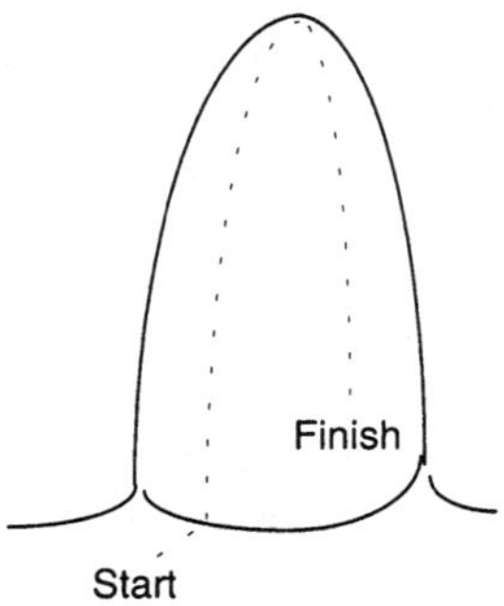

A straight path is not always the shortest.

The Pythagorean Theorem, one of the cornerstones of mathematics, has fascinated logical thinkers for centuries, and they have come up with over 300 different ways to prove it. Of course, only one proof is needed to classify a statement as a theorem. However, even though we may have a proof for a theorem, a different type of proof may deepen our understanding of why the theorem is true. A different proof may also suggest other possible theorems or provide us with new ideas on how to construct proofs for other theorems. As you search for a proof, do keep in mind that there may be more than one way to do it, and even after you find a proof, you may want to continue to search for a simpler path or a path with a different view of why the sentence is true.

Inductive vs. Deductive Reasoning

According to legend, a wise man used deductive reasoning to prove to an ancient Chinese Emperor that the volume of a sphere varies as the cube of its radius. The Emperor, not understanding the deductive argument, ordered his servants to bring in spheres of various sizes. He had the spheres filled with water and compared the volume of each sphere with its radius. At last, he was convinced of the wise man's assertion. Like the Chinese Emperor, most people are easily convinced by the inductive method. Deductive reasoning, on the other hand, is not an innate faculty, but a learned skill whose rules must be mastered in order to appreciate and understand its power. Both types of reasoning are essential in mathematics.

Inductive reasoning depends on the ability to recognize and describe patterns so that one can make a prediction as to what will continue to happen. The pattern might be the relationship between the numerical values of the volume of a sphere and its radius, or the relationship between the lengths of the sides of a right triangle. There are an amazing number of numerical and geometric patterns that seem to govern nature and the mechanics of the universe, so there are plenty of patterns to be discovered. Having discovered a pattern, we then use our powers of deductive reasoning to try to explain *why* such a pattern occurs.

The exercises in this chapter will have some inductive challenges, but most will focus on developing your skill in deductive reasoning. Please feel free, though, to inductively search for patterns and relationships and then try to certify them with the 100% guarantee that is provided by the deductive process of reasoning.

Exercise Set 2.1

1. Determine if the given argument is valid.

   a. s Therefore, r and s.

   b. s Therefore, r or s.

   c. r or s Therefore, s.

   d. If s, then t. t Therefore, s.

   e. If s, then t. $\sim s$ Therefore, $\sim t$.

   f. If s, then t. $\sim t$ Therefore, $\sim s$.

   g. s or t. $\sim t$ Therefore, s.

   h. Assume c. Hence, b. Therefore, if c, then b.

   i. c Hence, b. Therefore, c and b.

2. Determine if the given argument is valid.
   a. p and q. Therefore, p or q.
   b. p and q. Therefore, $p \Rightarrow q$.
   c. p or q. Therefore, $p \Rightarrow q$.
   d. $p \Rightarrow q$. Therefore, $\sim p$ or q.
   e. For every real number x, $p(x)$ is true. Therefore, $p(5)$ is true.
   f. For some real number x, $p(x)$ is true. Therefore, $p(5)$ is true.
3. Determine if the given argument is valid.
   a. x is in A or x is in B. Therefore, x is in A
   b. x is in A and x is in B. Therefore, x is in A.
   c. x is in A. So, x is in A or x is in B.
   d. x is in A. So, x is in A and x is in B.
4. Let x be a real number. Determine if the given argument is valid. If the argument is not valid, give a counterexample.
   a. If $x^2 < 1$, then $x < 1$. $x^2 \nless 1$. Therefore, $x \nless 1$.
   b. If $x^2 < 1$, then $x < 1$. $x \nless 1$. Therefore, $x^2 \nless 1$.
   c. If $x^2 < x$, then $x < 1$. $x^2 \nless x$. Therefore, $x \nless 1$.
   d. If $x^2 < x$, then $x < 1$. $x \nless 1$. Therefore, $x^2 \nless x$.
5. Let x be a point in a plane. Determine if the given argument is valid. If it is not valid, draw a sketch of circular sets and a point x for which the hypotheses are true and the conclusion is false.
   a. If x is in B, then x is in A. x is not in B.
      Therefore, x is not in A.
   b. If x is in B, then x is in A. If x is in A, then x is in D.
      x is not in D. Therefore, x is not in A.
   c. If x is in C, then x is in D. x is not in D.
      Therefore, x is not in C.
   d. If x is in B, then x is in C. If x is in C, then x is in D.
      x is not in B. Therefore, x is not in D.
6. Determine if the argument is valid. Don't get nervous about any words that you do not understand. For validity, it is only the structure that matters.
   a. If f is a differentiable function, then f is continuous.
      f is not continuous. Therefore, f is not differentiable.
   b. If f is a differentiable function, then f is continuous.
      f is not differentiable. Therefore, f is not continuous.
   c. For all x, if $|x - 1| < \delta$, then $|f(x) - f(1)| < \varepsilon$.
      $|f(x_0) - f(1)| \geq \varepsilon$. Therefore, $|x_0 - 1| \geq \delta$.

7. If possible, make a valid deduction from the given information. You may want to write some of the sentences in an equivalent form.

   a. If $x \in A$, then $x \in B$.
      If $x \notin C$, then $x \notin B$.
      Therefore, _____

   b. If $x \in C$ and $x \in D$, then $x \in$ F.
      $x \notin$ F.
      Therefore, _____

   c. If $x \in B$, then $x \in A$.
      If $x \in A$, then $x \in C$.
      $x \notin$ C.
      Therefore, _____

   d. x is in D or x is in E.
      x is not in E.
      Therefore, ________

   e. If $x \in A$, then $x \notin B$.
      $x \in B$.
      Therefore, _____

   f. If $x \in A$, then $x \in B$ and $x \in C$.
      $x \notin B$ or $x \notin C$.
      Therefore, _____

   g. If $x \in D$ or $x \in E$, then $x \in Z$.
      $x \notin Z$.
      Therefore, _____

   h. If x is in A, then x is in B.
      If x is in B, then x is in A.
      Therefore, ___________

8. Make the strongest deduction possible.

   a. Assume x is in C.
      So, x is in D.
      Therefore, _______

   b. Assume that x is not in C.
      So, x is not in R.
      Therefore, _______

   c. Assume that x is not in C.
      So, x is in R.
      Therefore, _______

   d. x is in C.
      So, x is in D.
      Therefore, _______

   e. x is in A or x is in C.
      Assume that x is in A.
      So, x is in D.
      Assume that x is in C.
      So, x is in E.
      Therefore, _______

9. a. What is an argument?
   b. What is a valid argument?
   c. What is a proof?
   d. What is a theorem?
   e. What is a conjecture?

10. a. How is the structure of a proof similar to a good essay?
    b. What are the 3 types of sentences that we use in a proof?

11. Is the given expression a sentence?

    a. $x + y$
    b. Assume $x + y$.
    c. Assume $x + y = 1$.
    d. $A \cup B$
    e. Assume $A \cup B$.
    f. Assume $x \in A \cup B$.
    g. $A \subseteq B$
    h. Assume $A \subseteq B$.
    i. $A \cap B$

12. If an argument is valid, does the conclusion have to be true? If not, give a counterexample.

13. Let n be an integer greater than 1.

    n is *prime* if and only if its only positive factors are n and 1.

    a. Suppose that n is not a prime number. What can you deduce? Write your deduction in terms of variables and quantifiers.

    b. Suppose that n is a prime number and $n = ab$ where a and b are positive integers. What can you deduce about a?

    c. Suppose that n is a prime number and $n = ab$ where a and b are integers. What can you deduce about a?

14. In this exercise, you are asked to prove one of the most important theorems in all mathematics, the Pythagorean Theorem.

    a. Draw 4 copies of the same right triangle. Label the legs as a and b, and the hypotenuse as c. Cut out the 4 triangles and try to arrange them so that somewhere you see c^2. (*There are different ways to do this.*)

    b. Using your picture, try to derive the Pythagorean Theorem. Expressing the same area in different ways may produce the desired result. You may use the formulas for the area of a square and the area of a triangle.

    c. Explain why any "squares" that you used in your proof are really squares. Comment on whether your proof is dependent on the size and shape of the right triangle that you use.

15. See if you can find a pattern in the given data that will enable you to predict the number of eggs on the 90th day and the number of eggs on the nth day.

    a. On the 1st day there were 2 eggs, on the 2nd day there were 4 eggs, on the 3rd day there were 6 eggs, and on the 4th day there were 8 eggs.

    b. On the 1st day there were 3 eggs, on the 2nd day there were 5 eggs, on the 3rd day there were 7 eggs, and on the 4th day there were 9 eggs.

    c. On the 1st day there were 2 eggs, on the 2nd day there were 4 eggs, on the 3rd day there were 8 eggs, and on the 4th day there were 16 eggs.

    d. On the 1st day there was 1 egg, on the 2nd day there were 3 eggs, on the 3rd day there were 7 eggs, and on the 4th day there were 15 eggs.

16. What type of reasoning did you use in the previous exercise, inductive reasoning or deductive reasoning?

Activity 2.2

Let x and y be integers. Do you think the given statement is true? (*Test it with a lot of examples before you jump to any conclusions.*) If so, try to prove it. If not, give a counterexample.

1. If x is even and y is even, then $x+y$ is even.
2. If x is odd and y is odd, then $x+y$ is odd.
3. If x is odd, then x^2 is odd.
4. If x^2 is even, then x is even.

≡ 2.2 Proving Implications ≡

Most sentences that we try to prove are phrased in terms of an implication. There are two methods for deriving an implication, a *direct proof* and an *indirect proof* (page 123). The method that we select gives us the beginning and end of the proof, as illustrated in the adjacent templates.

> *Derive*: If p, then q.
>
> *Direct Proof*
> Assume that p is true.
> . . .
> Therefore, q is true.
> So, if p, then q.

> *Derive*: If p, then q.
>
> *Indirect Proof*
> Assume that q is false.
> . . .
> Therefore, p is false.
> So, if $\sim q$, then $\sim p$.
> So, if p, then q.

With a direct proof, we assume the hypothesis is true.

With an indirect proof, we assume the conclusion is false.

An indirect proof is a direct proof of the contrapositive. If we assume q is false, we are assuming $\sim q$ is true. When we derive that p is false, we have derived that $\sim p$ is true, which means that we have proved the contrapositive: $\sim q \Rightarrow \sim p$. We can prove an implication by either doing a direct proof of the implication or a direct proof of its contrapositive.

Some implications are easier to prove in their contrapositive form, but usually we first try to do a direct proof. Let's work through the process of constructing a direct proof of the following theorem:

For every integer x, if x is even, then x^2 is even.

First, we set up the outside structure of the proof, writing our assumptions at the beginning and stating what we want to derive at the end, as illustrated on the next page.

With the outside structure in place, we then work on connecting the beginning with the end, with our focus always on what we are trying to derive. We may need to rephrase it in a form that helps us see how to structure the proof. Usually we work backwards from the end until we have a firm grasp of what we have to demonstrate. In the adjacent template, what do we need in Step 4 in order to derive Step 5? To answer this question, we substitute in the definition of even:

> *Theorem A*: For every integer x, if x is even, then x^2 is even.
>
> ---
>
> *Direct Proof*
> 1. Let x be an integer.
> 2. Assume that x is even.
> 3. ...
> 4. **There exists an integer k such that $x^2 = 2k$.**
> 5. Then x^2 is even.
> 6. So, if x is even, then x^2 is even.

x^2 is even if and only if
there exists an integer k such that $x^2 = 2k$.

The above substitution gives us Step 4, which is written in bold in the adjacent box to remind us that it is our focus. Our job now is to find an integer k such that $x^2 = 2k$. It is not yet apparent as to where we will find k, so let's start working down from the top and translate Step 2.

When we translate "x is even," we cannot use k again because we used it with x^2. So, we must use another letter, such as j: $x = 2j$. Since we're focused on our goal of finding a k so that $x^2 = 2k$, we know that we need to bring x^2 into the picture. One way to do this is to square both sides of the above equation. Then all we have to do is factor 2 from the right side and we have found the k that we were looking for. The complete proof of this theorem is given below in an outline style.

Theorem A For every integer x, if x is even, then x^2 is even.

Direct Proof – Outline Style

1. Let x be an integer.
2. Assume that x is even.
3. There exists an integer j such that $x = 2j$.
   So, $x^2 = 4j^2 = 2(2j^2)$.
   Define k as follows: $k = 2j^2$.
   Since j is an integer, k is an integer.
   Furthermore, $x^2 = 2k$
4. So, there exists an integer k such that $x^2 = 2k$.
5. Thus x^2 is even.
6. So, if x is even, then x^2 is even.

Using the above outline, we can convert the proof to the following paragraph style.

Theorem A

For every integer x, if x is even, then x^2 is even.

Direct Proof – Paragraph Style

Let x be an even integer. Since x is even, there exists an integer j such that $x = 2j$. Squaring both sides, we get $x^2 = 4j^2$. Define k as follows: $k = 2j^2$. Since j is an integer, k is an integer. Furthermore, $x^2 = 2k$. Thus, x^2 is even. Therefore, if x is even, then x^2 is even.

The converse of the above theorem is also a theorem, but look what happens when we try to prove it with a direct proof.

Theorem B

For every integer x, if x^2 is even, then x is even.

Attempted Direct Proof

1. Let x be an integer.
2. Assume that x^2 is even.
3. There exists an integer j such that $x^2 = 2j$.
   So, $x =$ ___
4. Thus, there exists an integer k such that $x = 2k$.
   Hence, x is even.
5. Therefore, if x^2 is even, then x is even.

In Step 3, we run into a brick wall. From the equation $x^2 = 2j$, we can deduce that $x = \sqrt{2j}$; however, this equation does not help us find an integer k such that $x = 2k$. The strategy we used to prove Theorem A does not work here. So, let's try another strategy and see if we can prove the contrapositive.

If x is not even, then x^2 is not even.

> *Theorem B*: For every integer x, if x^2 is even, then x is even.
>
> *Indirect Proof*
> 1. Let x be an integer.
> 2. Assume that x is not even.
> 3. ...
> 4. ...
> 5. So, x^2 is not even.

First, we set up the outside structure, as illustrated on the left. If an integer is not even, it has to be odd, so we can work backwards and translate Step 5 as follows:

4. There exists an integer k such that $x^2 = 2k + 1$.

Our goal now is to find an integer k such that $x^2 = 2k + 1$. Keeping this new goal in mind, we're ready to jump back to the beginning and work our way down. On the next page is a proof in outline style, followed by a proof in paragraph style. The paragraph style is longer because we have more explanation in it.

Theorem B For every integer x, if x^2 is even, then x is even.

Indirect Proof – Outline Style

Let x be an integer.
Assume that x is not even. Then x must be odd.
So, there exists an integer j such that $x = 2j + 1$.
$x^2 = (2j+1)^2 = 4j^2 + 4j + 1$
$= 2\,(2j^2 + 2j) + 1$
Define k as follows: $k = 2j^2 + 2j$.
Since j is an integer, k is an integer.
Furthermore, $x^2 = 2k + 1$. So x^2 is odd.
Thus, x^2 is not even.
We have shown that if x is not even, then x^2 is not even.
So, if x^2 is even, then x is even.

Theorem B For every integer x, if x^2 is even, then x is even.

Indirect Proof – Paragraph Style

Let x be an integer. Assume that x is not even. Since every integer is either even or odd, x must be odd. So, there exists an integer j such that $x = 2j+1$. Squaring both sides, we get

$$x^2 = (2j+1)^2 = 4j^2 + 4j + 1 = 2\,(2j^2 + 2j) + 1.$$

Define k as follows: $k = 2j^2 + 2j$. Since products and sums of integers are also integers, k is an integer. So there exists an integer k such that $x^2 = 2k + 1$, which means that x^2 is odd. An odd integer cannot be even, so x^2 is not even. Hence, if x is not even, then x^2 is not even. Therefore, if x^2 is even, then x is even.

Proving Equivalences

Derive: p if and only if q
$p\Rightarrow q$: . . .   So, if p, then q.
$q\Rightarrow p$: *Conversely,* . . .   So, if q, then p.
Therefore, p if and only if q.

The standard technique for proving an equivalence of the form *p if and only if q* is to prove the following two implications:

$$p \Rightarrow q \text{ and } q \Rightarrow p$$

The structure of this type of proof is actually that of two separate proofs, as illustrated in the adjacent template. What we assume in each part depends on whether we use the direct or indirect method to prove the implication. The beginning of the second part of the proof is usually flagged by saying "conversely." Otherwise, the reader might be confused by the two assumptions. Since the two parts of the proof are separate, derivations from the first assumption cannot be used in the second part of the proof.

The following proof illustrates the structure for proving an equivalence. Note how each part is a separate proof. In fact, the two proofs are merely copies of the previous proofs for Theorem A and Theorem B.

Theorem For every integer x, x is even if and only if x^2 is even.

Proof Let x be an integer.

$p \Rightarrow q$ Assume that x is even. So, there exists an integer j such that $x = 2j$. Squaring both sides, we get $x^2 = 4j^2$. Define k as follows: $k = 2j^2$. Since j is an integer, k is an integer. Furthermore, $x^2 = 2k$. Thus, x^2 is even. Therefore, if x is even, then x^2 is even.

$q \Rightarrow p$ Assume that x is not even. Since every integer is either even or odd, x must be odd. So, there exists an integer j such that $x = 2j+1$. Squaring both sides, we get

$$x^2 = (2j+1)^2 = 4j^2 + 4j + 1 = 2\,(2j^2 + 2j) + 1.$$

Define k as follows: $k = 2j^2 + 2j$. Since products and sums of integers are also integers, k is an integer. So there exists an integer k such that $x^2 = 2k + 1$, which means that x^2 is odd. An odd integer cannot be even, so x^2 is not even. Hence, if x is not even, then x^2 is not even. So, if x^2 is even, then x is even.

Thus, x is even if and only if x^2 is even.

Proving an Or-Sentence

Derive: p or q
Assume p is false. . . . So, q is true. Therefore, p or q.

To prove an or-sentence, we can assume that one of the parts is false and then derive the other part (page 127). In the adjacent template, we assume that p is false and then derive that q is true.

When we studied the logical operators in Chapter 1, we examined why *p or q* has the same meaning as $\sim p \Rightarrow q$. Note that the adjacent structure is the same as the form for proving $\sim p \Rightarrow q$.

On the other hand, we could assume that q is false and then derive that p is true. With this format, we are proving $\sim q \Rightarrow p$, which also has the same meaning as *p or q*. The structure of this type of proof is illustrated in the following example.

Theorem For every real number x, $x > 4$ *or* $x < 5$.

Proof Assume that $x \not< 5$.
Then $x \geq 5$.
Since $x \geq 5$ and $5 > 4$, $x > 4$.
So, $x < 5$ or $x > 4$.

Exercise Set 2.2

1. Set up the outside structure for a direct proof of the given implication. Then set up the outside structure for an indirect proof of the implication. Which do you think would be easier to prove?
   a. If m and n are odd integers, then mn is odd.
   b. If mn is odd, then m and n are odd integers.
   c. If a is not a factor of $b + c$, then a is not a factor of b or a is not a factor of c.
   d. If x is rational and y is rational, then $x + y$ is rational.
2. Set up the outside structure for proving the given equivalence:
   a. For all x, $x \in A$ if and only if $x \in B$
   b. m and n are odd integers if and only if mn is odd.
3. Set up the outside structure for proving the given or-sentence.
   a. For all x in U, $x \in A$ or $x \in B$.
   b. For every integer x, x is even or x^2 is odd.
4. Set up the outside structure for an indirect proof of the following statement. Then set up the inside structure for deriving the conclusion.

   If x is rational and y is irrational, then $x + y$ is irrational.
5. Let x and y be arbitrary integers. Prove the following. You may use previous results in a proof.
   a. If x is even and y is even, then $x+y$ is even.
   b. If x is even, then xy is even.
   c. If $x+y$ is odd, then x is odd or y is odd.
   d. If xy is odd, then x is odd.
   e. If xy is even, then x is even or y is even.

Activity 2.3

Do you think the given statement is true? Justify your answer.

1. 3 is a factor of x if and only if 3 is a factor of x^2.
2. 4 is a factor of x if and only if 4 is a factor of x^2.
3. 5 is a factor of x if and only if 5 is a factor of x^2.

≡ 2.3 Writing a Proof ≡

---

4 Stages in Writing a Proof

1. Analyze the structure.
2. Write the end and beginning of the proof.
3. Connect the beginning with the end.
4. Polish the proof.

Writing a proof is very different from reading a proof. Normally we read a proof from the beginning to the end in order to make sure that each step follows from the steps before it. However, when we write a proof, we usually start at the end and then work from both the end and beginning as we try to connect the two pieces in a logical framework.

Writing a proof is a four-stage process. First, we analyze the structure of the sentence that we want to prove in order to see how to structure the proof. Then we write the end and beginning of the proof. The challenging part is the third stage when we try to connect the beginning with the end. Last, but not least, we polish our proof until it emerges as a well-written argument. Let's examine each of these four stages.

1. Analyze the structure

Before we can write a proof, we need to understand what we are attempting to prove. We should look up the definitions of any words or terms that we cannot verbalize. There is no way that we can hope to prove a theorem if we cannot write the meaning of the terms that are involved.

After we have all relevant definitions at our fingertips, we should interpret the theorem for specific examples, which may give us some ideas on how to attack the proof. The examples will certainly give us a firmer footing for working with the abstractions in a meaningful way.

We should take advantage of our powers of visual reasoning and try to visualize the theorem by drawing general pictures or pictures of specific examples. The pictures need not be descriptive in a detailed way. As long as they give some hint of the flavor of the theorem, they can be an invaluable resource.

When we have a good understanding of the verbal and visual meaning of the sentence that we want to prove, we should then analyze its outside structure. For example, to prove $A \subseteq A \cup B$, we must see the outside structure of this sentence as $A \subseteq Y$. We would then use the definition of subset to set up the outside structure of our proof.

Introduce Variables

If the sentence that we want to prove has no variables in it, we usually translate the sentence in terms of variables so that we have something tangible to manipulate. For example, to prove that the sum of every two even numbers is even, we would introduce variables x and y to represent the even numbers.

We should scan the proposed theorem for hidden quantifiers, such as "there is" or "any." For each quantifier that we find, we introduce a variable, as illustrated in the following example.

Theorem C There is no largest real number.

Stage 1 First, notice the phrase "there is," which is flagging the existential quantifier. We will use x to go with this quantifier. Now, we must sort out where the negation goes. A little contemplation should convince us that it goes before the quantifier:

$$\sim (\exists x, x \text{ is the largest real number.})$$

Next, we translate "x is the largest real number":

$$\forall y, x \geq y$$

Then we apply the rules for negating quantifiers:

$$\sim(\exists x\ \forall y, x \geq y)$$

$$\forall x\ \exists y, x < y$$

Now we translate back to word form:

For every real number x, there is a real number y such that $x < y$.

Now we have the theorem in a form that is easy to prove.

Translate with Logical Operators

In addition to translating in terms of quantifiers, we should further translate the proposed theorem in terms of the logical operators so that we have a clear understanding of how to structure the proof. If the sentence contains words such as "only if," "whenever," or "necessary," we should translate it in terms of an implication. When we translate to an implication, we should then compare the implication with its contrapositive and pick the one that seems the easiest to prove.

2. Write the end and beginning

What could we put on Line 4 to derive Line 5?

What could we put on Line 3 to derive Line 4?

When planning a car trip across the country, we have to focus on where we're going in order to figure out how to get there. The same strategy is needed for constructing a proof. If a proof is not obvious, we should first write the end of the proof, which is the sentence that we want to prove, and then work backwards from the end as far as we can. If Line 5 is the last line of the proof, we should contemplate what is needed on Line 4 to derive Line 5. On the next-to-last line of a proof, we usually translate the theorem in terms of a definition, as illustrated in the following proof.

Theorem: For all sets A and B, $A \subseteq A \cup B$.

Proof: Let A and B be sets.

. . .

Definition → For every x, if $x \in A$, then $x \in A \cup B$.

So, $A \subseteq A \cup B$.

The above definition of a subset shows us how to structure the next layer of the proof. We jump back to the beginning of the proof where we introduce the necessary variables and write any assumptions needed for whatever we've written on the next-to-last line. Next, we jump down across the gap and list the sentence that we hope to derive, as illustrated below. We now have a simplified goal to work towards.

Theorem: For all sets A and B, $A \subseteq A \cup B$.

Proof: Let A and B be sets.

Let x be an element in A.

. . .

Then $x \in A \cup B$.

For every x, if $x \in A$, then $x \in A \cup B$.

So, $A \subseteq A \cup B$.

Using a similar technique, we can set up the outside structure of Theorem C, which we analyzed on page 142.

Theorem C: There is no largest real number.

1. Let x be a real number.
2. . . .
3. There exists a real number y such that $x<y$.
4. For every real number x, there is a real number y such that $x<y$.
5. So, there is no largest real number.

---

3. Connect the beginning with the end

In this stage of the construction process, we work our way from the bottom up and from the top down so that we can bridge the gap. We usually back up as far as we can from the bottom in order to see the direction needed for the beginning of the proof. We then jump back to the beginning and try to work our way down, focusing always on the line we are trying to derive.

If we don't see how to connect the beginning with the end, we should again review definitions of all the involved terms and look for previous theorems that might be useful. We should write this information off to the side. When we see where we can use any of the information, we can then move it into the proof in whatever translated form is needed.

If we're still stumped, now might be the time to switch to another type of proof. Perhaps the theorem can be rephrased in a different way, such as the contrapositive, that is easier to prove. Or perhaps we may want to try a proof by contradiction or use cases or try mathematical induction. We will discuss techniques for structuring these types of proofs in the following sections.

Constructing a proof may take a while. The serious thinkers do not give up. For Carl Gauss, one of the greatest mathematicians of all time, a year was not an unreasonable amount of time to spend on an interesting problem. Of course, you must have a great passion for what you are doing in order to invest that amount of time in a proof.

When the proof process drags on and on, there are techniques to help us get over the humps. If it looks hopeless, take a break, and come back to the proof in a couple of hours or

---

For a whole year, this theorem tormented me and absorbed my greatest efforts. . . . Finally, two days ago, I succeeded, not on account of my painful efforts, but by the grace of God. Like a sudden flash of lightning, the riddle happened to be solved. I myself cannot say what was the conducting thread which connected what I previously knew with what made my success possible.

Carl Friedrich Gauss
1777–1855

---

even the next day. Meanwhile, we may find that our subconscious has continued to work on it. We may get a sudden flash of insight as we wait in the cafeteria line. The more times we go back and think about the problem, the more likely we are to find a solution. A common attribute of all logical thinkers and all good athletes and, in general, all successful people, is perseverance. In an extreme sport like mathematics, perseverance is essential. Without perseverance, we will not spend the necessary time to even get in the game.

Sometimes we find ourselves frustrated at having spent so much time on a proof, especially if we have no happy ending to proudly display. When this happens, it is important to realize that we have not wasted our time. Like lifting weights at the health club, the time that we spend thinking about a proof develops our mental muscles.

Theorem C: There is no largest real number.

Proof

1. Let x be a real number.
2. Define y_o as follows: $y_o = x + 1$
   Since x is a real number,
   y_o is a real number.
   Since $x < x + 1$,
   $x < y_o$.
3. So, for every x, there is a real number y such that $x < y$.
4. So, there is no largest real number.

The proofs assigned in the exercises are designed to help you develop your own technique for constructing proofs. Templates and tips will be given and writing styles will be suggested, but the whole goal is for you to figure out the way that works best for you. As long as you come up with a proof, it doesn't matter how you got there. If you can't get there, though, use the techniques presented here. More then likely, they will save you a lot of time and frustration.

With Theorem *C*, the gap that we left on the previous page is very easy to bridge. We have to demonstrate that there exists a y such that $x < y$. One way to show something exists is to construct it. In the adjacent proof, x is introduced on Line 1, so we can use it in our construction of y_o on Line 2. We then show that y_o satisfies the conditions and we have a proof. It's that simple.

4. Polish the Proof

After we logically connect the beginning of our proof with the end, we are ready for the final stage of the proof process, which is to polish and refine it. We should remove any information that is not essential to the line of reasoning that connects the beginning with the end. A proof should be as simple as possible and easy for other people to follow. The reader should know that everything we have previously written still holds, so we do not need to keep repeating it, unless we are using it to explain our reasons. We sometimes leave it to the reader to fill in reasons that we consider obvious, but as you probably know from your own experience, what an author considers obvious is not always obvious to the reader. The amount of justification we insert for the steps in our proof depends on the audience for whom we are writing.

The proof for Theorem C on the previous page is written in an outline style, similar to an outline for an essay. The outline style of proof is often used in teaching high school geometry, but the paragraph style is always used in mathematics journals and in most textbooks. The outline style simplifies the appearance of the line of reasoning so that we can quickly scan it and spot any logical errors or missing steps. The visual organization of having each new thought start on the left side seems to make it easier for the brain to see the connections. The reasoning goes straight down, from top to bottom, with no wraparounds. The extra white space gives breathing room so that we are not overwhelmed by the denseness of the text. The numbering of lines in the outline style lends an organizational structure to the proof, grouping related sentences together. If we number each line in a long proof, though, the organizational impact of the numbering is lost. Leaving space on the right side for reasons reminds us that we do need a reason to support each line, regardless of whether or not we feel it necessary to write it down.

The same proof is written below in a more conversational paragraph style.

Theorem C There is no largest real number.

Proof – Paragraph Style Let x be a real number. Define y_o as follows: $y_o = x+1$. Since y_o is the sum of 2 real numbers, y_o is a real number. Furthermore, $x + 1 > x$, so $y_o > x$. So, for every real number x, there exists a real number y such that $y > x$. Therefore, there is no largest real number.

Writing Style The style that we use when writing a proof should be the best format for presenting it to the intended audience. In math textbooks, authors usually write proofs in paragraph form, but students often find it difficult to follow the reasoning. If this happens to you, rewrite the author's proof in outline style and you may be surprised at how much easier it is for you to see the connecting steps and fill in the missing reasons that were obvious to the author.

The proof styles used in this book were selected for an audience who wants to develop their ability to reason logically. Most proofs are written in a relaxed outline format, with new

Well-written homework:

Find the absolute minimum value of $f(x) = x^2 - 4x$.

$f'(x) = 2x - 4$

Set $f'(x) = 0$.

$2x - 4 = 0$

So, $x = 2$.

Take the derivative of $f'(x)$:

$f''(x) = 2$

Since $f''(x) > 0$, the graph is always concave up. So f has an absolute minimum at $x = 2$.

$f(2) = (2)^2 - 4(2) = -4$

So, the minimum value of f is -4.

A man who is always able to present his subject in such a way that it is readily understood, is a man who understands it himself.

A. G. Drachmann

thoughts starting on a new line. By seeing the relevant steps lined up on the left side, we are programming our brain with the essential details needed for logical reasoning. When we have our technique down, we will let our brain do more of the work in the background and switch over to a more conversational wrap-around paragraph format.

Whether writing in outline or paragraph style, formatting equations and inequalities so that equal and inequality signs line up is an immense help to the reader, as illustrated in the adjacent homework. The format for writing the solution to a calculus problem is essentially the same as the format for writing a proof. When we solve a problem, we use deductive reasoning. If we get confused in our homework, more than likely it is because of the way we are writing our reasoning.

If we write sentence fragments, such as $2x - 4$
instead of the complete sentence, $f'(x) = 2x - 4$,
we are setting ourselves up for major confusion. When we jot down the derivative, we know at that moment that we are computing the derivative, but if we have no written record of it, we cannot check our reasoning to see if we did it correctly, and we may end up substituting in the wrong equation.

In addition to using "equals" to form complete sentences, we should supplement our equations with leading words such as "set" and "so" and write sentences to explain what is going on, as illustrated in the adjacent homework. Review your current writing style by looking at homework from last week or last semester. Can you figure out what you did? If not, perhaps you need to work on how you write for yourself. Writing is one of the most important learning tools you have. By improving your writing, you are improving your understanding.

Exercise Set 2.3

1. The domain for x, y, a, b, and c is the set of integers. Translate the following sentences.
   a. x is odd.
   b. xy is odd.
   c. a is a factor of b.
   d. a is a factor of $b + c$.
   e. 7 is a factor of x^2.

2. Do you think the given sentence is true?
(*Test it with examples before jumping to a conclusion.*)
If so, try to construct a proof using the method described in this section. First write the end and beginning of the proof, leaving a space between the two. Then work backwards as far as you can, using your definitions in the previous exercise. Next, jump to the beginning and use the definitions to fill in the missing steps. If the direct method seems difficult, set it up for the indirect method.
   a. For all integers a, b and c, if a is a factor of b and a is a factor of c, then a is a factor of $b + c$.
   b. Let x be an integer.
   If x has 7 as a factor, then x^2 has 7 as a factor.
   c. For every integer x, x is even or x^2 is odd.
   d. For every integer x, x^2 is odd if and only if x is odd.
   e. For all integers x and y, x is odd and y is odd if and only if xy is odd.
3. Prove the following statement. First set up the outside structure, then set up the inner structure. Work backwards as far as you can. Complete the steps by focusing on what you want to derive.

   For all real numbers a and b, if $a < b$, then $a < \frac{a+b}{2} < b$.

Activity 2.4

1. When you disprove a statement p, what statement do you prove?
2. Let a and b be integers. The notation $a \mid b$ is read as "a divides b."
a divides b if and only if there exists an integer k such that $b = ak$.
Prove or disprove the given statement.
   a. $3 \mid 0$
   b. $0 \mid 3$
   c. $a \mid a$
   d. If $a \mid b$, then $b \mid a$.
   e. If $a \mid b$ and $b \mid a$, then $a = b$.
   f. If $a \mid b$ and $b \mid c$, then $a \mid c$.
   g. If $a \mid b$ and $a \mid c$, then $a \mid b + c$.

≡ 2.4 Working with Quantifiers ≡

To prove that a sentence is true for all x, we must demonstrate that it is true for each element in the domain of x. We cannot prove it by examples unless we give an example for each element in the domain. This method of proof is called the *method of exhaustion.* If we have 1000 elements in the domain, this technique will certainly be exhausting, that is, unless we're using a computer. If the domain is an infinite set and an infinite list of examples must be individually checked, not even a computer can handle the task. So we must find a general way to prove the result.

Universal Quantifier

When we construct a proof, for each variable x that is universally quantified, we should have a sentence of the following form:

Let x be an element in the domain.

After this type of introduction, x represent a generic member of the domain. Later in the proof we cannot assign x a specific value, such as "set $x = 2$," because x would no longer represent an arbitrary element. However, based on the assumptions in our proof, we may be able to derive that $x = 2$. We could also assume that $x = 2$, but we would have to use that assumption in any stand-alone deductions that we make.

We can change the domain or meaning of x when we have a subproof in which the reader is fully aware that we are no longer using any threads from our previous usage of x. As a good writer, we must plant the necessary signposts so that the reader knows when a separate part of the proof starts and stops.

Existential Quantifier

To prove that a sentence is true for some x, one single example will suffice. The existence of such an x can be proved by constructing or defining an x_o that satisfies the condition. We use the subscript in x_o to remind us that x_o is a specific element and not an arbitrary element, as is the case when x is universally quantified.

Constructing x_o

In a proof, we indicate that we are constructing x_o by writing a sentence of the following form:

Define x_o as follows: $x_o =$____

When we construct an x_0, we should point out that it satisfies the required conditions, as illustrated in the following proofs.

Theorem: There exists an integer x such that for every integer y, $x + y = y$.

Proof: Let $x_o = 0$. Note that 0 is an integer.
Let y be an integer. Then $x_o + y = 0 + y = y$.
Therefore, there exists an integer x such that
for every integer y, $x + y = y$.

Theorem: For every integer y, there exists an integer x such that, $x + y = 0$.

Proof: Let y be an integer.
Define x_o as follows: $x_o = -y$.
Since y is an integer, $-y$ is an integer.
$x_o + y = (-y) + y = 0$.
Therefore, for every integer y, there exists an integer x
such that, $x + y = 0$.

Order of the Quantifiers

Note how the order of the first two lines in the above proofs follows the order of the quantifiers in the theorem. The order of mixed quantifiers makes a difference in the meaning (page 30), so we must always introduce mixed quantifiers in the order in which they appear in the theorem. If y has already been introduced, we may use y in the construction of x_o. However, if y has not been introduced, we cannot use y in the construction of x_o, as illustrated in the following two templates:

Theorem: $\forall y\ \exists x,\ p(x,y)$

Proof: Let y be an element in the domain.
Let x_o = __ (*We can use y to construct x_o.*)

. . .

So, $p(x_o,y)$ is true.

Theorem: $\exists x\ \forall y,\ p(x,y)$

Proof: Let x_o = __ (*We cannot use y to construct x_o.*)
Let y be an element in the domain.

. . .

So, $p(x_o,y)$ is true.

Onto Functions

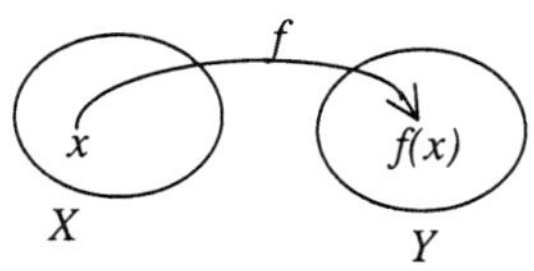

To practice the technique of working with mixed quantifiers, we will work with the definition of an onto function. If a function f maps the set X *into* the set Y, then each element x in the set X is mapped to an element $f(x)$ in the set Y, as illustrated on the left. However, an element in Y does not necessarily have an x that maps to it. If each element in Y does have an x that maps to it, we say that f maps X *onto* Y.

f maps X *onto* Y
if and only if
for every y in Y,
there exists an x in X
such that $f(x) = y$.

f maps X *onto* Y
if and only if
for every y in Y, there exists an x in X such that $f(x) = y$.

Because of the mixed quantifiers, many students find it difficult to write proofs that use the definition of onto. However, if we condition ourselves to structure our thinking in the correct order, the proofs will seem fairly simple.

✣ *Example*

Let $f(x) = 2x + 1$ where x is a real number. Let $\mathbb{R}$ represent the set of real numbers. Prove that f maps $\mathbb{R}$ onto $\mathbb{R}$.

f maps $\mathbb{R}$ *onto* $\mathbb{R}$
if and only if
for every y in $\mathbb{R}$,
there exists an x in $\mathbb{R}$
such that $f(x) = y$.

First we translate the previous definition of an onto function for the set $\mathbb{R}$, which gives us the adjacent definition. Next we set up the outside structure of our proof by working through the quantifiers from left to right. The first quantifier gives us the first sentence in our proof; the second quantifier gives us the second sentence. The remaining part of the definition gives us the third part of our proof.

1. Let y be an element in $\mathbb{R}$.
2. Let $x_o =$ ___
3. So, $f(x_o) = y$.

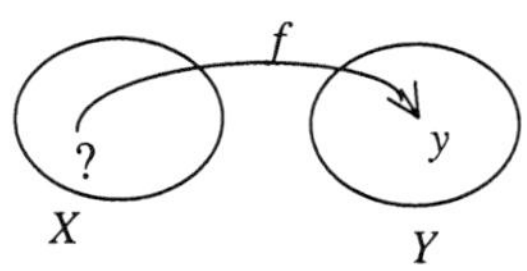

Our job now is to find x_o. Since y is introduced in the first step of our proof, we can use y to define x_o. The only clue we have is that $f(x_o) = y$. Working backwards from this equation, we can figure out how to express x_o in terms of y.

$$
\begin{aligned}
f(x_o) &= y \\
f(x_o) &= 2x_o + 1 && \text{.... } \textit{Definition of } f \\
y &= 2x_o + 1 && \text{.... } \textit{Substitution} \\
2x_o &= y - 1 \\
x_o &= \tfrac{y-1}{2} && \text{...... } \textit{Algebra}
\end{aligned}
$$

We have found x_o. However, we will not insert the above steps in our polished proof. Instead, we will first define x_o. Then, we will arrange the previous steps in the reverse order to demonstrate that $f(x_o) = y$. We must also check to see if x_o is a real number as required in the definition of onto. In the following proof, the steps are numbered to correspond to the 3 steps we used to structure the proof.

Theorem Let $f(x) = 2x + 1$ and let $\mathbb{R}$ represents the set of real numbers. Then f maps $\mathbb{R}$ onto $\mathbb{R}$.

Proof

1. Let y be a real number.
2. Define x_o as follows: $x_o = \frac{y-1}{2}$.
   x_o is a real number. ...... *Since y is real, $y-1$ is real.*
   .... *So $\frac{y-1}{2}$ is a real number.*
3. $f(x_o) = 2x_o + 1$ .... *Definition of f*
   $= 2\left(\frac{y-1}{2}\right) + 1$ .... *Substitute for x_o*
   $= (y-1) + 1$
   $= y$

   So, for every real number y, there exists a real number x such that $f(x) = y$. Therefore, f maps $\mathbb{R}$ onto $\mathbb{R}$.

⌖ *Example* Let $f(x) = 2x$ where x is in $\mathbb{N}$. $\mathbb{N} = \{1, 2, 3, \ldots\}$
Prove that f does not map $\mathbb{N}$ onto $\mathbb{N}$.

First, we negate the definition of onto:

$$\sim(\forall y \text{ in } \mathbb{N}, \exists x \text{ in } \mathbb{N},\ f(x) = y)$$
$$\exists y \text{ in } \mathbb{N}, \forall x \text{ in } \mathbb{N},\ f(x) \neq y$$

Using the above translation, we set up the outside structure of our proof by working with the quantifiers from left to right:

1. Let $y_o =$ ______
2. Let x be an element in $\mathbb{N}$.
3. So, $f(x) \neq y_o$.

Our job now is to find a specific natural number y_o so that for every natural number x, $f(x) \neq y_o$. Because of the order of the quantifiers, we cannot use x in our construction of y_o.

To find a y_o that will do the described task, let's think about the range of the function f. Since $f(x) = 2x$, $f(x)$ is even for every natural number x. Thus, we can pick y_o to be any natural number that is not even. In the following proof, we let y_o be 3.

Theorem Let $\mathbb{N} = \{1, 2, 3, \ldots\}$. Let $f(x) = 2x$ where x is in $\mathbb{N}$. The function f does not map $\mathbb{N}$ onto $\mathbb{N}$.

Proof

1. Let $y_o = 3$. $3 \in \mathbb{N}$.
2. Let x be an element in $\mathbb{N}$. Then $f(x) = 2x$.
3. Since x is an integer, $2x$ is an even number. But 3 is not even. So, $f(x) \neq 3$.

Thus, by the definition of onto, f does not map $\mathbb{N}$ onto $\mathbb{N}$.

Disproving a Statement

To disprove a statement, we prove its negation. First, we negate the sentence, bringing the negation across any quantifiers or logical operators so that we can see how to structure the proof. We then use the standard techniques for constructing a proof of the negated statement.

⊕ *Example* Disprove the following: For all real numbers, a, b, and c, if $a < b$, then $ca < cb$.

First, we negate the above sentence so that we can see how to structure the proof:

$$\sim(\forall a\ \forall b\ \forall c,\ a < b \Rightarrow ca < cb)$$
$$\exists a\ \exists b\ \exists c,\ a < b \text{ and } ca \nless cb$$

Our job now is to find real numbers a, b, and c so that:

$$a < b \text{ and } ca \nless cb.$$

We can write our proof as follows:

Proof Set $a = 2$, $b = 3$, and $c = -1$. $2 < 3$, but $(-1) \cdot 2 > (-1) \cdot 3$. So, there exists real numbers a, b, and c such that $a < b$ and $ca \nless cb$.

We can disprove a universally quantified sentence by finding a counterexample. However, we cannot disprove an existentially quantified sentence with a counterexample because its negation will be universally quantified.

Example Disprove the following: There exists an integer x such that x is odd and x^2 is even.

First we translate the negation.

$$\sim(\exists x,\ x \text{ is odd and } x^2 \text{ is even.})$$
$$\forall x,\ x \text{ is not odd or } x^2 \text{ is not even.}$$

Next we translate the or-sentence as an implication:

$$\forall x,\ x \text{ is odd} \Rightarrow x^2 \text{ is not even.}$$

For every integer x, if x is odd, then x^2 is odd.

We set up the outside structure of our proof in Steps 1 and 5 of the following proof. Then we move to Step 4 and use the definition of odd to translate Step 5.

> Step 4 is our focus as we figure out how to bridge the gap. Our job is to find j. If we stay focused on this task, the connecting steps become fairly obvious.

From Step 2, we square both sides of the equation to get x^2 into the picture. Then guided by our goal, which is to find j, we do algebraic manipulations and juggle the expression into the desired form:

Proof

1. Let x be an integer.
   Assume that x is odd.
2. There exists an integer k such that $x = 2k + 1$.
3. $x^2 = (2k+1)^2 = 4k^2 + 4k + 1 = 2(2k^2 + 2k) + 1$.
   Define j as follows: $j = 2k^2 + 2k$.
   Since k is an integer, j is an integer.
   Next we substitute j in the above expression for x^2.
   $x^2 = 2j + 1$.
4. So, there exists an integer j such that $x^2 = 2j + 1$.
5. Therefore, x^2 is odd.
6. So, there exists does not exist an integer x such that x is odd and x^2 is even.

To Prove or Disprove?

Before we get to the stage of writing a proof, a considerable amount of detective work may be needed to decide whether to prove or disprove the statement. With some statements, our choice may be fairly obvious, but if we're not sure, we should contemplate both the sentence and its translated negation. If we think a universally quantified sentence is true, we should test a wide range of examples. If we do not come across a counterexample, we may be ready to try to write a general proof of the statement. However, we should keep in mind that our examples do not guarantee that the sentence is always true. If we are unable to construct a proof for the general case, we should look further for counterexamples.

For an existentially quantified sentence, we should diligently search for a particular example that makes the sentence true. Sometimes, we may be able to derive the existence of an element without actually constructing it, as illustrated in the proof on page 157.

Layers of Quantifiers

When a proof has layers of quantifiers, we can easily become confused unless we structure our thinking. To work through several layers of quantifiers, we need to carefully peel away the layers with our focus always on what we're trying to derive.

For example, let's try to prove the following theorem, which has three layers of mixed quantifiers embedded in the definitions of the three onto functions.

Theorem If f maps X onto Y and g maps Y onto Z, then $g \circ f$ maps X onto Z.

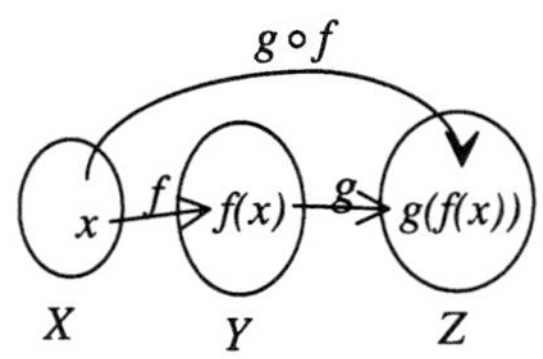

---

f maps X *onto* Y
if and only if
for every y in Y, there exists an x in X such that $f(x) = y$.

Before we start structuring the proof, we should review the definitions of the involved terms and try to visual what the theorem says. First, we draw the adjacent sketch to help us keep in mind the relation between these three functions. The function f maps a point in X to a point in Y, whereas g maps a point in Y to a point in Z. Since $f(x)$ is an element in Y and g maps Y into Z, $g(f(x))$ is located in Z. The function $g \circ f$, which maps X into Z, is defined as follows:

$$g \circ f(x) = g(f(x))$$

We will examine $g \circ f$ in more detail on page 358. The definition of an onto function is given on the left. Now that we have the definitions in front of us, we are ready to write the outside structure of the proof:

Assume that f maps X onto Y and g maps Y onto Z.

. . .

So $g \circ f$ maps X onto Z.

Most students want to start at the top and translate Step 1, but we are not going to do that because we are logical thinkers and we are going to focus on what we want to derive. An assumption in a proof is similar to money in a savings account where the owner does not draw on the money until there is a real need for it. We have no idea yet as to where we will need to use the two assumptions in our first step, so we will leave them in the bank for future use. Now, let's focus on what we want to derive and translate the last step, "$g \circ f$ maps X onto Z."

For every z in Z, there exists an x_o in X such that $g \circ f(x_o) = z$.

The above translation tells us how to set up the inner structure of the proof. As always, we start with the first quantifier at the beginning of the sentence, which gives us Step 2 in our proof.

1. Assume that f maps X onto Y and g maps Y onto Z.
2. Let z be an element in Z.

   . . .

   $g(f(x_o)) = z$

   So, $(g \circ f)(x_o) = z$.
6. **There exists an x_o in X such that $g \circ f(x_o) = z$.**
7. For all z in Z, there exists an x_o in X such that $g \circ f(x_o) = z$.
8. So $g \circ f$ maps X onto Z.

We put the remainder of the sentence in Step 6, as illustrated on the left. Step 6 is highlighted in bold to remind us that it is our goal.

When we find x_o, we must compute $g \circ f(x_o)$ and show that $g \circ f(x_o) = z$, which gives us Step 5 in our proof.

Now we focus on Step 6. Our job is to find x_o. Unlike our previous proofs, though, we will not construct it; instead, we will deduce its existence from the fact that both f and g are onto functions.

Let's visualize what's going on. At this stage, we have z over in the set Z, as illustrated in the adjacent sketch. We are looking for an x_o over in X. In the sketch, we can see that g gives us a way to back up closer to the set X. Since g maps Y onto Z:

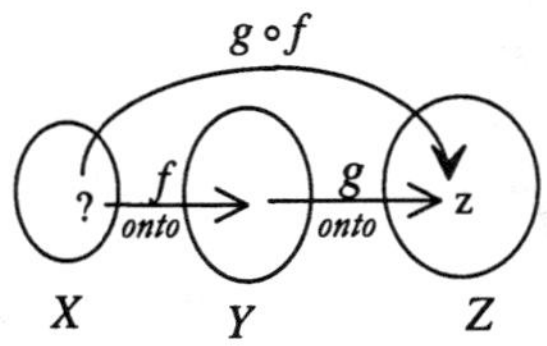

3. There exists a y_0 in Y such that $g(y_0) = z$.

Now, we focus on y_0. Since f maps X onto Y:

4. There exists an x_0 in X such that $f(x_0) = y_0$.

At last, we have found x_0. All we have left to do is justify Step 5 by using substitution. In the following polished version of this proof, we leave out Step 7 since the reader should be able to see that we have satisfied the definition of onto.

Theorem If f maps X onto Y and g maps Y onto Z, then $g \circ f$ maps X onto Z.

Proof Assume that f maps X onto Y and g maps Y onto Z.

Let z be an element in Z.

Since g is onto, there exists a y_o in Y such that $g(y_o) = z$.

Since f is onto, there exists an x_o in X such that $f(x_o) = y_o$.

By the definition of composition: $g \circ f(x_o) = g(f(x_o))$

Substitute y_o for $f(x_o)$: $= g(y_o)$

Substitute z for $g(y_o)$: $= z$

So, there exists an x_o in X such that $g \circ f(x_o) = z$.

Therefore, $g \circ f$ maps X onto Z.

If you understand the reasoning process for structuring the above proof, congratulations, for you have reached a new level in the development of your reasoning skills. We must resist the temptation to start working from the top down. Instead, we should focus on what we want to achieve and then structure our thinking so that we can reach our goal.

Exercise Set 2.4

1. Define the following using variables and quantifiers.
   a. x is a rational number.
   b. $x+y$ is rational.
   c. $\frac{x}{y}$ is rational.
   d. x is irrational.
2. Let x and y be arbitrary real numbers.
   Prove or disprove each statement.
   a. If x is rational, then $-x$ is rational.
   b. If x is irrational, then $-x$ is irrational.
   c. If x is rational and y is rational, then $x+y$ is rational.
   d. If x is irrational and y is irrational, then $x+y$ is irrational.
   e. If x and y are rational numbers, then $\frac{x}{y}$ is rational.
   f. If x and y are rational numbers and $y \neq 0$, then $\frac{x}{y}$ is rational.

g. If x is rational and y is irrational, then $x+y$ is irrational.

Hint: If you are clever, you can set up your structure so that you don't have to deal with irrationals. See (4) on page 140.

3. Prove or disprove each statement.
   a. For every real number y, there exists a real number x such that $x+2y=7$.
   b. There exists a real number x such that for every real number y, $x+2y=7$.
   c. There exists a real number x such that for every real number y, $x<y$.
   d There exists a real number x such that for every natural number y, $x<y$.
4. Let $f(x)=5x+2$. Prove or disprove each statement. $\mathbb{R}$ is the set of real numbers. $\mathbb{N}$ is the set of natural numbers.
   a. f maps $\mathbb{R}$ onto $\mathbb{R}$. b. f maps $\mathbb{N}$ onto $\mathbb{N}$.
5. Let $f(x)=x^2$. Prove or disprove that f maps $\mathbb{R}$ onto $\mathbb{R}$.
6. *Theorem*: If g maps A onto B and h maps B onto C, then $h \circ g$ maps A onto C.
   a. Draw a sketch that illustrates the above theorem.
   b. Set up the outside structure for a proof of the above theorem.
   c. Complete your proof in part (b).
7. Prove or disprove each statement.
   a. For every real number x, $2x<3x$.
   b. There is no smallest positive real number.
   c. The interval (3,5) does not have a smallest element.
8. Between any two distinct rational numbers, regardless of how close together they are, can you always find another rational number between them? If so, how would you do it?
   a. Consider the question for specific examples. Can you find a rational number between $\frac{3}{1000}$ and $\frac{4}{1000}$?
   b. Consider the question visually. In the adjacent sketch, let a and b be rational numbers. Can you find a rational number between a and b? If so, what is it?

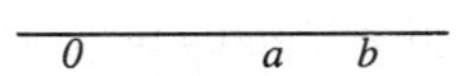

   Hint: Let x_o be halfway between a and b. Does x_o have to be a rational number? Find a formula for x_o in terms of a and b. Use visual reasoning to justify your formula.

9. Complete the proof of the following theorem. You may cite previous theorems that you have proved about rational numbers. The word *"claim"* is used as a signpost to tell the reader that you are now going to do a subproof and prove this claim. It serves the same function as the *"theorem"* heading. You cannot make derivations from the claim.

   Theorem: Between every two distinct rational numbers, there is another rational number.

   Proof: Let a and b be rational numbers with $a<b$.
   Set $x_o =$ ______.
   x_o is a rational number because ______
   Claim: $a<x_o$.
   . . . Therefore, $a<x_o$.
   Claim: $x_o<b$.
   . . . Therefore, $x_o<b$.
   So, x_o is between a and b.

10. Prove or disprove each statement.
    a. Between every two real numbers, there is another real number.
    b. If $a<b+x$ for every positive number x, then $a\leq b$.

11. The theorem in exercise 9 tells us that the rational numbers are very densely distributed along the number line. Between every two rational numbers there is another rational number.
    a. Between every two distinct rational numbers, are there an infinite number of rational numbers? If so, explain your reasoning.
    b. Do you think the rational numbers fill up the number line? Is the coordinate of each point on a number line a rational number? Justify your answer.

12. An algorithm is a procedure for accomplishing a specified task. An algorithm for finding a number y larger than a given number x could be stated as $y=x+1$. This algorithm is not unique.
    a. Give an algorithm that could be programmed into a computer so that for every two different rational numbers that we input, the computer will output a rational number that is between the two input numbers.
    b. Find another algorithm that outputs a rational number between two given rational numbers, but this time make the output different than the output for the algorithm in part (a).

Activity 2.5

1. Let x and y be real numbers.
   a. If $xy > 0$, what do you know about x and y?
   b. If $(x-2)(x+1) > 0$, what do you know about $x-2$ and $x+1$?
   c. Use part (b) to solve the following inequality: $(x-2)(x+1) > 0$
2. Let x and y be real numbers.
   a. If $xy < 0$, what do you know about x and y?
   b. If $(x-1)(x+1) < 0$, what do you know about $x-1$ and $x+1$?
   c. Use part (b) to solve the following inequality: $(x-1)(x+1) < 0$.

≡ 2.5 Using Cases ≡

When trying to construct a proof, we may sometimes feel as though our hands are tied because we don't have enough information. When this happens, we may want to introduce cases so that we have additional information to use within each case. For example, a proof about real numbers could be subdivided into the following cases:

Case 1: Assume that $x \geq 0$.

Case 2: Assume that $x < 0$.

Since "$x \geq 0$ *or* $x < 0$" is true, either Case 1 or Case 2 must be true. When we do our reasoning in Case 1, we have the additional information that $x \geq 0$, which may help us derive the desired result. Within Case 2, we can use that $x < 0$. If we are able to derive the desired result in both cases, we then know that the result is always true.

A proof can be subdivided into cases by using a true or-sentence, such as "$x \geq 0$ or $x < 0$." In the adjacent template, the or-sentence is represented as p *or* q. The letter r represents the sentence that we want to prove. Each part of the or-sentence determines a case. In Case 1, we assume that p is true and then derive r. In Case 2, we assume that q is true and then derive r. Since one of the two cases is true, we can conclude that r must always be true. The following proof illustrates the technique for using cases in a proof.

Derive: r

Proof with Cases:

p or q

Case 1: Assume p is true.

. . .

Therefore, r.

Case 2: Assume q is true.

. . .

Therefore, r.

Since one of the 2 cases must occur and r is true in both cases, r is always true.

Theorem Let x and y be integers. If x is even or y is even, then xy is even.

Proof Assume that x is even or y is even.

Case 1: Suppose that x is even.
Then $x = 2k$ for some integer k.
So, $xy = (2k)y = 2(ky)$
Since k and y are both integers, ky is an integer.
Therefore, xy is even.

Case 2: Suppose that y is even.
Then $y = 2j$ for some integer j.
So, $xy = x(2j) = 2(jx)$
Since j and x are both integers, jx is an integer.
So xy is even.

Since one of the two cases must occur, xy is even.
So, if x is even or y is even, then xy is even.

If we have an or-sentence in the middle of a proof, we can branch into cases. Since "p *or* $\sim p$" is always true, any sentence p can be used to set up cases in a proof:

Case 1: Assume p is true.
Case 2: Assume p is false.

> *Derive*: r or s
>
> *Proof with Cases*:
> p or q
> *Case* 1: Assume p is true.
> ...
> Therefore, r.
> *Case* 2: Assume q is true.
> ...
> Therefore, s.
> Therefore, r or s.

The trick is to find cases that will help us derive the desired conclusion. We can use as many cases as we like, as long as we know that at least one of the cases must be true. If we are able to derive the desired conclusion for each case, we can then conclude that the conclusion must always be true. Sometimes when we use cases, though, we derive different conclusions. When this happens, we can conclude that either the conclusion in Case 1 or the conclusion in Case 2 must be true, as illustrated in the adjacent template.

When we use cases to solve an inequality, we often get different conclusions as illustrated in the next example. Cases are introduced by using the following property of real numbers:

If a product of two numbers is positive,
both factors are positive or both are negative.

⊕ *Example*

Solve the inequality: $(x-1)(x+3)>0$.

Assume that $(x-1)(x+3)>0$.

Both factors must be positive or both must be negative.
So ($x-1>0$ and $x+3>0$) or ($x-1<0$ and $x+3<0$).

Case 1: Assume that $x-1>0$ and $x+3>0$.
Then $x>1$ and $x>-3$.
In order for both of these inequalities to hold, x must be greater than 1.
So the solution for this case is: $x>1$.

Case 2: Assume that $x-1<0$ and $x+3<0$.
Then $x<1$ and $x<-3$.
In order for both of these inequalities to hold, x must be less than -3.
So the solution for this case is: $x<-3$.

Since either Case 1 or Case 2 must occur, the solution to the original inequality is: $x>1$ or $x<-3$.

Unlike a proof, when we solve an equation or inequality, our derivations have to go both ways. In a proof, we could conclude in Case 1 that $x>-3$. However, that condition does not reverse and give a solution to the inequality.

Inequalities

We often use cases when we work with inequalities, so let's briefly review the basic rules for inequalities. The set of all real numbers can be divided into three nonoverlapping sets, the positive numbers, the negative numbers and 0. We do not classify 0 as either positive or negative; 0 serves as the boundary, separating the positive numbers from the real numbers.

The negative numbers are a mirror image of the positive numbers. When we look in a mirror, left becomes right and right becomes left. Consequently, when we work with positive and negative numbers, these mirror images cause a lot of switching, which can be confusing if we do not keep the rules straight.

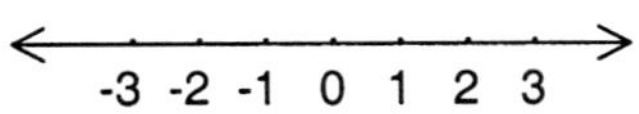

The set of real numbers is ordered from left to right with no regard for the mirror images. On a number line, $a<b$ if and only if a is to the left of b. Since "left" is a relative term that depends on the point of view, we need a more objective definition for "less than."

Using the subtraction operation and the concept of a positive number, we can define "less than" as follows:

$$a<b \text{ if and only if } b-a \text{ is positive.}$$

$a<b$
if and only if
$b-a$ is positive

We can now define "greater than" in terms of "less than."

$$a>b \text{ if and only if } b<a.$$

Since the > symbol is a mirror image of the < symbol, our eyes may sometimes play tricks on us when both symbols are present. If we translate to only one of the inequality symbols the sentence may seem simpler. For example, we may want to rewrite "$a<b$ and $c>b$" as "$a<b$ and $b<c$."

Multiplication was extended to the negative numbers in a way that would preserve the existing properties of multiplication on the positive numbers. Consequently the product of two negative numbers is defined to be a positive number, whereas the product of a positive number and a negative number is defined to be a negative number. Translating in terms of variables gives the following sentences:

$ab>0$ if and only if
$(a>0$ and $b>0)$ or $(a<0$ and $b<0)$.

$ab<0$ if and only if
$(a>0$ and $b<0)$ or $(a<0$ and $b>0)$

Using the above two properties, we can derive the following rules for multiplying an inequality by a real number:

Theorem Let a, b, and c be real numbers and let $a<b$.
If $0<c$, then $ac<bc$.
If $c<0$, then $ac>bc$.

Proof Let $a<b$. By the < definition, $0<b-a$.
Assume that $0<c$.
Since the product of two positive numbers is positive,
$0<(b-a)c$. Thus, $0<bc-ac$.
So, by the < definition, $ac<bc$.
Assume that $c<0$.
Since the product of a positive and a negative number is negative, $(b-a)c<0$.
Thus, $bc-ac<0$. By the < definition, $bc<ac$.
So, by the > definition, $ac>cb$.

If we multiply an inequality by a positive number, the inequality does not change, but if we multiply by a negative number, the inequality is reversed. Students sometimes forget to use the latter rule because we do not have to make this type of distinction when working with equations:

If $a = b$ and c is a real number, then $ac = bc$.

With inequalities, though, we have 3 cases:

Case 1: If $a < b$ and $c > 0$, then $ac < bc$.
Case 2: If $a < b$ and $c < 0$, then $ac > bc$.
Case 3: If $a < b$ and $c = 0$, then $ac = bc$.

We will use Case 2 to prove the following two theorems.

Theorem D Let x be a real number. If $-1 < x$, then $1 > -x$.

Proof Assume that $-1 < x$.
Multiplying the inequality by -1 reverses the inequality:

$$(-1)(-1) > -x$$

Therefore, $1 > -x$.

Theorem E Let x be a real number. If $-1 < x < 0$, then $-x > x^2$.

Proof Assume that $-1 < x$ and $x < 0$.
Since $x < 0$, multiplying the first inequality by x reverses it:

$$-1 < x$$

$$x \cdot (-1) > x \cdot x$$

Thus, $-x > x^2$.

We will use the above two theorems in Case 2 of the following proof. In the construction of this proof, we first set up the outside structure, and then we will work from the top down:

Assume that $-1 < x < 1$.

. . .

Then $x^2 < 1$.

We will multiply an inequality by x, so we must split the following proof into three cases. Since we assume $-1 < x < 1$ in the beginning of the proof, we can use it in each case.

Theorem Let x be a real number. If $-1 < x < 1$, then $x^2 < 1$.

Proof Assume that $-1 < x < 1$.
Then $-1 < x$ and $x < 1$.
Since x is a real number, $x > 0$ or $x < 0$ or $x = 0$.

Case 1: Assume that $x > 0$.
In our original assumption, $x < 1$.
Since $x > 0$, $x \cdot x < x \cdot 1$
Thus, $x^2 < x$.
We now have that $x^2 < x$ and $x < 1$.
Since $<$ is transitive, $x^2 < 1$.

Case 2: Assume that $x < 0$.
In our original assumption, $-1 < x$.
From Theorem E on the previous page, $-x > x^2$.
From Theorem D on the previous page, $1 > -x$.
We now have that $1 > -x$ and $-x > x^2$.
Since $>$ is transitive, $1 > x^2$.
Thus, $x^2 < 1$.

Case 3: Suppose that $x = 0$.
Then $x^2 = 0$. So $x^2 < 1$.

In each of the 3 cases, $x^2 < 1$.
Furthermore, one of these cases must occur.
Therefore, if $-1 < x < 1$, then $x^2 < 1$.

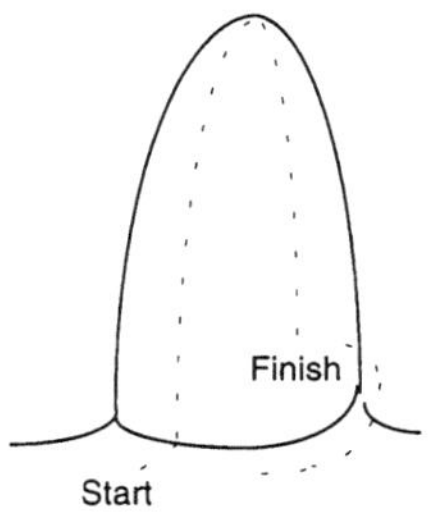

In Case 2, note the clever way that transitivity is used to produce the desired result. This proof is a little tedious, but it does show why the statement is true.

The method used to construct the above proof was to first set up the outside shell, then go to the top and work straight down, plowing through all the detail until we finally saw the light at the end of the tunnel. However, we could have walked around the mountain if we had focused more on *working backwards* from the end, as illustrated on the next page.

Assume that $-1<x<1$.

. . .

So, $0<(1+x)(1-x)$

$0<1-x^2$

Then $x^2<1$.

This more thoughtful analysis gives us a simpler goal to work towards, namely, show that the product of $1+x$ and $1-x$ is positive.

Theorem Let x be a real number. If $-1<x<1$, then $x^2<1$.

Proof Assume that $-1<x<1$.

Then $-1<x$ and $x<1$.

So, $0<1+x$ and $0<1-x$.

Since the product of two positive numbers is positive:

$0<(1+x)(1-x)$.

$0<1-x^2$

Therefore, $x^2<1$.

After working through the detail of the former proof, the beauty in the simplicity of the above proof shines like polished crystal.

Impossible Cases Sometimes when we use cases, we discover that one of the cases cannot occur. If a case cannot occur and we derive the desired result in each of the other cases, then our theorem is true. This happens in the following proof of the converse of the previous theorem.

Theorem For every real number x, if $x^2<1$, then $-1<x<1$.

Proof Assume that $x^2<1$. Then $x^2-1<0$.
So $(x-1)(x+1)<0$. Since the product is negative, one factor must be positive and the other negative.

Case 1: Assume that $x-1>0$ and $x+1<0$.

Then $x>1$ and $x<-1$.

There is no x that satisfies both inequalities, so this case cannot occur.

Case 2: Assume that $x-1<0$ and $x+1>0$
Then $x<1$ and $x>-1$.
So, $-1<x<1$.

Since only Case 2 can occur, $-1<x<1$.

Therefore, for every real number x, if $x^2<1$, then $-1<x<1$.

Exercise Set 2.5

1. Use cases to solve each inequality.
   a. $(x+2)(x-4)>0$
   b. $(x-3)(x-4)<0$
   c. $x^2-9>0$
   d. $x^2-4<0$
2. Prove each statement. Let x and a be real numbers. $|x|$ denotes the absolute value of x.

   Definition: $|x| = x$, if $x \geq 0$.
   $|x| = -x$, if $x<0$.

   Definition: $x \leq a$ if and only if $x<a$ or $x=a$.
   $x \geq a$ if and only if $x \leq a$.

   a. $x^2 \geq 0$
   b. $x \leq |x|$
   c. $|x| = |-x|$
   d. If $x>1$ or $x<-1$, then $x^2>1$.
   e. If $x \geq 1$ or $x \leq -1$, then $x^2 \geq 1$.
   f. Let $a>0$. $x^2<a^2$ if and only if $-a<x<a$.
3. Let n be an integer. Then n is even or n is odd.
   Use this statement to set up cases and prove the following:
   a. n^2-n is even.
   b. $\frac{n(n+1)}{2}$ is an integer.

Activity 2.6

1. Using only a straightedge and a compass, describe how to locate $\sqrt{2}$ on a number line.
2. Assume that $\sqrt{2}$ is a rational number. Then try to derive a contradiction.

≡ 2.6 Proof by Contradiction ≡

Do I contradict myself?
Very well then .... I contradict myself.
I am large .... I contain multitudes.

Walt Whitman
Leaves of Grass

Contradictions have a place in other systems of thought, such as Walt Whitman's poetic view of the world, or in a Buddhist's meditation on a contradiction to reach a higher level of spiritual experience. However, in the house of mathematics, we do not allow contradictions. As we will see on page 205, one little contradiction wipes out our whole logical system. If we make an assumption that produces one of these lethal contradictions, we deduce that the assumption had to be false, which is the basis for a proof by contradiction.

The structure of a proof by contradiction is illustrated in the adjacent template; r represents the sentence we want to prove and c represents the contradiction that we find. First, we assume the negation of what we want to prove. Then we search until we find a sentence c that we can derive as true and also derive as false, which makes it a contradiction. Since a contradiction cannot exist, our original assumption that produced the contradiction must be false. Therefore, r is true.

> *Derive:* r
>
> *Proof by Contradiction*
> Assume that r is false.
> . . .
> Therefore, c is true.
> . . .
> Therefore, c is false.
> Contradiction!
> So, r must be true.

A proof by contradiction is rather negative in spirit. It shows us why a sentence can't be false, but it doesn't really show us why it is true. We deduce that the sentence is true because it can't be false. Nevertheless, this method of proof is a powerful tool that enables us to prove some theorms that we might not be able to prove with a more positive outlook.

When we prove the contrapositive of a sentence, we also switch into negative mode, assuming that the conclusion is false. Our goal then is to show that the hypothesis is false. However, when we do a proof by contradiction, we have no idea as to what our goal is, other than to find a contradiction. We have no clue as to where to look for it. Like Sherlock Holmes, we must be very clever as we search for a deadly contradiction lurking somewhere.

Validity of Contradiction Proofs

To justify the validity of a proof by contradiction, we can argue as follows. When we assume $\sim r$ is true and derive c *and* $\sim c$, we have proved the following implication:

$$\sim r \Rightarrow (c \text{ and } \sim c)$$

Next we translate the above implication as an or-sentence.

$$r \text{ or } (c \text{ and } \sim c)$$

Since (c and $\sim c$) is false, r must be true.

A Very Famous Proof

One of the most famous proofs by contradiction is the proof that $\sqrt{2}$ is an irrational number. If you tried to prove this theorem in Activity 2.6, you understand the challenge in finding a contradiction with no clear goal to guide us. If you were not successful, you may be surprised to learn that someone did prove it way back in the 4th century B.C.E. Before we prove that $\sqrt{2}$ is irrational, let's take a historical look at why there was a great interest in this question 2400 years ago.

Deep thinkers have always sought out other thinkers, for ideas get a nurturing cross-fertilization in a community of thinkers. In the 5th century B.C.E., Pythagoras founded one of the earliest known schools of thinkers, known as the Pythagorean brotherhood. The ideas developed by the Pythagoreans had a lasting impact on the developing cultures of the western world. Consumed with a desire to explain *why* things happen, the Pythagoreans were among the first thinkers to use deductive reasoning. They developed it into a fine art, proving the Pythagorean theorem and many others as well. Using the deductive process to analyze the sounds of music, they asked the question as to *why* some combinations of sounds are more pleasing to the ear than other combinations. When they discovered that harmonic tones of music came from plucking strings whose lengths were simple ratios of natural numbers, they then created the musical scale from which western music evolved. The beautiful simplicity of this mathematical relation between ratios of natural numbers and musical harmony led the Pythagoreans to a spiritual belief that all nature originated from the natural numbers, causing them to endow these numbers with mystical properties.

After the counting numbers were developed to measure sizes of sets, the next giant step was to develop numbers for measuring lengths and distances. For this purpose, the ruler was created by evenly spacing the counting numbers on a ruler. The unit interval was then subdivided into n equal subintervals whose right endpoints were labeled as $\frac{1}{n}, \frac{2}{n}, \frac{3}{n}, \ldots, \frac{n}{n}$, and thus were the rational numbers created. Like the harmonic tones of music, the Pythagoreans believed that all lengths could be represented as a ratio of natural numbers, which certainly seems visually plausible. The Pythagoreans knew that between any two rational numbers, no matter how close together they are, there is always another rational number (page 159), and so it does appear as if the rational numbers should fill up the number line. So, it appeared as though any length could be represented as a ratio of natural numbers.

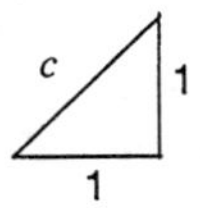

$$c^2 = 1^2 + 1^2$$
$$c^2 = 2$$
$$\text{So, } c = \sqrt{2}$$

New knowledge sometimes brings disturbing revelations for it forces us to reexamine our previous beliefs in terms of the new insight. Such was the case with the Pythagorean Theorem. Through this theorem, the Pythagoreans could easily construct a length that represents $\sqrt{2}$, as illustrated on the left. However, try as they might, the Pythagoreans could not find two natural numbers whose quotient was $\sqrt{2}$. According to legend, while sailing on the sunny Mediterranean, a Pythagorean brother came up with a proof by contradiction that $\sqrt{2}$ could not be expressed as a ratio of natural numbers. His companions, greatly distressed by the fatal impact his proof had on their religious beliefs, ungratefully threw the author overboard and drowned him. Of course, this story may not be true, but it is certainly true that whoever came up with the following proof was a very deep thinker.

Theorem $\sqrt{2}$ is not a rational number.

Proof by Contradiction Suppose that $\sqrt{2}$ is a rational number.

So there exist integers a and b such that $\sqrt{2} = \frac{a}{b}$.
If a and b have any common factors, cancel them so that we are left with integers c and d such that that $\sqrt{2} = \frac{c}{d}$, and c and d have no common factors.

We will now contradict that c and d have no common factors.

First, we will do some simple algebra:
$$\sqrt{2} = \frac{c}{d}$$
$$2 = \frac{c^2}{d^2}$$
$$2d^2 = c^2$$

Since c^2 has 2 as a factor, c^2 is even. By a previous theorem (page 138), if c^2 is even, then c is even. So, c is even, and hence, 2 is a factor of c. Thus, $c = 2k$ for some integer k.

Now we will do some more algebra: $c^2 = 4k^2$

Since $2d^2 = c^2$, $2d^2 = 4k^2$

So, $d^2 = 2k^2$

Since d^2 is even, d is even. So 2 is a factor of d.
Thus, 2 is a factor of both c and d.
But c and d have no common factors. *Contradiction!*

So our original assumption is false.
Thus, $\sqrt{2}$ is not a rational number.

Derive: If p, then q.
[*Assume that p is true.*]
Assume q is false.
...
Therefore, p is false.
[*Contradiction!*
Therefore, q is true.]
So, if p, then q.

As you admire the ingenuity in the previous chain of reasoning, please observe how unrelated the contradictory sentence is to the original assumption. The great challenge in constructing a proof by contradiction is to actually find a contradiction, for there are no pointers as to where to look for it like we have in a proof by contraposition.

A contrapositive proof can be construed as a contradiction, as illustrated in the adjacent template. After we assume p is true, we want to derive that q is true, so we switch into contradiction mode and assume q is false. We then derive that p is false, which contradicts our original assumption. However, if we did not use the fact that p was true, we could remove the first line and the other two italicized lines, which leaves us with a proof of the contrapositive, $\sim q \Rightarrow \sim p$. Using the word 'contradiction' for a proof of the contrapositive makes the proof longer and the reasoning more complex.

Exercise Set 2.6

1. Prove or disprove the following. Since we have already proved $\sqrt{2}$ is irrational, you may cite this result.
   a. $1+\sqrt{2}$ is an irrational number.
   b. For every rational number x, $x+\sqrt{2}$ is an irrational number.
   c. Let x be a rational number and y an irrational number. Then $x+y$ is irrational.
   d. Let x and y be irrational numbers. Then $x+y$ is irrational.
2. A *prime number* is a natural number n, greater than 1, whose only positive factors are 1 and n. The Fundamental Theorem of Arithmetic gives a very important property of natural numbers.

Fundamental Theorem of Arithmetic

Every natural number, other than 1, can be represented in a unique manner as a product of prime numbers, with smaller factors written to the left of larger factors.

---

   a. Let x be a natural number greater than 1. In the prime factorization of x, can a factor appear only once? 2 times? 3 times?

b. In the prime factorization of x^2, can a factor appear only once? 2 times? 3 times? 4 times? 5 times?

3. Let n be a natural number. Consider the following sentence:

   For every natural number x, if n is a factor of x^2,
   then n is a factor of x.

   a. Is the above sentence true for $n = 2$? For $n = 3$? For $n = 4$?
   b. For what values of n is the above sentence true?
      Hint: Reflect on your work in the previous exercise.

4. Prove that $\sqrt{3}$ is an irrational number.
   Hint: You can write a proof similar to the one for $\sqrt{2}$ and cite your results from (3). Or be creative and use your answers from (2) to construct a different type of proof.

5. Generalize your proof in (4) and prove the following.
   Theorem: If p is a prime number, then $\sqrt{p}$ is irrational.

6. Prove that there are an infinite number of prime numbers.
   Hint: Suppose that there are only a finite number of prime numbers. Label them as $p_1, p_2, \ldots, p_n$. Set $x = p_1 p_2 \ldots p_n + 1$. Now demonstrate that none of the prime numbers in the list can divide x. Then explain why this gives a contradiction.

7. In Euclidean geometry, the sum of the angles in a triangle is 180°. Using this result, prove the given statement with a proof by contradiction. All points and lines are in the same plane.
   a. Through a point P not on a line t, there is only one line through P that is perpendicular to t.
   b. If line t is perpendicular to two distinct lines s and r, then s and r cannot intersect.
   c. If line t intersects two distinct lines s and r and the corresponding angles formed are congruent, then s and r cannot intersect.

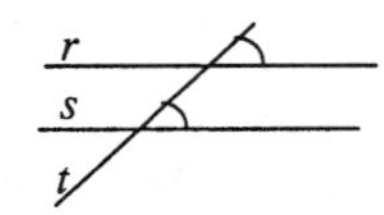

Activity 2.7

1. Let n be a positive integer. Write the sum of the following in the style given in part (a), but also list the last term.
   a. The first n positive integers: $1 + 2 + 3 + \ldots + \_\_\_$
   b. The first n even positive integers: $2 + 4 + 6 + \ldots + \_\_\_$
   c. The first $n + 1$ even positive integers.
   d. The first n odd positive integers.

2. For each n in the adjacent table, list the sum of the first n even positive integers in the sum-column.

n	Sum
2	
3	
4	
5	
6	
7	
8	
n	?

   a. What is the pattern in the numbers as you read down the sum-column?

   b. What is the pattern that goes *across* the list? How is n related to the adjacent number in the sum-column? Try writing the numbers in the sum-column in different forms until you see a pattern emerge across the row.

   c. What is the sum of the first 1000 even positive integers? Did you use your pattern from part (a) or from part (b)?

   d. Use your work from (1b) and (2b) to write an equation that gives a formula for the sum of the first n even positive integers.

3. Find a formula for the sum of the first n positive integers.

n	Sum
2	
3	
4	
5	
6	
7	
8	
n	?

   a. You may want to do some detective work similar to your work in (2). Fill in the adjacent table and look for a pattern that goes *across* the list. Write the numbers in the sum-column in different forms until you see a pattern emerge. You may want to factor the numbers in the sum-column, or multiply by $\frac{2}{2}$.

   b. You may want to consider the relation of this formula to the formula that you found in (2d).

   c. You may want to use visual reasoning. Imagine the numbers as increasing columns of black dots, then fill in with white dots, as illustrated on the right. In a similar representation for $1+2+3+\ldots+n$, how many columns and rows will there be? What will the total number of dots be? How will the number of black dots compare with the number of white dots? Use this data to determine the total number of black dots.

1 + 2 + 3 + 4

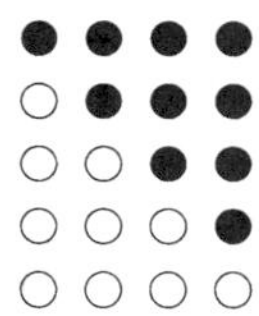

4. Demonstrate the *Domino Theory*. Stand a bunch of dominoes vertically, positioning them so that if you push over a domino, it knocks over the next domino.

   a. If you push over the 1st domino, what happens? What happens if you push over the 3rd domino?

   b. Suppose that you have an infinite sequence of dominoes, $p_1, p_2, p_3, \ldots$, positioned so that the following is true:

   For every natural number n, if p_n falls, then p_{n+1} falls.

   If you push over the 1st domino, what happens? What happens if you push over the 3rd domino?

≡ 2.7 Mathematical Induction ≡

Mathematical induction is a technique for proving an infinite sequence of sentences, analogous to the Domino Theory. Suppose that the sentences are labeled as illustrated on the left. If we can demonstrate that each sentence implies the one after it, then we know that:

Is $p(n)$ true for every n?

$p(1)$

$p(2)$

$p(3)$

$p(4)$

.
.
.

$p(n)$

.
.
.

$$p(1) \Rightarrow p(2)$$
$$p(2) \Rightarrow p(3)$$
$$p(3) \Rightarrow p(4)$$
$$\ldots$$
$$p(n) \Rightarrow p(n+1)$$
$$\ldots$$

The above implications tell us absolutely nothing about the truth values of the individual sentences. After all, $p(1)$ could be false and then $p(1) \Rightarrow p(2)$ is automatically true. However, this chain of implications does position the sentences like a row of dominoes, so that if we are able to verify one of them, then all the others *after* that one will have to be true. For example, if we establish that $p(1)$ is true, we can work down the list, using the Law of Detachment to deduce that each successive sentence is true:

$p(1)$ is true.

Since $p(1) \Rightarrow p(2)$, we can deduce that $p(2)$ is true.

Since $p(2) \Rightarrow p(3)$, we can deduce that $p(3)$ is true.

And so on, down the list.

Principle of Mathematical Induction

Let $p(n)$ be an open statement. Suppose the following are true:

1. For all integers $n \geq 1$,
   $p(n) \Rightarrow p(n+1)$.
2. $p(1)$ is true.

Then $p(n)$ is true for all integers n where $n \geq 1$.

The Principal of Mathematical Induction guarantees that the Law of Detachment can be applied an *infinite* number of times. In the adjacent description of mathematical induction, Part (1) is equivalent to arranging the dominoes so that if any particular domino falls, the next one must also fall. Part 1 is called the *inductive step*. Part (2) is equivalent to pushing over the first domino. The conclusion is that all the dominoes will fall. The induction game is for master players because we're playing with an *infinite* set of dominoes. Most anyone can do Part 1 for a finite set of dominoes. Doing the same task for an infinite set is far more challenging for we must rely on our powers of deductive reasoning rather than eye and hand coordination.

General Form

Before we look at some examples, let's state the principle of mathematical induction in a more general form. When a row of dominoes are positioned according to the inductive step, they do not all fall when we push over the third domino. The first and second dominoes will be left standing, but all the others will come tumbling down. In a similar manner, suppose that a sequence of sentences satisfies the inductive step. In other words, we have been able to prove the following:

For every integer n, $p(n) \Rightarrow p(n+1)$.

However, when we check $p(1)$, we find that it is not true. We then check $p(2)$, but, alas, it is also not true. Do we give up? Not yet, because our infinite sequence of sentences are positioned just like those dominoes. All we have to do is find one that is true, and from that point on, each one will be true. Suppose that we check $p(3)$ and find that, yes, it is true. We can then proclaim that $p(n)$ is true for all $n \geq 3$:

$p(3)$ is true.

Since $p(3) \Rightarrow p(4)$, we can deduce that $p(4)$ is true.

Since $p(4) \Rightarrow p(5)$, we can deduce that $p(5)$ is true.

And so on, down the list.

Principle of Mathematical Induction

Let $p(n)$ be an open statement.
Let c be a fixed integer.
Suppose the following are true:

1. For all integers $n \geq c$,
   $p(n) \Rightarrow p(n+1)$.
2. $p(c)$ is true.

Then $p(n)$ is true for all integers n where $n \geq c$.

Actually, we did not need to know that $p(n) \Rightarrow p(n+1)$ for every positive integer n. It does not matter whether or not $p(1) \Rightarrow p(2)$ is true nor does it matter for $p(2) \Rightarrow p(3)$. If we can establish that $p(n) \Rightarrow p(n+1)$ for every integer $n \geq 3$ and we can also verify that $p(3)$ is true, we can then deduce that $p(n)$ is true for all $n \geq 3$, which is the following list:

$$p(3),\ p(4),\ p(5),\ p(6), \ldots,\ p(n), \ldots$$

Hence, we can generalize the previous statement of mathematical induction as stated on the left, using c to represent the point in our sequence beyond which we're going to demonstrate that every sentence is true. This version is identical to the previous version except that we replace 1 with c. The conclusion guarantees that the following sentences are each true:

$$p(c),\ p(c+1),\ p(c+2),\ \ldots,\ p(n), \ldots$$

In the above example, c was positive, but it could also be negative. Let's interpret the theorem for $c = -2$. Suppose that we have a sequence of open statements about an integer n and we can prove the inductive step in Part 1:

$$\text{For every integer } n \geq -2, \; p(n) \Rightarrow p(n+1).$$

Suppose also that $p(-2)$ is true. Using mathematical induction, we can deduce that $p(n)$ is true for every $n \geq -2$, which is the following list of sentences:

$$p(-2),\; p(-1),\; p(0),\; p(1), \ldots,\; p(n),\; \ldots$$

Grammatical Structure

Mathematical induction is a simple concept, but its grammatical structure is quite complex, as you can see in the following symbolic translation, where $c = 1$:

$$[(\forall n,\, p(n) \Rightarrow p(n+1)) \text{ and } p(1)] \Rightarrow \forall n,\, p(n)$$

Theorem: For all positive integers n, $p(n)$ is true.
Induction Proof Let $p(n)$: ____ where n is a positive integer.
Part 1 – Inductive Step Let n be a positive integer. Assume $p(n)$ is true. . . . So, $p(n+1)$ is true. Thus, $p(n)$ implies $p(n+1)$.
Part 2 – Verification Step . . . So $p(1)$ is true.
Therefore, $p(n)$ is true for all positive integers n.

The outside structure of the above sentence is an implication, but its hypothesis is a conjunction that has a quantified implication within it. It is no wonder that students sometimes get confused by the structure of an induction proof, especially when they assume that $p(n)$ is true in Part 1. They feel that they are assuming what they want to prove, which is never allowed. We sometimes assume the negation of what we want to prove, hoping to find a contradiction, but we never ever assume what we want to prove.

To eliminate any confusion, let's examine the adjacent template for a proof by mathematical induction. At the beginning of an induction proof, we identify the open sentence on which we will do the induction and express it in terms of the variable n. After we identify $p(n)$, the proof has two separate parts. We can prove them in whichever order we choose.

In the inductive step, we must prove $p(n) \Rightarrow p(n+1)$ for every positive integer n. First, we let n be a positive integer. Then we assume $p(n)$ is true and derive $p(n+1)$. It may look like we are assuming what we want to prove; however, $p(n)$ is not a stand-alone statement and neither is $p(n+1)$. Our stand-alone conclusion is that $\forall n,\, p(n) \Rightarrow p(n+1)$. In essence, we have proved the adjacent list of implications. As observed earlier, these implications tell us nothing about the truth value of the individual sentences. They do position the sentences like a row of dominoes, though, so that all we have to do is verify one of them and all the others after that one will have to be true.

Part 1: $p(1) \Rightarrow p(2)$
$p(2) \Rightarrow p(3)$
$p(3) \Rightarrow p(4)$
. . .
$p(n) \Rightarrow p(n+1)$
. . .

In the verification step, we demonstrate that $p(1)$ is true. This step is usually very easy. We finish the proof by applying the Principle of Mathematical Induction. Having proved Part 1 and Part 2, we can then deduce by mathematical induction that $p(n)$ is true for all positive integers n.

An Induction Example

$1+2+3+\ldots+n=\frac{n(n+1)}{2}$

Let's put our theory into action and look at an induction example. If you did the detective work in (3) on page 173, you probably discovered the adjacent pattern based on a few examples. Examples do not guarantee that this formula always holds, but we can get a guarantee using mathematical induction. First, we let $p(n)$ represent the following open sentence:

$$p(n)\text{: } 1+2+3+\ldots+n=\frac{n(n+1)}{2}$$

The above colon means "represents."

$p(n)$ represents the sentence: $1+2+3+\ldots+n=\frac{n(n+1)}{2}$.

Students sometimes erroneously set $p(n)=\frac{n(n+1)}{2}$.

If we do not write the correct representation for $p(n)$ as a sentence, we cannot set up the proper structure for an induction proof. Another error is to use an equals sign instead of a colon in the above representation:

$$p(n)=1+2+3+\ldots+n=\frac{n(n+1)}{2}$$

From the above form, we would surmise that $p(n)$ is equal to the left side of the equation and also equal to the right side of the equation. With this misconception, our thinking process is completely derailed. The equals verb is reserved for sets and numbers, so when we specify $p(n)$, we cannot replace the colon with an equals sign.

In this example, $p(n)$ represents a sentence whose predicate is "equals" and whose subject is "the sum of all the positive integers from 1 to n." In the symbolic representation,

$$1+2+3+\ldots+n$$

the three dots indicate that we are to continue in the same pattern until we reach n. However, when $n=1$ or $n=2$, we do not include +3 in the translation. To interpret the meaning of $p(1)$, we read the left side of the equation as the sum of all positive integers from 1 to 1 and we write $p(1)$ as follows:

$$1=\frac{1(2)}{2}$$

We usually check several examples, such as those given on the left, to make sure that the formula works before we attempt the induction step.

$p(1)$: $1=\frac{1(2)}{2}$

$p(2)$: $1+2=\frac{2(2+1)}{2}$

$p(3)$: $1+2+3=\frac{3(3+1)}{2}$

$p(4)$: $1+2+3+4=\frac{4(4+1)}{2}$

The Inductive Step To set up the induction part of the proof, we translate $p(n+1)$ by substituting $n+1$ for n.

$$p(n)\colon 1+2+3+\ .\ .\ .\ +n=\frac{n(n+1)}{2}$$

$$p(n+1)\colon 1+2+3\ +\ .\ .\ .\ +(n+1)=\frac{(n+1)(n+2)}{2}$$

We usually list the next-to-last term on the left side of $p(n+1)$ because it may give us a connection with $p(n)$:

$$p(n+1)\colon 1+2+3+\ .\ .\ .\ +n+(n+1)=\frac{(n+1)(n+2)}{2}$$

> *Part 1 – Inductive Step*:
>
> Let n be a positive integer.
>
> Assume that $p(n)$ is true.
>
> $1+2+3+...+n=\frac{n(n+1)}{2}$
>
> . . .
>
> $1+2+3+...+(n+1)=\frac{(n+1)(n+2)}{2}$
>
> Hence, $p(n+1)$ is true.
>
> Therefore, $p(n) \Rightarrow p(n+1)$.

Next we set up the structure to prove $p(n) \Rightarrow p(n+1)$, as illustrated on the left. This structure gives us a clear focus on what we need to do. How can we derive the second equation from the first equation? Let's simplify this task by focusing on how we can derive either the left side or the right side of the second equation. We can easily derive the left side by adding $(n+1)$ to both sides of the first equation. Then we will try to manipulate the right side of the new equation to obtain the desired result, as illustrated in the following proof.

Theorem For every positive integer n, $1+2+3+\ .\ .\ .\ +n=\frac{n(n+1)}{2}$.

Induction Proof Let $p(n)\colon\ 1+2+3+\ .\ .\ .\ +n=\frac{n(n+1)}{2}$.

Part 1 Let n be a positive integer.
Assume that $p(n)$ is true.

$$1+2+3+\ .\ .\ .\ +n=\frac{n(n+1)}{2}$$

Add $n+1$ to both sides:

$$1+2+3+\ .\ .\ .\ +n+(n+1)=\frac{n(n+1)}{2}+(n+1)$$

$$1+2+3+\ .\ .\ .\ +n+(n+1)=\frac{n^2+n+2n+2}{2}$$

$$1+2+3+\ .\ .\ .\ +n+(n+1)=\frac{(n+1)(n+2)}{2}$$

So $p(n+1)$ is true.

Therefore, $p(n) \Rightarrow p(n+1)$ is true.

Part 2 $p(1)\colon 1=\frac{1(2)}{2}$ Note that $p(1)$ is true.

Conclusion Therefore, by mathematical induction, $p(n)$ is true for every positive integer n.

Summation Notation

The sum in the previous theorem can be written in a more concise form using the Greek letter Σ, which corresponds to our letter *S*.

$$\sum_{i=1}^{n} i = 1 + 2 + 3 + \ldots + n$$

We read the above notation as "the summation of i as i goes from 1 to n." When we see the sigma notation, we should mentally view it as the sum on the right side of the above equation. Using sigma notation, we can state the previous theorem as follows:

For every positive integer n, $\sum_{i=1}^{n} i = \frac{n(n+1)}{2}$.

Factorial Notation

In the next example, we have a sequence of sentences where $p(1)$ is not true, but at some point, it kicks into gear and from then on out, all the sentences are true. This example involves n factorial, which is notated as $n!$. If n is a natural number, $n!$ is the product of all natural numbers from 1 to n:

$$n! = n \cdot (n-1) \cdot (n-2) \cdot \ldots \cdot 3 \cdot 2 \cdot 1$$
$$5! = 5 \cdot 4 \cdot 3 \cdot 2 \cdot 1$$

⊕ *Example*

Let $p(n)$ represent the following sentence: $2^n < n!$
For which positive integers is $p(n)$ true? Prove your answer.

$p(1)$:	$2^1 < 1$	$p(1)$ is not true.
$p(2)$:	$2^2 < 2!$	$p(2)$ is not true.
$p(3)$:	$2^3 < 3!$	$p(3)$ is not true.
$p(4)$:	$2^4 < 4!$	$p(4)$ is true. (16<24)
$p(5)$:	$2^5 < 5!$	$p(5)$ is true. (32<120)

Part 1 – Inductive Step:

Let n be an integer where $n \geq 4$.

Assume that $p(n)$ is true.

$2^n < n!$

. . .

So, $2^{n+1} < (n+1)!$

Hence, $p(n+1)$ is true.

Thus, $p(n) \Rightarrow p(n+1)$ for $n \geq 4$.

Since the right side of the inequality grows more rapidly than the left side, we suspect that $p(n)$ is true for $n \geq 4$. Let's try to prove it by mathematical induction.

First, we set up the structure for the inductive step, as illustrated on the left. Then we work on bridging the gap. We can derive the left side of the second inequality by multiplying both sides of the first inequality by 2:

$$2^n < n!$$
$$2 \cdot 2^n < 2 \cdot n!$$
$$2^{n+1} < 2 \cdot n!$$

We now have the left side the way we want it, so let's work on the right side. One way to bridge the gap is to show that:

$$2(n!) < (n+1)!$$

We could then use transitivity to get the desired conclusion:

$$2^{n+1} < 2(n!) \text{ and } 2(n!) < (n+1)! \quad \text{Therefore, } 2^{n+1} < (n+1)!$$

We will now work backwards to try to see if $2(n!) < (n+1)!$

$$2(n!) < (n+1)!$$

Divide both sides by $n!$: $2 < \frac{(n+1)!}{n!}$

Write out the factorials: $2 < \frac{(n+1)\cdot n\cdot(n-1)\cdot\ldots\cdot 3\cdot 2\cdot 1}{n\cdot(n-1)\cdot\ldots\cdot 3\cdot 2\cdot 1}$

Most factors cancel: $2 < n+1$

If we can establish this last sentence, we have a proof. But wait, $4 < n$, so 2 must be less than $n + 1$. Yes, we do have a proof. We will now polish it and write it in the correct order.

Theorem For every integer n, if $n \geq 4$, then $2^n < n!$.

Induction Proof Let $p(n)$ represent the following sentence: $2^n < n!$

Part 1 Let n be an integer such that $n \geq 4$.
Assume that $p(n)$ is true: $2^n < n!$

Multiply both sides by 2: $2 \cdot 2^n < 2 \cdot n!$
$2^{n+1} < 2n!$

Since $n \geq 4$, $2 < n+1$
Multiply by $n!$: $2(n!) < (n+1)n!$
Thus, $2(n!) < (n+1)!$

Since $2^{n+1} < 2(n!)$ and $2(n!) < (n+1)!$,
by transitivity, $2^{n+1} < (n+1)!$
This last sentence is $p(n+1)$. So, $p(n+1)$ is true.

Thus, $p(n) \Rightarrow p(n+1)$ when $n \geq 4$.

Part 2 $p(4)$ is the following sentence: $2^4 < 4!$
$2^4 = 16$ and $4! = 24$. So, $p(4)$ is true.

Conclusion Therefore, by mathematical induction, $2^n < n!$ for every integer n where $n \geq 4$.

The Natural Numbers

The Principle of Mathematical Induction is intimately connected with the set $\mathbb{N}$ of natural numbers. Suppose that $S \subseteq \mathbb{N}$ and $1 \in S$. Suppose also that the following is true:

For every integer n, if $n \in S$, then $n+1 \in S$.

From this data, we can deduce that $S = \mathbb{N}$. This fundamental property of the set of natural numbers is the Principle of Mathematical Induction. We will now prove that this version is equivalent to the first version given on page 174. This proof will test our ability to structure our thinking, for the grammatical structures of what we assume and what we derive are quite complex. When we assume that the first version is true, we are not assuming individual parts of it are true; we are assuming that the complete statement is true:

Principle of Mathematical Induction

Let S be a subset of $\mathbb{N}$ that has the following properties:

1. For every positive integer n, if $n \in S$, then $n+1 \in S$.
2. $1 \in S$.

Then $S = \mathbb{N}$.

1st Version: $[(\forall n, p(n) \Rightarrow p(n+1))$ and $p(1)] \;\Rightarrow\; \forall n, p(n)$

We then must derive that the complete statement of the second version is true:

2nd Version: $[(\forall n, n \in S \Rightarrow n+1 \in S)$ and $1 \in S] \;\Rightarrow\; S = \mathbb{N}$

1st Version ⇒ 2nd Version

Assume that the first version of the Principle of Mathematical Induction is true.

We will now prove that the second version must be true. Since its outside structure is an implication, we assume that its two hypotheses are true. Let S be a subset of $\mathbb{N}$ that has the following properties:

1. For every positive integer n, if $n \in S$, then $n+1 \in S$.
2. $1 \in S$.

Let $p(n)$: $n \in S$. Now translate the above two statements:

1. For every positive integer n, $p(n) \Rightarrow p(n+1)$.
2. $p(1)$ is true.

Since $p(n)$ satisfies the hypotheses of the first version, we can deduce that $p(n)$ is true for all positive integers n. Thus, for every positive integer n, $n \in S$. So, $\mathbb{N} \subseteq S$.

Since S is also a subset of $\mathbb{N}$, $S = \mathbb{N}$.
Thus, the second version is true.

2nd Version ⇒ 1st Version

Conversely, assume that the second version of the Principle of Mathematical Induction is true.

We will now show that the first version has to be true. We start by assuming the hypotheses in the first version. Let $p(n)$ be an open sentence. Assume the following:

1. For every positive integer n, $p(n) \Rightarrow p(n+1)$.
2. $p(1)$ is true.

Define S as follows: $S = \{\, n \mid p(n) \text{ is true}\}$
Translate the above two statements in terms of S:

1. For every positive integer n, if $n \in S$, then $n+1 \in S$.
2. $1 \in S$.

Since the second version is true, we can deduce that $S = \mathbb{N}$.
So, for every positive integer n, $p(n)$ is true.
Thus, the first version is true.

Some textbooks use the second version for the Principle of Mathematical Induction and some use the first version. As the above proof shows, the two versions are equivalent.

Inductive Definitions

We sometimes define a sequence in an inductive manner. A *sequence* is a function whose domain is the set of natural numbers. Using the notation s_n to indicate $s(n)$, we can represent a sequence in the following manner:

$$s_1,\ s_2,\ s_3,\ \ldots,\ s_n,\ \ldots$$

In an inductive (or recursive) definition, we define the first term of a sequence, then we define each successive term using previous terms, as illustrated in the following example.

⊕ Example

Inductively define the sequence s_n as follows:

$$s_1 = 5 \qquad s_n = s_{n-1} + 2$$

We compute the terms in the sequence as follows:

$$s_2 = s_1 + 2 = 5 + 2 = 7$$
$$s_3 = s_2 + 2 = 7 + 2 = 9$$
$$s_4 = s_3 + 2 = 9 + 2 = 11$$
$$\ldots$$

Using the method in this example, it would take a while to compute the 1000th term in the sequence. To find a faster method, let's redo the previous computations, but not do any simplifications:

$$s_2 = s_1 + 2 = 5 + 2$$
$$s_3 = s_2 + 2 = (5 + 2) + 2$$
$$s_4 = s_3 + 2 = (5 + 2 + 2) + 2$$

Since multiplication is repeated addition, we can write these terms as follows:

$$s_2 = 5 + 2$$
$$s_3 = 5 + (2 \cdot 2)$$
$$s_4 = 5 + (3 \cdot 2)$$

The above pattern indicates that we can compute s_n as follows:

$$s_n = 5 + (n - 1) \cdot 2$$

Now we have a closed formula for s_n that enables us to quickly compute any term of the sequence: $s_n = 2n + 3$

To prove that our closed formula is correct, we can use mathematical induction. First, we set up the outside structure, as indicated by the brackets. With our focus on what we want to derive, we use the definition of s_{n+1} and work our way down.

Theorem If $s_1 = 5$ and $s_n = s_{n-1} + 2$, then $s_n = 2n + 3$.

Induction Proof Let $p(n)$: $s_n = 2n + 3$

Part 1 Let n be a positive integer.

Assume that $p(n)$ is true: $s_n = 2n + 3$

Definition of s_{n+1}: $s_{n+1} = s_n + 2$

Substitute for s_n: $s_{n+1} = (2n + 3) + 2$

So $p(n + 1)$ is true: $s_{n+1} = 2(n + 1) + 3$

Therefore, $p(n) \Rightarrow p(n + 1)$ is true.

Part 2 $p(1)$: $s_1 = 2 \cdot 1 + 3$ Since $s_1 = 5$, $p(1)$ is true.

Conclusion By mathematical induction, $p(n)$ is true for every positive integer n. So our closed formula is correct.

Stronger Form

The Principle of Mathematical Induction can be stated in a stronger form by adding more hypotheses in Part 1. Instead of assuming only that $p(n)$ *is true*, we assume that $p(n)$ and all the sentences that precede it are true:

Assume that $p(1)$, $p(2)$, $p(3)$, . . . , *and* $p(n)$ *are each true.*

The conclusion $p(n + 1)$ is sometimes easier to derive with the extra assumptions.

The Stronger Principle of Mathematical Induction is stated on the left. Note that it is identical to the version on page 174 except for the extra hypotheses in Part (1). If we prove Part 1, we have proved the following infinite list of implications:

$$n = 1: \quad p(1) \Rightarrow p(2)$$
$$n = 2: \quad [p(1) \text{ and } p(2)] \Rightarrow p(3)$$
$$n = 3: \quad [p(1) \text{ and } p(2) \text{ and } p(3)] \Rightarrow p(4)$$
$$\cdots$$

Stronger Principle of Mathematical Induction

Let $p(n)$ be an open statement. Suppose the following are true:

1. For every integer $n \geq 1$, if $p(1) \wedge p(2) \wedge \ldots \wedge p(n)$, then $p(n+1)$.
2. $p(1)$ is true.

Then $p(n)$ is true for all integers n where $n \geq 1$.

If we verify that $p(1)$ is true, we can then work our way down the above list, making deductions on each line:

$p(1)$ is true.
Since $p(1) \Rightarrow p(2)$, we can deduce $p(2)$.

Now we know that $p(1)$ *and* $p(2)$ is true.
Since $p(1)$ *and* $p(2) \Rightarrow p(3)$, we can deduce $p(3)$.

Now we know that $p(1)$ *and* $p(2)$ *and* $p(3)$ is true.
Since $p(1)$ *and* $p(2)$ *and* $p(3) \Rightarrow p(4)$, we can deduce $p(4)$.

And so on, down the list. The Stronger Principal of Mathematical Induction guarantees that the Law of Detachment can be applied an infinite number of times.

The stronger version can be generalized by starting at an integer other than 1. We will use this form in the following example. A *prime number* is a positive integer greater than 1 whose only positive factors are 1 and itself. For example, 2, 3, 5, 7, and 11 are prime numbers. 12 is not a prime number, but 12 can be factored as a product of primes: $12 = 2 \times 2 \times 3$. It seems fairly obvious that every integer greater than 1 is a prime number or a product of prime numbers. However, to prove this statement, we need the stronger assumption from the stronger induction version.

Theorem Every integer greater than 1 is a prime number or a product of prime numbers.

Induction Proof Let $p(n)$: n is a prime or n is a product of primes.

Part 1 Let n be a integer where $n \geq 2$.

Assume that $p(2)$ and $p(3)$ and $p(4)$ and . . . and $p(n)$ are true.
We can translate this assumption as follows:

For every integer j where $2 \leq j \leq n$,
j is a prime or j is a product of primes.

We want to derive $p(n+1)$, which we can translate as:

$n+1$ is a prime or $n+1$ is a product of primes.

Assume that $n+1$ is not prime.
Then there exists positive integers a and b such that:

$$n+1 = ab, \text{ where } 1 < a < n+1 \text{ and } 1 < b < n+1.$$

So, $2 \leq a \leq n$ and $2 \leq b \leq n$.
By the induction hypothesis, a is prime or a product of primes. Similarly, b is a prime or a product of primes.
So $n+1$ can be expressed as a product of primes, using the prime factors from a and b.

→ Thus, $p(n+1)$ is true.

Therefore, $p(n) \Rightarrow p(n+1)$.

Part 2 $p(2)$: 2 is a prime or 2 is a product of primes.
Since 2 has no positive factors other than 1 and itself,
2 is a prime number. So $p(2)$ is true.

Conclusion Therefore, by mathematical induction, $p(n)$ is true for all positive integers n where $n \geq 2$.

Deductive Reasoning Mathematical induction is a type of deductive reasoning, even though its name sounds more like inductive reasoning. Inductive reasoning is when we suspect that something is true based on examples, experiments, or experiences. Before we prove a statement by mathematical induction, we normally verify it for various values of n, which may be the reason that Augustus De Morgan gave it the name of mathematical induction back in 1838. However, inductive reasoning provides no guarantee that our conjecture is true. On the other hand, a proof by mathematical induction does provide a 100% guarantee that the sentence is true for the specified integers. Deductive and inductive reasoning each have an important function in the reasoning process, but the proof method of mathematical induction is in the camp of deductive reasoning, not inductive reasoning.

Exercise Set 2.7

1. Consider the following formula.

$$2 + 4 + 6 + \ldots + 2n = n(n+1)$$

   a. Check to see if the formula works for various values of n.

   b. If your examples indicate that the formula is correct, try to use mathematical induction to prove that the formula is true for every positive integer n. State what $p(n)$ represents in your proof, state the conclusion for the first part of the proof, and also state the final conclusion.

   c. Let s_i denote the ith term in the sum: $2 + 4 + 6 + \ldots + 2n$ Give a formula for s_i:

   d. Rewrite the sum in part (c) in summation notation: $\sum_{i=1}^{n} s_i$

2. In the adjacent square, we have a picture of 1+3+5+7.

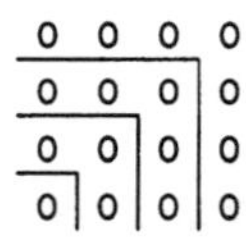

   a. Expand this sketch to get a picture of 1+3+5+7+9.

   b. Use visual reasoning to conjecture a formula for the sum of the first n odd positive integers.

   c. Translate "the sum of the first n odd positive integers" into the following form: $1 + 3 + 5 + \ldots$ _ ?

   To see the pattern, write the sum of the first 2 odd positive integers, the sum of the first 3 odd positive integers, etc., until you see how to express a formula for the last term.

   d. Use mathematical induction to prove that your formula from part (b) is true for every positive integer n. State what $p(n)$ represents in your proof, state the conclusion for the first part of the proof, and also state the final conclusion.

e. Give a formula for the ith term in the sum: $1+3+5+\ldots$ ___

f. Rewrite the sum in part (e) in summation notation: $\sum_{i=1}^{n} s_i$

3. Inductively define a sequence as follows: $s_1 = 3,\ s_n = s_{n-1} + 8$
   a. Compute the first 5 terms of this sequence.
   b. Find a closed formula for s_n so that we can compute s_{100} without computing s_{99}. *Hint*: Redo part (a) but do not simplify your computations and a closed formula will be easy to spot.
   c. Use mathematical induction to prove that your formula in part (b) always works.

4. Let a and d be fixed real numbers.
   Inductively define a sequence as follows: $s_1 = a,\ s_n = s_{n-1} + d$
   This type of sequence is called an *arithmetic sequence.*
   a. Find a closed formula for s_n.
   b. Use mathematical induction to prove that your formula is correct.

5. Inductively define a sequence as follows: $s_1 = 3,\ s_n = s_{n-1} \cdot 8$
   a. Compute the first 5 terms of this sequence.
   b. Find a closed formula for s_n so that s_{100} can be computed without computing s_{99}.
   c. Use mathematical induction to prove that your formula in part (b) always works

6. Let a and d be fixed real numbers.
   Inductively define a sequence as follows: $s_1 = a,\ s_n = s_{n-1} \cdot d$
   This type of sequence is called a *geometric sequence.*
   a. Find a closed formula for s_n.
   b. Use mathematical induction to prove your formula is correct.

7. $\sum_{i=1}^{n} s_i = s_1 + s_2 + s_3 + \ldots\ s_n$. Rewrite each with summation notation.
   a. $2 + 2^2 + 2^3 + \ldots + 2^n$
   b. $\frac{1}{2} + \frac{1}{4} + \frac{1}{8} + \ldots + \frac{1}{2^n}$
   c. $\frac{1}{1\cdot 2} + \frac{1}{2\cdot 3} + \ldots + \frac{1}{n(n+1)}$

8. Rewrite each sum in expanded notation.
   a. $\sum_{i=1}^{n} 5^i$ b. $\sum_{i=1}^{n} 5^{i-1}$ b. $\sum_{i=0}^{n} 5^i$ c. $\sum_{i=0}^{n} r^i$

9. Let $S_n = 2 + 2^2 + 2^3 + \ldots + 2^n$.
   a. Find a closed formula for S_n. *Hint*: Multiply both sides of the above equation by 2, then subtract the two equations and simplify.

b. Use mathematical induction to prove the closed formula for S_n in part (a).

c. Let $S_n = 3 + 3^2 + 3^3 + \ldots + 3^n$. Find a closed formula for S_n. Use mathematical induction to verify your formula.

d. Generalize your formula in part (c). Then use mathematical induction to verify your formula.

10. Compute the given sum for $n = 1, 2, 3,$ and 4.
Try to find a possible formula for the sum.
Make sure that your formula works for $n = 1, 2, 3,$ and 4.
Then try to verify your formula using mathematical induction.

a. $\frac{1}{2} + \frac{1}{4} + \frac{1}{8} + \ldots + \frac{1}{2^n} =$ ___

b. $\frac{1}{1 \cdot 2} + \frac{1}{2 \cdot 3} + \ldots + \frac{1}{n(n+1)} =$ ___

11. Use Mathematical Induction to prove the following:

a. $1^2 + 2^2 + 3^2 + 4^2 + \ldots + n^2 = \frac{n(n+1)(2n+1)}{6}$

b. For every positive integer n, $2^{n-1} \leq n!$.

c. For every positive integer n, $n < 2^n$.

12. Conjecture: $n^2 \leq 2^n$
Is the above conjecture true for every positive integer n?
If not, can you find an integer c so that it is true for every integer $n \geq c$? If so, prove your result by mathematical induction.

13. Prove Part 1 of the following theorem in a manner different from the proof given on page 180. Instead of multiplying the inequality by 2, multiply both sides by $n + 1$, which will give you the desired right side. Then you must figure out how to derive the left side.

Theorem For every integer $n \geq 4$, $2^n < n!$.

Induction Proof Let $p(n)$ represent the following sentence: $2^n < n!$

Part 1 Let n be an integer such that $n \geq 4$.
Assume that $p(n)$ is true: $2^n < n!$

Multiply both sides by $n+1$: _______

So, $2^{n+1} < (n+1)!$

Therefore, $p(n+1)$ is true. So, $p(n) \Rightarrow p(n+1)$ for $n \geq 4$.

15. Use mathematical induction and cases to prove that every integer is even or odd.

Theorem	Every integer is even or odd.
Induction Proof	Let $p(n)$: n is even or n is odd.
Part 1	Let n be an integer. Assume that $p(n)$ is true. Then n is even or n is odd.

Case 1: Assume ________
Therefore, ________
Case 2: Assume ________
Therefore, ________
Therefore, $p(n+1)$ is true.
So, $p(n) \Rightarrow p(n+1)$.

16. The complex number $a + bi$ can be identified with the point (a,b) (page 15). The complex number $\cos\theta + i\sin\theta$ can be identified with the point $(\cos\theta, \sin\theta)$.

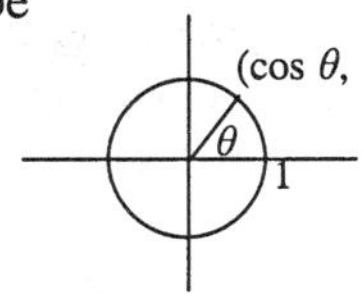

a. Us the Pythagorean Theorem and the definitions of $\cos\theta$ and $\sin\theta$ to explain why the point in the adjacent sketch is $(\cos\theta, \sin\theta)$.

b. Use mathematical induction to prove the following:

DeMoivre's Theorem: Let n be a positive integer and let θ be a real number. Then $(\cos\theta + i\sin\theta)^n = \cos n\theta + i\sin n\theta$.

Hint: $\cos(\alpha + \beta) = \cos\alpha\cos\beta - \sin\alpha\sin\beta$
$\sin(\alpha + \beta) = \sin\alpha\cos\beta + \cos\alpha\sin\beta$

The rules for multiplying complex numbers are the same as for real numbers, with the additional rule that $i^2 = -1$.

c. How far is the point, $\cos\theta + i\sin\theta$, from the origin? Describe the location of the point, $\cos 45° + i\sin 45°$.

d. Use *DeMoivre's Theorem* to describe the location of the point, $(\cos 45° + i\sin 45°)^3$.

17. Let n be an integer greater than 1. Use mathematical induction to prove the following: The product of n odd integers is odd.

18. Derive a formula for the sum of the first n counting numbers as follows. Let $S = 1 + 2 + 3 + \ldots + n$. Write the sum in reverse order: $S = n + (n-1) + (n-2) + \ldots + 3 + 2 + 1$

Add the two equations. On the right side, compute the sum by adding the first 2 terms together, the second 2 terms together, etc. Solve for S. Compare your derivation with the proof on page 178.

Activity 2.8

In a monastery, 64 disks, with each a different size, were placed on one of three posts, with the largest disk on the bottom, the next largest on top of it, and continuing in this manner with the smallest disk on top, as illustrated below.

The monks were ordered to move the disks, one at a time, to one of the other 2 posts, subject to the condition that a disk must always be set on top of a larger disk. The task must be continued until all 64 of the disks are transferred to one of the other two posts. When the monks complete the task, the people were told that the world would end. Naturally, many people were concerned as to how long it would take the monks to finish.

The monks move one disk each second, and they always make the least number of moves to accomplish the goal. How long will it take them to move all 64 disks to another post?

1. Before you start on this fascinating problem, make a wild guess as to what you think the answer might be in years.
2. Make a model and practice the movements.
3. This counting task will be beyond our grasp unless we analyze the process in small steps and look for a pattern. How long does it take the monks to move the two disks on top to another single post?
4. How long does it take the monks to move the three disks on top to another single post? *(Use your previous answer)*
   To move the four disks on top? *(Use your previous answer.)*
   To move the five disks on top? To move n disk?
5. Verify your last answer using mathematical induction.
6. So, how many years will it take the monks to move the 64 disks? Is it longer than you initially imagined?

≡ 2.8 Axiomatic Systems ≡

Axiomatic systems provide a structure in which we logically order our reasoning about some area of interest, such as geometry or set theory or number theory. To construct an axiomatic system, we first work backwards through what we personally understand about the subject and identify what we want to use as the foundation for the system. For example, what is the foundation for what we accept as true?

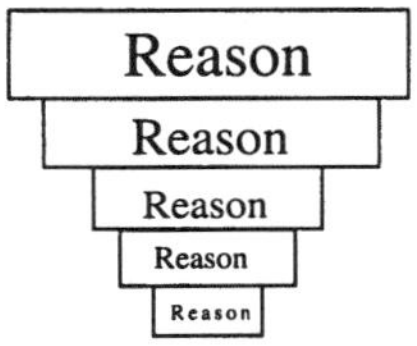

To be a creative thinker, we must ask good questions which will lead us on a path of inquiry. Along this path, suppose that we discover an interesting relation. To authenticate our discovery, we must validate it with a proof. As we write our proof, we justify each step with a reason that is either simpler or refers to previous results.

Having finished our proof, suppose that we now go through the same process for each reason that we used in our proof. We prove each reason, justifying it with simpler reasons. If we continue this process with each new reason that we use, we will back up, step by step, until we cannot find any simpler reasons to use in our explanation. At this point, we will have reached an impasse. We cannot explain or prove these simplest of sentences because there is nothing simpler to use to explain them.

Axioms

An axiomatic system has a few sentences, called *axioms*, which we consider true but cannot prove. It is not possible to prove everything because we would have no previous knowledge to use when we try to construct our first proof. Consequently, we must accept some sentences as true so that we have something to use in our first proof.

An *axiom* for a system is a statement assumed true in that system, requiring no proof.

In the first proof that we write in a system, the only tools we can use are the axioms and definitions. In the second proof, we can also use the first theorem if we so choose. The more theorems that we prove, the more tools that we have to use in future proofs. To get started with the first proof, though, we must have axioms. The axioms form the foundation for truth in an axiomatic system. They provide us with simple truths from which we construct more sophisticated truths.

Axioms
Theorem 1
Theorem 2
Theorem 3
...

Someone else going through the same process of explaining each reason with simpler reasons may back up in a different way, ending up with different axioms, so the axioms selected are not necessarily unique. Different sets of axioms can produce the same system.

If Statement A is equivalent to Statement B, we may choose to call Statement A an axiom and then prove Statement B, in which case Statement B would be a theorem. On the other hand, we could call Statement B an axiom and then prove Statement A. As a matter of form, though, we always choose the axioms to be as simple as possible. If Statement A sounds simpler than Statement B, we would use it as an axiom. The ultimate goal in logical reasoning is to make things as simple as possible. When we have a complex situation, we look for a simple explanation as to why things happen the way that they do. Perhaps we are guided by the same laws that drive the universe, which make objects traveling through space always seek the simplest path.

Let's go back in time 2300 years and imagine Euclid pacing through the magnificent rooms of the great library of Alexandria, rooms filled with the greatest collection of knowledge known to mankind. Having just started his own school in Alexandria, Euclid is thinking about the best way to teach his students how to reason in a logical manner, for he knows that only through reasoning can humanity continue to expand the knowledge housed in the greatest library of antiquity. He also know that nowhere can the pure structure of reasoning be demonstrated in a brighter light than in the field of geometry, for there students can use their visual reasoning to develop their skills in deductive reasoning. As he ponders the vast body of geometric knowledge, possibly thinking of how he will teach his students the ingenious proofs of Pythagoras or Eudoxus, Euclid realizes that the whole body of geometric knowledge needs to be organized in a simpler format with the theorems arranged in a linear order. This arrangement would make it easier to prove the more complex theorems with a sequence of simple steps, referencing previous theorems. Furthermore, by making the tools used to prove the more complex theorems stand alone on their own right as theorems, these tools would become available for general use thereafter.

This wonderful book [Euclid's *Elements*], with all its imperfections, which are indeed slight enough when account is taken of the date it appeared, is and will doubtless remain the greatest mathematical textbook of all time.

Thomas L Heath
1861–1940

Euclid then organized the work of the brilliant thinkers before him into a series of textbooks whose influence on western civilization has been ranked as second only to the Bible. Of course, Euclid had to work backwards to find out where to start. As he tried to explain each reason with a simpler reason, he arrived at a surprisingly small core of only five simple truths from which he could derive all the other results. His five axioms, or *postulates* as they were called at

Postulates of Euclidean Geometry

1. Through any two points, *there exists* a unique straight line.
2. For any two points on a line, *there exists* another point on the line beyond the other two.
3. For every point C and length r, *there exists* a circle of radius r with C as the center.
4. All right angles are congruent.
5. In the plane determined by a line ℓ and a point P not on ℓ, *there exists* a unique line through P that is parallel to ℓ.

that time, can be translated into the equivalent forms in the adjacent list. Notice the beautiful simplicity of these postulatès. Notice also that four of these five postulates have an existence clause.

Most axioms focus on existence questions, as do the axioms of faith in religious systems. A basic axiom in most religions is whether or not god exists, and, if so, how many gods are there? The atheists postulate 0 gods, the Christians and Moslems postulate 1 god, and the Hindus postulate more than 1 god. True believers in these religious groups have spiritual experiences that sustain their belief, spiritual experiences that cannot be backed up with deductive reasoning. Many religious people believe that their axioms of faith are the only *true* axioms of faith, which often generates conflict and wars with those who do not see the world from their religious perspective.

Until the 19th century, a similar view was shared by intellectuals concerning the truth of Euclid's postulates (page 198). It was believed that Euclid's postulates were absolute truths, truths that were not to be questioned for they described inherent truths of reality. From this unquestioning faith in the five axioms of Euclid, a geometric picture of our universe emerged with straight lines traveling across cosmic distances, straight lines that behaved as they were perceived to behave here on Planet Earth. Current scientific evidence, though, suggests that our view of the universe as Euclidean may be as outdated as our earlier view that the earth was flat. However, regardless of the future scientific verdict on the geometry of our universe, the axioms of Euclidean geometry will still be true in Euclidean geometry. We cannot disprove an axiom for there is nothing to disprove. The axioms serve as the truth foundation of that particular system. If that system does not suit our needs for a particular purpose, such as creating a mathematical model of our cosmic universe, then we can use another slate of axioms and build another axiomatic system. Thus, the axioms are *relative* truths. An axiom in one system is not necessarily true in another system.

After the tremendous paradigm shift caused by the great debate over the absolute truth of axioms, mathematicians found it necessary to impose more rigorous standards on axiomatic systems. By the new standards, Euclidean geometry needed a few more axioms to make it completely rigorous. The new standards also required a closer inspection of how we work with sets. Sets seem so simple that they had always been worked with in an intuitive manner. The intuitive bubble burst, though, when Bertrand Russell developed a contradiction from

Axiom of Set Theory
There exists an empty set.

an "intuitively obvious" construction of a set (page 206), and so it became necessary to construct an axiomatic foundation for sets that would eliminate Russell's paradox. The first axiom, "there exists an empty set," is the Big Bang of Set Theory, giving us the existence of an initial set from which we construct all other sets (page 276). The other set theory axioms (page 277–285) give us ways to construct other sets from this one set, until we build a rather miraculous universe of sets reaching into the mysterious realms of infinity.

Undefined Terms

The process of proving a theorem and then proving each reason that we use leads us back to the problem of how we write the first proof in a linearly ordered system where we can only use previous theorems. To solve this dilemma, axioms were introduced. We run into a similar problem when we start analyzing our definitions. To understand a definition, we must understand the definition of each word in it. If we look up the definition of a word in the dictionary, along with the definition of each word used in that definition, and continue this process, sooner or later we will find definitions that are circular. For example, in the adjacent definitions, a set is defined in terms of a collection, which is defined in terms of a group, which is defined in terms of a set. If we do not know the meaning of either *set* or *collection* or *group*, these definitions will have no meaning for us.

Set – a collection of objects.
Collection – a group of objects.
Group – a set of objects.

Definitions in a dictionary are by necessity circular for it is impossible to define every word in a linear manner. Otherwise, we could arrange the definitions in an ordered list with our first definition at the top. If we wish to make our definitions non-circular, each time we define a new word, we must use only previously defined words. The dilemma is the construction of the first definition since there are no previously defined words that we can use in it. The only way out of this dilemma is to admit that it is not possible to define every word, and select certain words as *undefined.*

Definition 1. ________
Definition 2. ________
Definition 3. ________
Definition 4. ________
. . . . . .

Undefined terms are the basic words from which we construct the vocabulary for an axiomatic system. The terms selected to be undefined represent the simplest concepts, concepts that cannot be explained by simpler concepts. However, the notion of what is simplest is a matter of personal preference. Different starting points can lead to the same results. Using different undefined terms and different axioms, mathematicians have produced different axiomatic systems that generate the theorems of Euclidean geometry.

When we select the undefined terms for a system, we need both subjects and verbs so that we have the necessary ingredients to specify our axioms, which are sentences. In the axiomatic system of set theory, the undefined terms are *set,* which is a noun, and *is an element of*, a verb that gives a relation between two nouns. Using only these two undefined terms, along with the five logical operators and two quantifiers, we can build definitions for all the concepts of set theory. Logical reasoning is a linear process and it is rather amazing how little we need to get us started. Like the Big Bang Theory that our whole universe miraculously originated from one little point, it is no less spectacular that the vast universe which the theory of sets encompasses originates from one little noun and one little verb.

Undefined Terms of Set Theory

set
is an element of

We all know what a straight line is, but can we put it into words – using no visuals – so that an intelligent being from another galaxy could understand what we meant? If we do come up with a definition, we will then be faced with the task of defining each word in the definition for these intergalactic friends do not yet understand our language. Even though we personally know what a straight line is and what a point is, it is extremely difficult to define them. Euclid defined a straight line as "a line which lies evenly with the points on itself," but he did not define what it meant to "lie evenly." He defined a point as "that which has no part," but he did not define what "part" meant. By today's standards, Euclid's definitions are not considered axiomatic definitions. Since there are no simpler concepts with which to explain them, we classify "point" and "line" as undefined. In this context, "line" means "straight line." Using these two undefined nouns and the three undefined verbs listed in the adjacent box, we can build the vocabulary for Euclidean geometry. Notice how the undefined verbs give a relation between two nouns:

Undefined Terms of Euclidean Geometry

point
line
is on
is between
is congruent to

The point A *is on* the line ℓ.

The point A *is between* the points B and C.

Like the axioms selected for a system, the undefined terms that are selected are not unique. Other axiomatic systems have been constructed for Euclidean geometry that use different undefined terms.

Axiom: Through every two points, there exists a unique straight line.

Translation: For every two *points* A and B, there exists a unique *line* ℓ such that A *is on* ℓ and B *is on* ℓ.

If a term is undefined, though, how can we work with it? How do we know what it represents? We work around this dilemma by using the axioms to state the properties that an undefined term must have. In the translation of the adjacent axiom, note how the axiom gives a property of the undefined terms, *point*, *line*, and *is on*.

The axioms give the basic properties of the undefined terms; they are the basic assumptions that we make about the undefined terms. On the other hand, the *undefined terms* are the basic building blocks, similar to atoms, from which we build the vocabulary in the system. The undefined terms form the foundation for the vocabulary of an axiomatic system, whereas the axioms form the foundation for what is true in the system. Axioms are sentences, but undefined terms are not sentences; they are terms.

Definitions

When we define a term, we phrase it in a sentence and then establish the meaning of that sentence with a string of words that also form a sentence. For example, to define the term, $\subseteq$, we define the meaning of the sentence $A \subseteq B$ with the following sentence: for every x, if $x \in A$, then $x \in B$. Since we can use the sentence being defined and its definition interchangeably, we always use the "if and only if" connective in a definition.

Using the undefined terms, we make our first *definition*, then our second definition, and so on. At each stage, we can only use the undefined terms and previously defined terms in our new definition, which then gives us another term that we can use in future definitions. Slowly and carefully we build the language of the system.

Grammar

Since a definition must be stated as a sentence, we must have a clear understanding of the grammar of the system before we make our first definition or state our first axiom. The *grammar* for an axiomatic system is a clearly defined set of rules for forming sentences, specifying the required syntactical structure that an expression must have in order to be classified as a sentence. For example, the expression $A \cup B$ does not have the proper syntax to be classified as a sentence. The rules for grammar in our everyday language are very complex, which makes it rather difficult to program computers to translate from one natural language to another without some distortions. However, the limited vocabulary of mathematics makes it possible to have a clearly defined set of grammar rules, which are a subset of the grammar rules for everyday language. If an expression that we write does not form a sentence in our everyday language, it is not a sentence in the standard mathematical systems. $A \cup B$ is not a sentence in our everyday language, so we cannot use it as a sentence in a proof.

Using the undefined terms and defined terms in conjunction with the proper grammar, we build the sentences that can be formed within our system. Now, we look out over our vast universe of sentences and we wonder – which of those sentences are true and which are false? This question brings us to the most important component of an axiomatic system, the component that sets the rules for making logical deductions – the proof procedure.

Proofs

Proofs provide a deductive procedure for deciding what is true in a system and what is false. If we prove a sentence, we classify it as *true* and label it as a theorem; if we disprove a sentence, we classify it as *false*. Using the proof procedure, we build on the foundation of axioms, finding other statements that we can prove are true. A sentence in a proof must be either:

- a sentence that we assume is true
- a sentence that we already know is true
- a sentence that we can derive from previous sentences by a valid argument

To form a proof, the sentences must be connected as described in the adjacent box. A proof is a linearly ordered structure of interwoven valid arguments. As we saw in Section 2.1, the rules for the outside construction of a valid argument are essentially the rules for how we use the logical operators and quantifiers. The definition of *implies* gives different ways to set up the structure to derive an implication; the definition also gives the structure for making derivations from an implication. These are two different proof techniques:

- how we derive an implication
- what we can derive from an implication

Similarly, the definition of *or* gives the rules for setting up the structure to derive an or-sentence; it also gives the structure for making derivations from an or-sentence. In a proof, it is essential to remember that deductions made from an assumption are not stand-alone conclusions. An assumption used in a valid argument must always be included in the final conclusion. For example, if we assume p is true and then derive q, our final conclusion is $p \Rightarrow q$.

A *proof* is a linearly ordered structure of interwoven valid arguments where each sentence is one of the following:

- An assumption used in a valid argument
- An axiom, previous theorem, or definition
- A sentence that can be derived from previous sentences by a valid argument

The final stand-alone conclusion is the theorem that has been proved.

An Axiomatic System

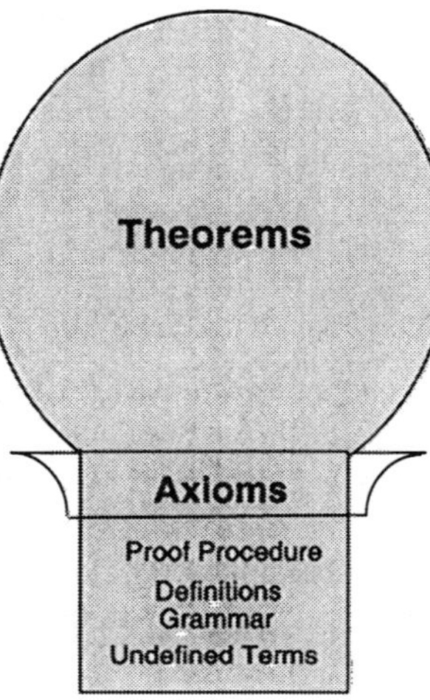

In summary, an axiomatic system is composed of the following:

- *Undefined terms* from which we construct the vocabulary of the system.
- A *grammar* which gives the rules for forming sentences.
- *Definitions* built from the undefined terms.
- A *proof procedure* which give a deductive method for deciding what is true in the system and what is false.
- *Axioms,* sentences assumed to be true.
- *Theorems*, sentences that we can derive from the axioms.

The early Greeks devised the axiomatic method for deductive reasoning. It is rather surprising that since the time of Euclid, the deductive method of reasoning and constructing proofs has remained essentially the same. However, our view of the axioms has drastically changed, causing one of the most profound paradigm shifts in all of human history.

Absolute Truths

For over 2000 years, axioms for mathematical systems were considered to be self-evident truths. They were considered as *absolute truths*, statements whose truth could not be questioned because it was thought that they described basic properties of physical space. This view was held until the 19th century when the quest for establishing the independence of Euclid's axioms forced a radical change in our perception of truth, a change so radical that its reverberations affected all areas of intellectual thought.

Independent Axioms

A set of axioms is *independent* if none of the axioms can be derived from the others. If we can derive a statement from the axioms, it is not necessary to list it as an axiom for we can list it as a theorem instead of an axiom. We prefer not to make more assumptions than are necessary. When the a set of axioms are independent of each other, we have a minimal list of assumptions about that system.

Many mathematicians regarded Euclid's Fifth Postulate with suspicion, not because its truth was in question, but because it was more complex than his other four axioms. They thought that it should be a theorem instead of an axiom. Euclid stated his Fifth Postulate in the following form:

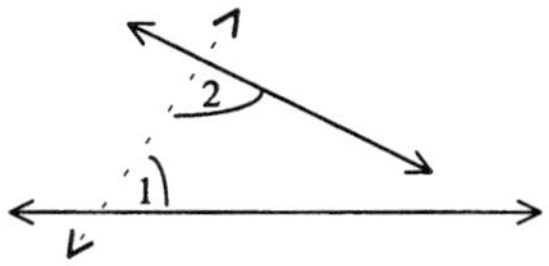

Euclid's Fifth Postulate: If two lines in a plane are cut by a transversal so that the sum of the interior angles on one side of the transversal are less than a straight angle, then the 2 lines must intersect on that side.

Euclid could have phrased his Fifth Postulate in one of the following equivalent forms:

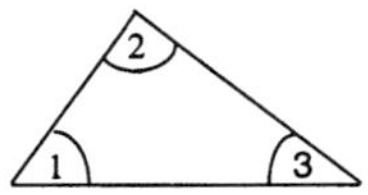

Triangle Sum Postulate: The sum of the angles in a triangle is 180°.

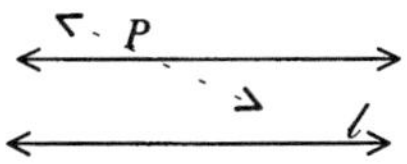

Parallel Postulate: In the plane determined by a line ℓ and a point P not on ℓ, there exists a *unique* line through P that is parallel to ℓ.

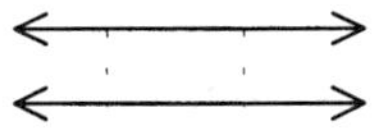

Equidistant Postulate: Parallel lines are everywhere equidistant.

If one structures their thinking in the way described in this chapter, it is not too difficult to prove that Euclid's Fifth Postulate is equivalent to each of the above statements, which certainly seem like obvious "truths." A proof of any one of these statements would also prove Euclid's Fifth Postulate since they are equivalent.

Through the centuries, the challenge of proving Euclid's Fifth Postulate attracted many great intellects, but with no success. In 1733, Girolamo Saccheri, a Jesuit mathematician, tried to prove the triangle sum version with a proof by contradiction. He assumed that the sum of the angles in a triangle is not 180° and tried to find a contradiction. We can imagine his great excitement when he thought that he had found one. He published his work, but after his death, a logical error was found in his reasoning. Although Saccheri did not demonstrate a contradiction, he did derive several interesting theorems that were logically correct.

Continuing the quest that had spanned two millennium, in the early 19th century, Carl Gauss, Janos Bolyai, and Nikolai Lobachevsky noticed independently that replacing Euclid's Fifth Postulate with its negation yielded a strange, but interesting, new set of theorems, including those that Saccheri had first proved. They began to suspect that a contradiction would not be found – that they were seeing a new axiomatic system that was logically valid. This new system, known today as non-Euclidean geometry, was extremely controversial in the 19th century because it defies our intuitive perception of reality. From our visual perception of the world around us, a point P, neither of which intersect a given line, with all three lines contained in the same plane?

Gauss, one of the greatest mathematicians of all time, did not publish his results, perhaps because he did not want to waste his time defending something that he knew was true, especially against people like Emanuel Kant and other philosophers who adamantly believed that space had to be Euclidean. Bold enough to risk the ridicule of academia, Bolyai (1833) and Lobachevsky (1829) each published the theorems they had derived in this strange new system. However, neither proved that their system had no contradictions. Thus, the naysayers continued to believe that someday a contradiction would surface in their bizarre system. On the other hand, though, no one had ever proved that Euclid's system has no contradictions.

It was not until 1868 that the question was finally settled. By constructing a model of non-Euclidean geometry within the framework of Euclidean geometry, Eugenio Beltrami proved that if Euclidean geometry has no contradictions, then non-Euclidean geometry also has no contradictions. In other words, non-Euclidean geometry is as logically sound as Euclidean geometry. This result proved that Euclid's Fifth Postulate could not be derived from the other four axioms.

Models

To understand Beltrami's method of proof, we need to understand what is meant by a model of an axiomatic system. To construct a model of an axiomatic system, we construct a model of each undefined term in the system. A model of an undefined term is some type of example that has those properties specified by the axioms. Consider the axiomatic system built from the following three undefined terms and three axioms.

Undefined Terms: *point*, *line*, and *is on*:

A1: For every two points, there is a line on them.

A2: Every line is on at least two points.

A3: Every point is on at least two lines.

Suppose that we use houses to model points and paths to model lines. For "is on," we will use the same interpretation. Doing a direct substitution in the above axioms gives the following translation of the axioms within the context of our model.

M1: For each two houses, there is a path on the two houses.

M2: Every path is on at least two houses.

M3: Every house is on at least two paths.

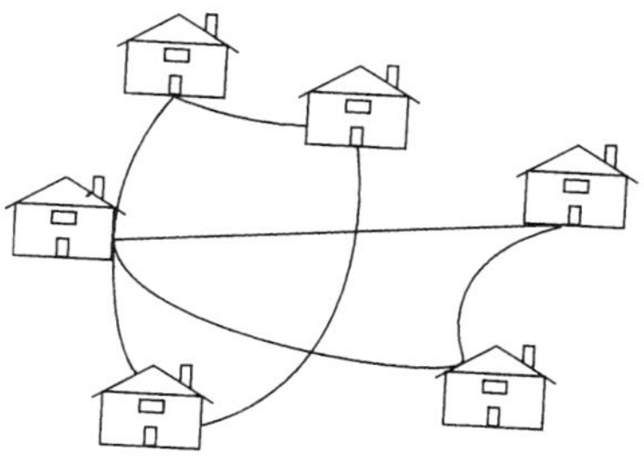

If we have a set of houses and paths that satisfy the previous three sentences, then we have a physical model of our axiomatic system. For example, in the adjacent illustration, we have six houses and various bicycle trails between them. By a path between two houses, we will mean a bicycle path where the paper boy can ride from one house to the next. If we check carefully, we will see that this interpretation of point and line satisfies the three axioms. So, we have a model of our axiomatic system.

Since a model of a system has all properties specified by the axioms, any theorem that can be derived from the axioms must also be true in the model. Consequently, the theorems of an axiomatic system are true in all models of the system. By proving a theorem in an abstract setting, we can then apply it to a host of different models with no further derivations required, which is the main reason that abstract thinking is such a powerful tool.

Beltrami settled the famous controversy over Euclid's Fifth Postulate by building a model of non-Euclidean geometry in a Euclidean plane. His interpretation of the meaning of the undefined term, "straight line," was not what we would traditionally call straight, but it did satisfy the first four axioms of Euclidean geometry. However, it did not satisfy Euclid's Fifth Postulate. If there were a contradiction that could be derived by negating Euclid's Fifth Postulate, then Beltrami's model would have to contain the contradiction, and, in turn, the Euclidean space in which his model was embedded would also contain the contradiction. Beltrami's model proves that if non-Euclidean space has a contradiction, so does Euclidean space.

Relative Truths

This startling new result – that non-Euclidean geometry is as valid as Euclidean geometry – forced scholars in all disciplines to change their view of truth. The announcement of this result had a momentous impact on the mindscape of 19th century intellectuals. It made scholars question their concept of truth and their faith in mathematics as the bastion of objective truth. The previously sacrosanct view of mathematical truths as *absolute* would no longer hold water. "Obviously true" sentences, such as the parallel postulate, were false in this new system. Mathematical truths could no longer be considered absolute; they now had to be viewed as true *relative* to a system.

The shock waves spread throughout the intellectual world. If truth cannot be absolute in mathematics, the most objective of all disciplines, how can it be absolute anywhere? Truth is relative, relative to the system that we construct for it. The concept of relativity then invaded the worlds of philosophy, theology, science, and the humanities. It is rather amazing that the seemingly isolated abstract reasoning of mathematicians – in their quest to establish the independence of Euclid's axioms – paved the way for Einstein's Theory of Relativity, new radical theories of literary criticism, and new directions in art, forcing the western dependence on absolute realism to fade into the background. Today, after all the dust has settled, we consider an *axiom* to be nothing more than a sentence that is *assumed* true for a particular system. The same sentence could possibly be false in another system.

As we discussed on page 200, a model of an axiomatic system is an example where each undefined term is assigned a specific meaning so that the corresponding interpretation of each axiom is true in the example. Since axiomatic systems, like Euclidean geometry, were originally constructed to model some aspect of reality, we also call an axiomatic system a mathematical model of its physical counterpart.

The axioms for Euclidean geometry were based on the human visual perception of straight lines, but our visual perception is limited to very small distances. Furthermore, we can only see very narrow bands of visibility in the radiation spectrum. Since light rays are the medium through which we see the physical world, their shape is how we see "straight." Fence posts appear to be lined up straight because of the way the light rays carry their image to our eyes. If light rays were sufficiently curved, we could look straight ahead and see the back of our head. Our intuitive notion of straightness is based completely on light rays.

At the turn of the century, Albert Einstein startled the scientific community with his prediction that a ray of light would be curved over large distances. Several years later his prediction was verified. If we try to explain this phenomenon by saying that the light ray was distorted by a gravitational field, then we have to ask where the gravitational field came from. If we say that it comes from all the mass hanging out in that vicinity, then we have to ask why is it that all the mass is huddled together in that particular location. If we say that it may be that space is curved and the mass is hanging out there because of the curvature of space, then we have stripped away the physical accouterments down to the abstract realm of mathematics. An axiomatic system that models space as curved is different from Euclidean geometry. Even though we have curves in Euclidean geometry, space itself is not curved.

The Euclidean truths that we were taught in high school, such as the sum of the angles in a triangle is 180°, do not necessarily transfer to cosmic space. Euclidean geometry is a good model of physical space when we are only concerned with small distances, such as those on planet Earth, but when cosmic distances are involved, non-Euclidean geometry may provide a better model.

Complete Systems

Given any well-formed sentence that can be built from the terms of an axiomatic system, we would like to be able to either prove it or disprove it. An axiomatic system that has this capability is called *complete*.

A sentence is called *decidable* if it is possible to decide its truth value with our proof procedure. If a sentence can be proved from the axioms, then the sentence is true; if the sentence can be disproved from the axioms, then the sentence is false. If it is not possible to either prove a sentence nor disprove it, then the sentence is *undecidable*. The axioms are decidable since they are true by virtue of being an axiom.

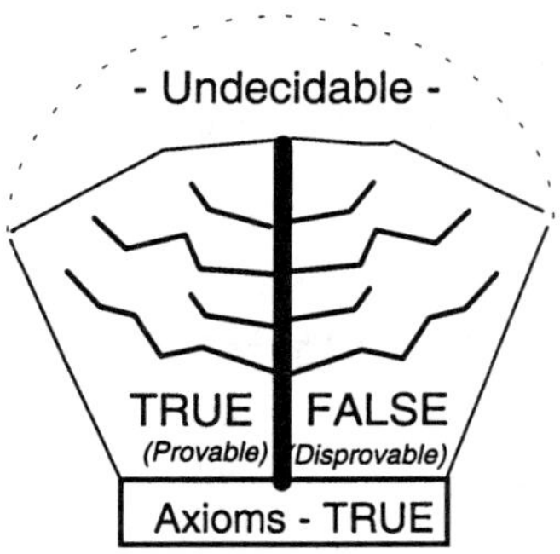

The sentences of an axiomatic system fall into four categories. The first category is the axioms, represented in the adjacent illustration at the foundation of the system, for they are the seeds from which all the other truths emerge. On the left side of our axiomatic Tree of Knowledge, we have the true sentences which can be proved. On the right side we have the false sentences which form a mirror image of the left side: the sentence p is on the TRUE side if and only if $\sim p$ is on the FALSE side. The fourth category, high above the provable and disprovable sentences, way beyond the reach of our axiomatic branches, are the undecidable sentences which can neither be proved nor disproved.

Since the ultimate goal of logical reasoning is to sort out what is true and what is false, we prefer to have a complete system, a system that has no undecidables floating beyond our grasp – a system whose axioms are rich enough to enable us to classify every sentence as either true or false. As an example of system that has an undecidable sentence, consider a system, called *neutral geometry*, whose axioms are Euclid's first four postulates. Now consider the following sentence:

For every triangle, the sum of its angles is 180°.

As we saw in our earlier discussion, this sentence can neither be proved nor disproved in neutral geometry, so it is undecidable. Thus, neutral geometry is not a complete system.

If we take this undecidable statement and add it as an axiom to our system, we produce a new axiomatic system which is

Euclidean geometry. The problem child has now been taken care of for we made it an axiom. Furthermore, by making it an axiom, we pulled a whole cadre of undecidables down into the axiomatic Tree of Knowledge. We can prove all the equivalent formulations of our new axiom, including the parallel postulate, and derive a host of our theorems as well. The question now is – did the addition of this new axiom eliminate all the undecidables, or are there still more floating around beyond the grasp of our tools of deductive reasoning?

Gödel's Theorem

In an axiomatic system that contains an infinite set, there will always exist sentences that cannot be proved or disproved from the axioms of that system. Furthermore, even if that statement is added as an axiom to the system, there will still be other sentences that cannot be proved or disproved.

Until 1931, scholars believed that it should be possible to construct a strong enough set of axioms so that every sentence would be decidable. This belief was shattered, though, when Kurt Gödel (1906–1978), a German mathematician working at Princeton, proved one of the most remarkable theorems of the 20th century, or perhaps of any century. Gödel proved that in an axiomatic system that contains an infinite set, such as the set of natural numbers, there will *always* be some well-formed sentences that can be neither proved nor disproved.

Consider the simple operations of arithmetic on the set of natural numbers. Since there are an infinite number of natural numbers, there will always be some sentences about natural numbers that we will not be able to prove or disprove. Perhaps the following famous conjecture, proposed by Christian Goldbach in the 18th century, is one of those sentences that can neither be proved or disproved:

Goldbach's Conjecture: Every even integer greater than 2 is the sum of two prime numbers.

$$4 = 2+2$$
$$6 = 3+3$$
$$8 = 3+5$$
$$10 = 3+7$$
$$\dots$$

Mathematicians have worked on Goldbach's Conjecture for over 200 years, but have not yet proved or disproved it. A proof may be found some day. After all, it did take over 2000 years to settle the independence question of Euclid's Fifth Postulate. More recently, a 300 year quest was laid to rest in 1995 when Andrew Wiles presented a proof of Fermat's last theorem. Even if a proof is never found for a conjecture like Goldbach's, the continued effort of trying to find a proof often leads to new, fertile areas for mathematical inquiry.

Contradictions

If we tell a lie and get caught, the outraged party may say, "You contradict yourself." We have a similar meaning in mathematics. When we lie, we represent a false sentence as true, so we are making the claim: *p and ~p*. This sentence form is a contradiction in mathematics. A *contradiction* is an abstract compound sentence that is always false. Since a law of logic is always true, its negation will be a contradiction. The contradiction form that we use most often is: *p and ~p*. If we can deductively derive a sentence and also derive its negation, we have a full-blown contradiction, which has a fatal impact on the system that contains it. One little contradiction makes every sentence a contradiction! Below is a simple proof of this amazing statement, which uses nothing more than the definition of *implies*.

Theorem If an axiomatic system has a contradiction, then *every* sentence in the system is a contradiction.

Proof Suppose p is a contradiction. Then p is true and p is false.

Let q be an arbitrary sentence in the system.
By the definition of implies, since p is false, $p \Rightarrow q$ is true.
But p is also true and since we now have that $p \Rightarrow q$ is true, we can deduce that q is true.

Since q represents an arbitrary sentence, *every* sentence is true.
$\sim q$ is a sentence, so $\sim q$ is also true.

Thus, every sentence is both true and false.
Therefore, every sentence is a contradiction.

A contradiction can never be ignored because it turns every sentence in the system into a contradiction! We no longer have to look for theorems, for every sentence is a theorem, and the negation of every sentence is a theorem. The system is completely trivial.

Logically speaking, if we tell one little lie, every statement that we make automatically becomes a lie. Even in personal matters where we are not bound by the strict rules for logical reasoning, we still find it difficult to believe someone who has been caught in a lie. Since the discovery of a contradiction is logically catastrophic, it may surprise you to learn that contradictions have been found in mathematical systems.

Russell's Paradox

In the 19th century, mathematicians thought that sets could be formed from any mathematical property. If one had an open statement $p(x)$, one should be able to form the set of all x such that $p(x)$ is true: $\{ x \mid p(x) \}$. However, this belief was dramatically shattered in 1902 when Bertrand Russell conceived the following paradoxical set:

$$V = \{ x \mid x \text{ is a set and } x \notin x \}$$

Is V an element of V?

Before we attempt to answer the above question, let's try to find some sets that are elements of V. For example, consider the following set:

$$A = \{4,5,7\}$$

A has 3 elements and A is not one of the elements. So, $A \notin A$. Thus, A is an element of V.

Continuing in a similar manner, we can find a lot of sets that are elements of V. Let's try to find an element that is not in V. What if we take the set A and throw A into it?

$$B = \{4, 5, 7, \{4,5,7\}\}$$

B has 4 elements and A is one of the elements, so $A \in B$. However, $B \notin B$. So B is also an element of V.

Let's try an infinite set:

$$\mathbb{N} = \{1, 2, 3, \ldots\}$$

$1 \in \mathbb{N}$, $2 \in \mathbb{N}$, $3 \in \mathbb{N}$, etc. However, $\mathbb{N} \notin \mathbb{N}$. Thus, $\mathbb{N}$ is an element of V.

We have found several elements of V, but we have not yet produced an element that is not in V. We are now ready to contemplate Russell's question: Is $V \in V$?

$V = \{ x \mid x \text{ is a set and } x \notin x \}$.

Either $V \in V$ or $V \notin V$.

Case 1: Suppose that $V \in V$.
Then by the definition of V, $V \notin V$.
Contradiction!

Case 2: Suppose that $V \notin V$.
Then by the definition of V, $V \in V$.
Contradiction!

The set V produces a contradiction. The impact of Russell's Paradox was devastating to Gottlob Frege, the founder of modern mathematical logic. After developing a theory for symbolic logic and quantifiers, Frege spent many years constructing a logical foundation for arithmetic. In 1893 he published Volume 1 of his *Foundations of Arithmetic*. After nine more years of work, he completed the second volume, but while it was being printed, Frege received a letter from Bertrand Russell which described his paradox. This news was, of course, logically catastrophic, but it was also personally catastrophic to Frege. He tried to revise his axioms to eliminate the contradiction, but became very bitter and despondent, making no more significant contributions to mathematics.

Because of Russell's Paradox, a new axiomatic structure for sets had to be constructed. With the new axioms (page 279), the rules for constructing sets were restricted so that Russell's description of V cannot be classified as a set, which eliminates his paradox.

Consistent Systems

A axiomatic system is *consistent* if it has no contradictions. We would like to be able to prove that the axiomatic systems used in mathematics are consistent. However, Kurt Gödel again surprised the mathematical community by proving that it is impossible to prove that a system that contains an infinite set has no contradictions. Since the set $\mathbb{N}$ of natural numbers is infinite, we cannot prove that arithmetic on $\mathbb{N}$ has no contradictions; however, we operate under the belief that it is consistent. If someone does ever derive a contradiction in the system of arithmetic, the axioms will have to be revised in order to eliminate the contradiction, as was done with the axioms of set theory. A contradiction cannot be swept under the rug and ignored.

Exercise Set 2.8

1. What is an axiom? Why are they necessary?
2. Why do we need Axiomatic Systems?
3. Why are undefined terms necessary?
4. If a term is undefined, how can we work with it?
5. What is a proof?

6. What is a theorem?
7. Given a well-formed sentence in a mathematical system, is it always possible to prove that it is true or prove that it is false?
8. What is a contradiction?
9. What happens to a mathematical system if a contradiction is found. Explain why.
10. What is a consistent system?
11. Is the system of arithmetic consistent?
    Is the system of Euclidean geometry consistent?
12. What does it mean to say that a set of axioms are independent?
13. Suppose that you want to build an axiomatic system for the laws of logic. In the first stage of your construction, you must decide which terms to take as undefined. Select two of the five logical operators as undefined terms. Then use the undefined terms to build definitions for the other three operators. In each successive definition, you may use previous words that you have defined.
    Hint: If you use *not* and *or* for the undefined terms, how would you define *p and q*, using only *not* and *or*.
13. Discuss Euclid's Fifth Postulate and the impact that it had on our concept of truth. Is Euclid's Fifth Postulate always true?
14. What is a model of an Axiomatic System?

Review

Proof A linearly ordered structure of interwoven valid arguments where each sentence is one of the following:

- An assumption used in a valid argument
- An axiom, previous theorem, or definition
- A sentence that can be derived from previous sentences by a valid argument

The final stand-alone conclusion is the theorem that has been proved.

Theorem A statement that has been proved.

Conjecture A statement someone thinks is true, but no one has proved it.

Inductive reasoning Type of reasoning used when we discover a general relation from specific examples or experiences.

Deductive reasoning Type of reasoning used when we derive a conclusion through valid arguments from other sentences that we accept as true.

Argument A list of sentences called hypotheses followed by a sentence called the conclusion.

Valid argument An argument in which the conclusion follows from the hypotheses. Let $h_1, h_2, \ldots, h_n$ represent the hypotheses of an argument and c represent the conclusion. The argument is valid if and only if the following implication is a law of logic:

$$(h_1 \text{ and } h_2 \text{ and } h_3 \text{ and } \ldots \text{ and } h_n) \Rightarrow c.$$

Law of detachment A valid argument whose hypothesis has the form, $p \Rightarrow q$ and p, and whose conclusion is q. Also known as modus ponens.

Law of contraposition A valid argument whose hypothesis has the form, $p \Rightarrow q$ and $\sim q$, and whose conclusion is $\sim p$.

Transitive law A valid argument whose hypothesis has the form, $p \Rightarrow q$ and $q \Rightarrow r$, and whose conclusion is $p \Rightarrow r$.

Direct proof Proving an implication by assuming the hypothesis is true and then deriving that the conclusion must be true.

Indirect proof Proving an implication by assuming the conclusion is false and then deriving that the hypothesis must be false. An indirect proof of an implication is a direct proof of its contrapositive.

Proof by cases Subdividing a proof into special cases, one of which must be true. The conclusion in a proof by cases is the disjunction of the subconclusions within each case.

Proof by contradiction A method of proof in which we assume the negation of what we want to derive and then derive a contradiction.

Principle of mathematical induction Let $p(n)$ be an open statement. Let c be a fixed integer. If for every integer $n \geq c$, $p(n) \Rightarrow p(n+1)$, and $p(c)$ is also true, then $p(n)$ is true for all $n \geq c$. *Stronger Version*: If for every positive integer n, $[\, p(1) \wedge p(2) \wedge p(3) \wedge \ldots \wedge p(n)\,] \Rightarrow p(n+1)$, and $p(1)$ is also true, then $p(n)$ is true for all positive integers n.

Disprove a statement Prove its negation.

Axiom A statement that is assumed true in an axiomatic system, requiring no proof.

Undefined terms The basic words from which we construct the vocabulary for an axiomatic system. It is impossible to define every word without being circular. The terms selected to be undefined are chosen to represent the simplest concepts possible, concepts that cannot be explained by simpler concepts.

Axiomatic system A list of undefined terms, a list of axioms, and a proof procedure for deriving theorems in the system. Definitions are built from the undefined terms and previously defined terms. Theorems are derived from the axioms, previous theorems, and definitions using the proof procedure. The axioms, definitions, and theorems must be sentences in accord with the grammar for the system.

Independent axioms A set of axioms in which none of the axioms can be derived from the others.

Contradiction An abstract compound statement that is always false, like *p and ~p*. A negation of a law of logic is a contradiction.

Consistent system An axiomatic system that contains no contradictions.

Decidable A sentence that can be either proved or disproved.

Undecidable A sentence that is not decidable. It is not possible to derive the sentence or its negation from the axioms.

Complete system An axiomatic system in which every well-formed statement can be either proved or disproved. Every sentence is decidable.

Model An example of an undefined term that has the properties specified by the axioms. A model of an axiomatic system contains a model of each undefined term in the system.

Sequence	A function whose domain is the set of natural numbers. The notation s_n indicate $s(n)$, the nth term in the sequence.
Even	a is *even* if and only if $a = 2n$ for some integer n.
Odd	a is *odd* if and only if $a = 2n + 1$ for some integer n.
Prime	a is *prime* if and only if a is an integer greater than 1 whose only positive factors of a are a and 1.
Divides	Let a and b be integers. a divides b if and only if $b = ak$ for some integer k. a divides b if and only if a is a factor of b.
Fundamental theorem of arithmetic	Every natural number, other than 1, can be represented in a unique manner as a product of prime numbers, with smaller factors written to the left of larger factors.

Chapter Review

1. a. What is a mathematical proof? What is a theorem?
   b. What does it mean to say that an argument is valid?
   c. If an argument is valid, does its conclusion have to be true?
   d. What are the 4 stages involved in writing a proof?
   e. How do you disprove a statement?
   f. What is the difference between inductive reasoning and deductive reasoning?
2. Describe the basic structure of the following types of proof.
   a. A direct proof of an implication
   b. An indirect proof of an implication.
   c. A proof of an or-sentence.
   d. A proof of an equivalence.
   e. A proof of an existentially quantified sentence.
   f. A proof of a universally quantified sentence.
   g. A proof by cases.
   h. A proof by contradiction.
   i. A proof by mathematical induction.
3. You should be able to construct proofs involving the structures listed in exercise 2 in either outline form or paragraph form. On the next page are some examples from this chapter.

a. The sum (or difference or product) of two even numbers is an even number.
b. The sum (or difference or product) of two odd numbers is an ___ number.
c. The sum (or difference or product) of two rational numbers is a rational number.
d. For every real number x, if x is rational and y is irrational, then $x + y$ is irrational.
e. It is not true that the sum of every two irrational numbers is irrational.
f. Let x be a real number. $-3 < x < 3$ if and only if $x^2 < 9$.
g. $\sqrt{5}$ is an irrational number.
h. For every positive integer n,
$1 + 2 + 4 + 8 + 16 + . . . + 2^{n-1} = 2^n - 1$
$1 + 3 + 3^2 + 3^3 + . . . + 3^{n-1} = \frac{3^n - 1}{2}$
$n! \geq 2^{n-1}$.
i. For every positive integer n, n is even or n is odd.
j. The sum of the first n even positive integers is $n(n + 1)$.
k. The sum of the first n odd positive integers is n^2.
l. $n^2 \leq 2^n$ for every integer $n \geq 4$.

4. Use cases and the meaning of "and" and "or" to derive the solution of an inequality of the form $ab > 0$ or $ab < 0$. For example, $(x - 2)(x + 27) < 0$.
5. Discuss the following.
   a. Why are axioms necessary?
   b. Why are undefined terms necessary?
   c. If a term is undefined, how can we work with it?
   d. What is a contradiction?
   e. What happens to an axiomatic system if a contradiction is found in it?
   f. What is a consistent system?
   g. How has the concept of truth changed since the time of the ancient Greeks? Why did it change?
   h. What did Kurt Gödel prove and why was it important?
   i. What does it mean to say that a set of axioms is independent?
6. Comment on what is accepted as a proof in another discipline that you have studied. Compare and contrast their notion of a proof with the mathematical notion of a proof.

◈ Chapter 3

Sets – The Building Blocks

Sets play a fundamental role in the development of our reasoning faculties. To learn the meaning of a word such as *triangle*, a young child must learn how to identify those objects to which the word applies. To identify an object as a triangle, the child must be able to classify it as a member of a set of objects that share a certain property. If a child is asked how many triangles are on a tray of assorted objects, the child must mentally sort the objects into two sets: the set of objects that have the property of being a triangle and the set of objects that do not have that property. Psychologists use tests such as these to measure intelligence in very young children, as well as in birds, monkeys, and other animals. The ability to recognize sets lies at the very foundation of what we mean by intelligence. So it is no surprise that the concept of a set lies at the very foundation of mathematics.

All of the basic concepts in mathematics can be phrased in terms of sets. When we count, we are counting the number of elements in a set; when we analyze the form of a figure, we are analyzing a set of points; when we look at a function, we see a relation between two sets. Sets provide the framework for mathematical discourse; they are the building blocks for all quantitative and spatial concepts.

The basic operations for working with sets are so simple that we might think no formal training is needed. After all, a first grader knows how to combine sets, find elements that are common to several sets, and remove elements from a set. However, like the logical operators, the meaning can be easily misconstrued by the untrained mind when more than one set operation is used in the same sentence. It could be that the confusion comes from the logical operators, for they are used to define the set operations. Which came first, though, is a chicken/egg question. Our innate ability to combine, overlap and remove may be where the logical operators originated. At any rate, they are intimately related. Most young children can do the basic operations with sets, but to manipulate them and cross breed them in an abstract manner requires a higher level of intelligence.

As in any living organism, sets need a reproductive system. The four basic operations for making new sets from old sets are called union, intersection, set subtraction, and cross product. The first three operations are simple concepts that occur naturally in everyday life, but the cross product is a complete product of the imagination. Introduced by Renè Descartes in the 17th century as a coordinate plane for plotting ordered pairs, the cross product provides a powerful tool for combining the visual reasoning of geometry with the algebraic reasoning of numbers.

The set concept was used in mathematics in an intuitive manner until the turn of the 20th century when Russell's paradox (page 206) sent the intellectual community into a philosophical tailspin. Russell's paradox was produced from an intuitively obvious assumption on how sets could be formed, but his example involved an infinite set, which is where our intuition falls apart. In our physical experiences with sets, from early childhood on, we never encounter sets of an infinite magnitude.

The set of all natural numbers has the same size as the set of all even natural numbers.

Since ancient Greece, infinity had greatly troubled the deep thinkers, producing such disturbing results as the adjacent deduction about the size of the set of natural numbers. How could the set of all natural numbers have the same size as the set of all even natural numbers? From this paradoxical sounding statement, G. W. Leibniz (1646–1716), the great philosopher/mathematician, deduced that the number of all natural numbers implies a contradiction. What may seem like a contradiction, though, is inherited from a finite perspective of infinite sets. If we take the viewpoint that infinite sets do not behave in the same manner as finite sets, then it may sound

I protest above all against the use of an infinite quantity as a completed one, which in mathematics is never allowed. The Infinite is only a manner of speaking.

Carl Gauss
1777–1855

perfectly natural that the set of all natural numbers could have the same size as the set of all even natural numbers.

One way to deal with the apparent paradoxes of infinity was the stance taken by the great mathematician, Carl Gauss, in the adjacent quote. The natural numbers go on and on and on; we can never get to the end, there's always one more. So, how can we round them all up and neatly encage them in a set? This is precisely what we do when we write the following notation:

$$\mathbb{N} = \{1, 2, 3, \ldots\}$$

For the human mind to encompass infinity in this manner is a very audacious move. Georg Cantor (1845–1918) took on the challenge and developed a logical theory which captured the wild horses of infinity, making it possible to work with infinity in a perfectly logical manner. His courage in tackling such an enormous concept as that of infinity was no doubt supported by his belief that "*In mathematics the art of asking questions is more valuable than solving problems*" (page 19).

Because of Cantor's work, we can logically talk about not only infinite sets, but also infinite numbers. We can also talk about different sizes of infinity in a completely logical voice, and without batting an eyelash, we can logically agree that, yes, the set of all natural numbers does have the same size as the set of all even natural numbers (page 289).

After an axiomatic foundation was constructed for set theory, sets became a major unifying concept in 20th century mathematics. Its importance is eloquently described by the secret society of French mathematicians known as Bourbaki: "Nowadays it is known to be possible, logically speaking, to derive practically the whole of known mathematics from a single source, The Theory of Sets."

Nowadays it is known to be possible, logically speaking, to derive practically the whole of known mathematics from a single source, The Theory of Sets.

Bourbaki

This chapter covers the basic concepts of set theory, takes a brief excursion into the mysterious realm of infinite sets, and provides exercises to help you develop your reasoning ability, your ability to write proofs, and your understanding of the language used with sets.

Activity 3.1

1. Let $A = \{1,2,3\}$, $B = \{3,1,2\}$, and $C = \{1,2,3,5,6\}$.
   a. Is $A = B$? b. Is $A = C$? c. Is $A \subseteq C$?
2. Let A and B be arbitrary sets. Analyze the thought processes you went through to answer the questions in the previous exercise, then make up informal definitions for $A = B$ and $A \subseteq B$. Translate each definition in terms of quantifiers and logical operators.
3. a. Make a wild guess as to the number of different subcommittees that could be formed from a set of 27 students.
   b. Let S be a set with n elements. Make an educated guess as to how many subsets S has. Look at examples for $n = 1$, $n = 2$, $n = 3$, $n = 4$, etc., until you see a pattern.
   c. Make an educated guess as to the number of different subcommittees that could be formed from a set of 27 students.
   d. Make an educated guess as to the number of different subcommittees that could be formed from a set of 270 students. Is the number of subcommittees that can be formed from a group of 270 students greater than the number of atoms in our universe?

≡ 3.1 Sets & Elements ≡

The vocabulary for working with sets is built from one noun and one verb phrase: *set, is an element of.* Intuitively, a set is a collection of objects, and each object in a set is an element of the set. However, since there are no simpler concepts with which to define them, these two terms are undefined in the axiomatic construction of set theory (page 195). In the sentence, "*x exists,*" the verb gives no connection with other objects. In the sentence, "*x exists in A,*" we now have a connection. Connections are essential for the reasoning process.

Set Notation

The $\in$ symbol represents *"is an element of."*

$$x \in S\text{: } x \text{ is an element of } S.$$

A slash through the $\in$ symbol represents its negation.

$$x \notin S\text{: } \sim(x \text{ is an element of } S)$$

Listing Method

We sometimes notate a set by listing the elements and enclosing them in set braces.

$$S = \{2, 4, 6, 8, \ldots, 200\}$$

The ellipsis (. . .) indicates that the listing continues in the given pattern. We must list enough elements so that the reader can see the pattern. A listing of the set E of all even integers requires two ellipses:

$$E = \{ \ldots, -6, -4, -2, 0, 2, 4, 6, \ldots \}$$

Unlike the previous example, each of the above ellipses represents an infinite number of elements.

Property Method

When we use the property method to define a set, we specify a property that determines membership in the set. The property is stated in terms of an open statement $p(x)$ which we enclose in set braces:

$$\{ x \mid p(x) \}$$

For example, let $p(x)$: x is a real number.

$$\mathbb{R} = \{ x \mid x \text{ is a real number} \}$$

The vertical bar in the above notation is read as *"such that"*:

$\mathbb{R}$ is the set of all x such that x is a real number.

When we describe a set with the property method, it must be *well-defined*, which means that the property must clearly distinguish who is in the set and who is not. The set of all even numbers is a well-defined set, whereas the set of all lucky numbers is not well-defined.

$E = \{ x \mid x \text{ is an even number} \}$

$E = \{ y \mid y \text{ is an even number} \}$

$E = \{ x \mid x = 2k \text{ for some integer } k \}$

$E = \{ 2k \mid k \text{ is an integer} \}$

The set E of even numbers can be notated with the property method in different ways, as illustrated on the left. In the second form, we use y instead of x. In the third form, we use the definition of even, prominently displaying the existential quantifier. In the last form, the existential quantifier is hidden but we must use it when we translate $x \in E$:

$x \in E$ if and only if *there exists* an integer k such that $x = 2k$.

Let $S = \{ x \mid p(x) \}$.

$x \in S \Leftrightarrow p(x)$ is true.

$x \notin S \Leftrightarrow p(x)$ is not true.

When we make deductions about sets, we usually reason with the sentence $x \in S$. From a property description of a set, we translate $x \in S$ and $x \notin S$ as illustrated on the left and in the following examples.

⊕ *Example*

Translate $x \in S$ and $x \notin S$ for the given set.

1. Let $S = \{ x \mid x \in A \text{ and } x \in B \}$.
   $x \in S$ if and only if $x \in A$ and $x \in B$.
   $x \notin S$ if and only if $x \notin A$ or $x \notin B$).

2. Let $S = \{5^n \mid n \text{ is an integer}\}$.
   $x \in S$ if and only if there exists an integer n such that $x = 5^n$.
   $x \notin S$ if and only if for every integer n, $x \neq 5^n$.

3. Let $S = \{3, 8, 29\}$.
   $x \in S$ if and only if $x = 3$ or $x = 8$ or $x = 29$.
   $x \notin S$ if and only if $x \neq 3$ and $x \neq 8$ and $x \neq 29$.

Universal Sets

A universal set U is the universe for a particular discussion, setting the boundary for our considerations. In elementary algebra, we often use the set ℝ of real numbers as our universal set, but sometimes we use the set ℂ of complex numbers. In plane Euclidean geometry, the universal set is the set of all points in a plane, whereas in spherical geometry, the universal set is the set of points on a sphere.

If the reader understands what the universal set is, we do not have to mention it each time we describe a set. For example, if the reader knows that the universal set is the set of real numbers, we may define the closed interval [1,3] as follows:

$$[1,3] = \{ x \mid 1 \leq x \leq 3 \}$$

If the context is not clear, though, we should let the reader know that the domain for x is the real numbers.

$$[1,3] = \{ x \mid 1 \leq x \leq 3 \text{ and } x \text{ is a real number} \}$$

$S = \{ x \mid p(x) \}$

$S = \{ x \mid p(x) \text{ and } x \in U \}$

In a set description by the property method, the elements are always limited to the members of a universal set. When we write the property without mentioning the universal set, it is implicitly understood that each x must be in the current universal set. The restriction of sets that we define to a universal set that we already know exists eliminates the contradiction produced by Russell's Paradox (page 206, 279).

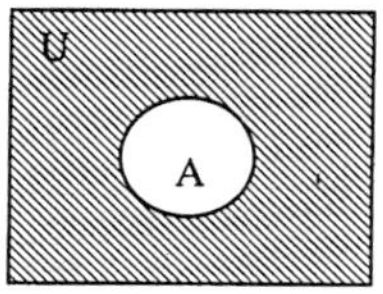

In the following definition of a set, it is implicitly understood that $x \in U$:

$$\text{Let } S = \{x \mid x \notin A\}.$$

In the adjacent illustration, the universal set U is represented as a rectangular region and the set A as a circular region. When we say $x \notin A$, we do not go outside of U.

The Empty Set

At the opposite extreme from the universal set, we have the empty set. The empty set is such a simple concept that we often gloss over its meaning. The *empty set* is a set that has no elements. We do not think of the empty set as nothing. The empty set is a container with nothing in it, similar to an empty box, which still exists even though it has nothing in it.

$\emptyset = \{\ \}$

We use the symbol $\emptyset$ to represent the empty set.

$\emptyset \notin \emptyset$

The empty set has no elements, so it is not an element of itself.

$\emptyset \in \{\emptyset\}$

If we put an empty box inside another box, the outside box is not empty. Likewise, the set $\{\emptyset\}$ is not empty. It has one element, the empty set.

$\emptyset \neq \{\emptyset\}$

The adjacent sets are not equal because the set on the left has no elements, whereas the set on the right has one element.

The empty set can be an element of a set, but it must be included in the list of elements or satisfy the property that determines the set. For example:

$$\text{Let } A = \{1, 5\} \text{ and } B = \{1, 5, \emptyset\}.$$

A has two elements, 1 and 5. The empty set is not one of these two elements, so $\emptyset \notin A$. On the other hand, B has three elements and $\emptyset \in B$.

Shapes as Sets

How do we describe what we mean by a shape? In mathematical language, we call a shape a geometric figure. Since a figure is composed of points, we can use the language of sets to define this visual concept:

A geometric figure.

A geometric figure is a set of points.

A set of points is a geometric figure.

A geometric figure.

With this definition, a figure does not have to be connected; it may be in more than one piece, as illustrated in the adjacent sketch. Note that the above definition uses the undefined concept of a point (page 195). Except for undefined terms, we can define almost all mathematical concepts in terms of sets.

One of the exceptions is when we deal with very large collections, such as the collection of all sets. Forming the set of all sets produces a contradiction (page 301), similar to Russell's Paradox. For this reason, the axioms of set theory do not allow a set to be a member of itself. So, we cannot call the collection of all sets a "set." Instead, we call it a "class," and there is an analogous theory for working with classes.

Equal Sets

Equality is one of the most important relations in mathematics. We use this relation in a stronger sense than is often used in everyday language. In the Declaration of Independence, the founders of the American democracy introduce their axioms with the sentence, "We hold these truths to be self-evident, that all men are created equal . . ." In a mathematical framework, however, two different objects cannot be equal to each other, not even identical twins.

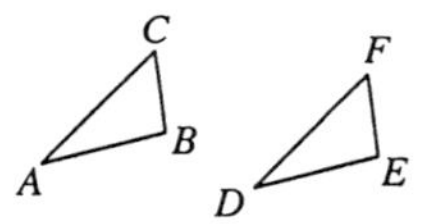

$\Delta ABC \neq \Delta DEF$

For example, in the adjacent illustration, we have triangles that are identical twins; corresponding sides and corresponding angles have the same measurements. These two triangles are congruent, but they are not equal because ΔABC is a different set of points than ΔDEF: $\Delta ABC \neq \Delta DEF$. Congruence is called an equivalence relation (page 326) because it has the same basic properties as equality, but it is important to note the distinction between the two.

When we say two sets are *equal*, we mean that they have identical elements. To say that sets A and B have the same elements can be translated as follows:

Two sets are *equal*
if and only if
they have the same elements.

Every element in A is in B, and every element in B is in A.

For every x, $x \in A \Rightarrow x \in B$, and for every x, $x \in B \Rightarrow x \in A$.

Since we can distribute the universal quantifier across an and-statement, we can translate the above statement as stated in the adjacent box. Notice how we are building the language of sets from the language of logic and the two undefined terms.

Let A and B be sets.
$A = B$
if and only if
for every x,
$(x \in A \Rightarrow x \in B)$ and $(x \in B \Rightarrow x \in A)$.

If $A = B$, then A and B are different names for the same set. We sometimes have different names for the same person, such as Jack and John, or Elizabeth and Beth. Parents do not normally give different children the same name because it destroys the function of a name, which is to identify the person. Similarly, when we name a set, such as A or B, we do not give the same name to another set that is included in the same discussion.

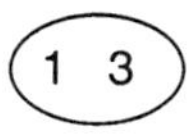

A set is completely determined by its elements; it does not matter how we arrange them. If we shake a box of dominoes, we still have the same set of dominoes.

$\{1,3\} = \{3,1\}$

Is $\{1,3\} = \{3,1\}$?

Is every element in the left set an element of the right set?

Is every element in the right set an element of the left set?

The answer to both questions is yes, so $\{1,3\} = \{3,1\}$.

Duplicate listings do not change a set. If we make a list of people to invite to a party and we list the same person twice, we have not changed the set of people being invited. Similarly, redundant listings in a set do not affect the set.

$\{1,3,3\} = \{1,3\}$

Is $\{1,3,3\} = \{1,3\}$?

Every element in the left set is an element of the right set and vice-versa. So, $\{1,3,3\} = \{1,3\}$.

When we count the number of elements in a set, we do not count duplicates. The set $\{1,3,3\}$ has only two elements.

Subsets

Another fundamental relation between sets is the subset relation. If every element in set A is also in set B, we say that A *is a subset of* B, which is notated as $A \subseteq B$.

Let A and B be sets.

$A \subseteq B$

if and only if

for every x, $x \in A \Rightarrow x \in B$.

The formal definition of subset is given in the adjacent box. Note that it is half of the condition used to define equal sets. We now have another verb phrase – *is a subset of* – that we can use to construct sentences. We are rapidly expanding our vocabulary. You may think that we are going at a rather slow pace; however, we have now covered the three fundamental verbs of set theory: *is an element of*, *is equal to*, and *is a subset of*. If we have a personal understanding of these concepts, all other definitions will be easy to interpret because all other definitions are built from these three verbs and the logical operators.

We can visualize the subset relation as illustrated on the left. This type of diagram, called a *Venn diagram*, was first used by John Venn to illustrate the laws of logic in a paper published in 1876. Note how the adjacent picture of the subset relation gives a picture of the implication operator:

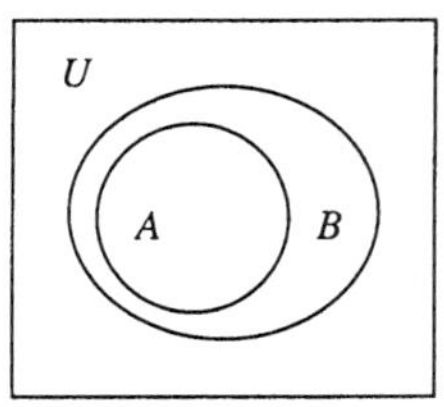

$A \subseteq B$

For every x, $x \in A \Rightarrow x \in B$.

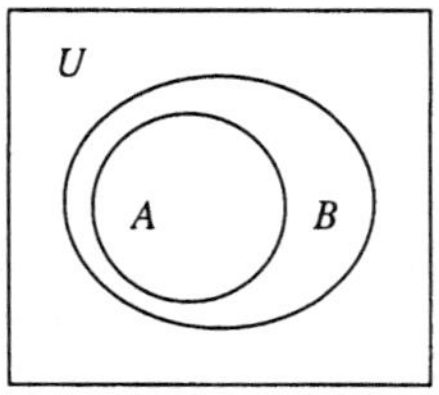

In the adjacent sketch, pick an x that is not in A. The implication, $x \in A \Rightarrow x \in B$, is true because the hypothesis is false. Thus, the implication is true for all x in the universal set. Students sometimes erroneously describe the subset relation in terms of *and*: for every x, $x \in A$ and $x \in B$. Note that this statement is not true in the adjacent picture.

In the hierarchy of number sets, the set $\mathbb{N}$ of all natural numbers is a subset of the set Z of all integers; the set Z of all integers is a subset of the set Q of all rational numbers.

$$\mathbb{N} \subseteq Z \subseteq Q$$

A verbatim reading of the above the symbols is grammatically incorrect. We use this notation as an abbreviation for:

$$\mathbb{N} \subseteq Z \text{ and } Z \subseteq Q$$

In a similar style, we use $3 < x < 5$ as an abbreviation for $3 < x$ *and* $x < 5$. When we negate an expression like $A \subseteq B \subseteq C$, we must be aware of the hidden *and*.

Subset Proofs To prove that $A \subseteq B$, we must prove the following:

For every x, if $x \in A$, then $x \in B$.

Derive: $A \subseteq B$

Direct Proof
Let x be any element.
Assume that $x \in A$.
...
So, $x \in B$.
Therefore, $A \subseteq B$.

We can prove this implication with either a direct or indirect proof. The outside structure for a direct proof is given in the adjacent template. At the end of the template, the implication is not explicitly stated in order to make the proof more concise. The amount of detail that we include in a proof is a delicate balance between putting in enough detail so that the reader can follow our reasoning, but not putting in too much detail and wasting the reader's time.

To prove that $\emptyset \subseteq A$, we must prove the following:

For every x, if $x \in \emptyset$, then $x \in A$.

Theorem For every set A, $\emptyset \subseteq A$.

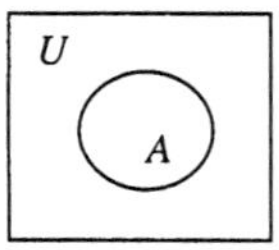

Proof Let A be a set. Let x be an element in the universal set.
Since the empty set has no elements, $x \notin \emptyset$.
Consider the implication: If $x \in \emptyset$, then $x \in A$.
Since the hypothesis is false, the implication is true.
Therefore, $\emptyset \subseteq A$.

Theorem For every set A, $A \subseteq A$.

Proof Let A be a set. Let x be an element in the universal set.

If $x \in A$, then $x \in A$. .... $p \Rightarrow p$

Therefore, $A \subseteq A$.

The above theorem states that any set is a subset of itself. When we count all the subsets of a given set, we include the empty set and the whole set. The set $\{1, 2\}$ has four different subsets: $\emptyset$, $\{1,2\}$, $\{1\}$, $\{2\}$

Transitivity

If we draw sets A and B such that $A \subseteq B$ and then we draw a set C such that $B \subseteq C$, we can clearly see that $A \subseteq C$. As demonstrated in the following proof, this property of sets is inherited from the transitivity property of *implies*.

Theorem For all sets A, B, and C, if $A \subseteq B$ and $B \subseteq C$, then $A \subseteq C$.

Proof Let A, B and C be sets. Assume that $A \subseteq B$ and $B \subseteq C$.
Let x be an element in the universal set.

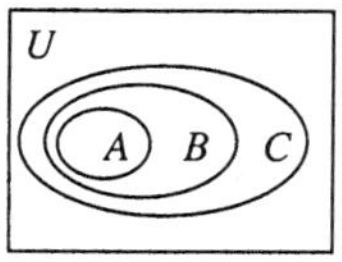

If $x \in A$, then $x \in B$. .... *Definition of* $A \subseteq B$

If $x \in B$, then $x \in C$. .... *Definition of* $B \subseteq C$

So, if $x \in A$, then $x \in C$. .... *Transitivity of implies*

So, $A \subseteq C$. .... *Definition of* $A \subseteq C$

Thus, for all sets A, B, and C, if $A \subseteq B$ and $B \subseteq C$, then $A \subseteq C$.

Set Equality Proofs

The definition of equal sets can be phrased in terms of subsets.

$$A = B$$

if and only if

$$A \subseteq B \text{ and } B \subseteq A.$$

Derive: $A = B$
Part 1. $A \subseteq B$ *Part* 2. $B \subseteq A$ Therefore, $A = B$.

To prove two sets are equal, we usually construct two different subset proofs, as illustrated in the adjacent template. We demonstrate that the left side is a subset of the right side and we demonstrate that the right side is a subset of the left side. From this, we conclude that the two sets are equal. We will have several examples of this type of proof in the next section.

Equivalence

The definition of equal sets can also be phrased in terms of an equivalence:

$$A = B$$

if and only if

for every x, $x \in A \Leftrightarrow x \in B$.

The equals relation between two sets is the analogue of the equivalence relation between two sentences. However, we do not use them interchangeably. We do not say that two sentences are equal; we say that they are equivalent. We only use the equals relation between sets and numbers.

Properties of the Equals Relation

The equals relation has three fundamental properties. Each of these properties is inherited from the corresponding property of equivalence.

Theorem

For all sets A, B and C, the following are true:

Reflexive Property: $A = A$

Transitive Property: If $A = B$ and $B = C$, then $A = C$.

Symmetric Property: If $A = B$, then $B = A$.

The reflexive property states that each set is equal to itself.

The transitive property gives us a way to deduce set equality when we have a middle set as a stepping stone. If the first set is equal to the second set and the second set is equal to the third set, we can then deduce that the first set is equal to the third set.

The symmetric property allows us to reflect sets about the equals sign; it doesn't matter which one we write first.

Properties of the Subset Relation

The subset relation also has the reflexive and transitive properties. However, it has the opposite extreme of the symmetric property, which is called *antisymmetric*. The only time that we can switch two sets around a subset sign is when the two sets are equal:

If $A \subseteq B$ and $B \subseteq A$, then $A = B$.

The subset relation is antisymmetric – the order makes a difference. We will demonstrate later (page 381) that the subset relation provides a model for any order relation.

Theorem For all sets A, B and C, the following are true:

Reflexive Property: $A \subseteq A$.

Transitive Property: If $A \subseteq B$ and $B \subseteq C$, then $A \subseteq C$.

Antisymmetric Property: If $A \subseteq B$ and $B \subseteq A$, then $A = B$.

Proper Subsets

Every set is a subset of itself. We use the term *proper subset* to distinguish those subsets that are not the whole set. "A is a proper subset of B" is notated as $A \subset B$.

> Let A and B be sets.
> $A \subset B$
> *if and only if*
> $A \subseteq B$ and $A \neq B$.

$$A \subset B \text{ if and only if } A \subseteq B \text{ and } A \neq B.$$

$$\{1,2\} \subset \{1,2,3\} \text{ since } \{1,2\} \subseteq \{1,2,3\} \text{ and } \{1,2\} \neq \{1,2,3\}$$

We have the same relation between $\subset$ and $\subseteq$ as between the symbols $<$ and $\leq$.

$$\{1,2,3\} \subseteq \{1,2,3\} \text{ but } \{1,2,3\} \not\subset \{1,2,3\}.$$

$A \supseteq B$

The reversal of the subset sign has the same meaning as reversing an inequality: $A \supseteq B$ if and only if $B \subseteq A$
$A \supseteq B$ is read as "A contains B." Mirror images like $\subseteq$ and $\supseteq$ can play tricks on the eye, so it may be less confusing to the reader to use the $\subseteq$ symbol whenever possible.

$A \not\subseteq B$

To translate the meaning of "is not a subset of," we write the negation as a prefix and then substitute the definition of subset:

> Let A and B be sets.
> $A \not\subseteq B$
> if and only if
> there exists an x such that
> $x \in A$ and $x \notin B$.

$$\sim (A \subseteq B)$$
$$\sim (\text{For every } x,\ x \in A \Rightarrow x \in B.)$$
$$\text{There exists an } x \text{ such that } x \in A \text{ and } x \notin B.$$

To prove that $A \not\subseteq B$, we must prove the above statement. Since this statement starts with an existential quantifier, we need only find one counterexample.

Theorem The set $\mathbb{R}$ of real numbers is not a subset of the set Q of rational numbers.

Proof $\sqrt{2}$ represents a length, so $\sqrt{2}$ is a real number.
However, $\sqrt{2}$ is not rational (page 170). So, $\mathbb{R} \not\subseteq Q$.

Variables and Constants

When more than one variable is used in the property description of a set, we must carefully observe which variables are changing and which are fixed. In the following definition of the closed interval $[a, b]$, both a and b are fixed.

$$[a,b] = \{ x \mid a \le x \le b \}$$

When a set is defined in the above form, $\{ x \mid p(x) \}$, we can see the changing variable x in the first field. However, if more than one variable is in the first field, we have to look in the second field to see what is changing.

⊕ *Example*

List the elements in the given set. Then translate $x \in S$.

1. Let $S = \{ y + n \mid n \text{ is a natural number} \}$.

   S is the set of elements of the form $y+n$ where n is a natural number. The variable n changes, but y does not.

   $$S = \{ y+1, y+2, y+3, y+4, \ldots, y+n, \ldots \}$$

   $x \in S$ if and only if
   there exists a natural number n such that $x = y+n$.

2. Let $S_n = \{ \frac{m}{n} \mid m \text{ is a natural number} \}$.

   As indicated in the second field, m changes but n is fixed.

   $$S_n = \{ \tfrac{1}{n}, \tfrac{2}{n}, \tfrac{3}{n}, \tfrac{4}{n}, \ldots, \tfrac{m}{n}, \ldots \}$$

   To list the elements in S_3, we substitute 3 for n:

   $$S_3 = \{ \tfrac{1}{3}, \tfrac{2}{3}, \tfrac{3}{3}, \tfrac{4}{3}, \ldots, \tfrac{m}{3}, \ldots \}$$

   $x \in S_3$ if and only if
   there exists a natural number m such that $x = \frac{m}{3}$.

   $x \in S_n$ if and only if
   there exists a natural number m such that $x = \frac{m}{n}$.

$\{a\} \neq a$

A set is a container for mathematical objects in the same way that a box is a container for physical objects. A box containing a butterfly is not the same entity as a butterfly that is not in a box. Similarly, set braces around an element change the meaning: $\{a\} \neq a$. The set whose only element is 1 is not equal to 1: $\{1\} \neq 1$.

Sets as Elements

We sometimes use sets as elements in another set. To work with these types of sets, we must read the notation carefully to identify the elements of the big set. For example, the following set S has a set as an element. At first glance we might think that $3 \in S$, but it is not:

Let $S = \{1, \{1,3\}\}$.
S has 2 elements: 1 and $\{1,3\}$.
$1 \in S$ and $\{1,3\} \in S$. However, $3 \notin S$.

We normally use uppercase letters for sets and lowercase letters for elements. However, when the elements of a set are also sets, we often use uppercase letters for the elements.

Let $S = \{ X \mid X \subseteq [0,1] \}$.
Let $A = [0, \frac{1}{2}]$. Then $A \in S$.

We sometimes use index notation to define elements of a set.

Let $A_i = [i, i+1]$.
Let $S = \{ A_i \mid i \text{ is an integer} \}$.
$[5,6] \in S$, but $[5,7] \notin S$.

The $\subseteq$ concept and the $\in$ concept are intimately related, but they have different meanings. When a set has sets as elements, we must carefully check to make sure that we are not confusing elements with subsets.

⊕ *Example*

1. Let $A = \{1,2,3\}$.

   $1 \in A$, but $1 \not\subseteq A$. On the other hand, $\{1\} \subseteq A$, but $\{1\} \notin A$.

2. Let $C = \{ \{1\}, \{2\}, \{5,9\} \}$.

   $\{1\} \in C$, but $1 \notin C$. Since $1 \notin C$, $\{1\} \not\subseteq C$.

3. Let $A = \{1,2\}$ and $S = \{ \{1,2\}, \{2,3,4\} \}$.

   $A \in S$. Since the elements in A are not elements of S, $A \not\subseteq S$.

4. Let $\mathbb{N}$ be the set of natural numbers and $\mathbb{R}$ the set of real numbers. Let $S = \{ \mathbb{N}, \mathbb{R} \}$.

   $\mathbb{N} \in S$ and $\mathbb{R} \in S$. However, $\mathbb{N} \not\subseteq S$ and $\mathbb{R} \not\subseteq S$.
   $\mathbb{N}$ and $\mathbb{R}$ each have an infinite number of elements,
   but S has only two elements.

Power Sets

The set of all subsets of a set S is called the *power set* of S, which is notated as $P(S)$. To be an element of $P(S)$, X must be a subset of S.

$$X \in P(S) \text{ if and only if } X \subseteq S.$$

Let S be a set.

$P(S) = \{ X \mid X \subseteq S \}$

Let $S = \{1, 2\}$. S has four subsets, so $P(S)$ has four elements:

$$P(S) = \{\emptyset, \{1\}, \{2\}, \{1,2\}\}$$

In the above example, $\{1\} \in P(S)$, but $1 \notin P(S)$. When working with a power set, we must carefully consider whether to use $\in$ or $\subseteq$ as the verb phrase. Let S be a set:

$$S \subseteq S, \text{ so } S \in P(S).$$
$$\emptyset \subseteq S, \text{ so } \emptyset \in P(S).$$

Number of Elements in P(S)

If S has n elements, how many elements does $P(S)$ have? To answer this question, let's start at the bottom with $n = 0$. The empty set has only 1 subset: $\emptyset \subseteq \emptyset$.

$$P(\emptyset) = \{\emptyset\}$$

If S has 1 element, say $S = \{a_1\}$, then S has two subsets:

$$P(S) = \{\emptyset, \{a_1\}\}$$

Suppose that S has 2 elements, say $S = \{a_1, a_2\}$.

$$P(S) = \{\emptyset, \{a_1\}, \{a_2\}, \{a_1, a_2\}\}$$

Note that the number of subsets doubled from the previous case when S had only 1 element. Let's see why it doubles at the next stage. As S gets larger, it becomes more difficult to find all the subsets. To make the task simpler, we will inductively build each successive stage on the previous work.

Suppose that $S = \{a_1, a_2, a_3\}$. If we remove a_3, we are back to the previous case. Let's divide the subsets of S into two categories: those that do not contain a_3 and those that do. First, we list all subsets of S that do not contain a_3, which we computed in the last stage:

$$\emptyset \quad \{a_1\} \quad \{a_2\} \quad \{a_1, a_2\}$$

To get the remaining subsets, we simply insert a_3 into each of these subsets.

$$\{a_3\} \quad \{a_1, a_3\} \quad \{a_2, a_3\} \quad \{a_1, a_2, a_3\}$$

Now we can see why the number of subsets doubled:

If S has 2 elements, $P(S)$ has 4 elements.
If S has 3 elements, $P(S)$ has 8 elements.

Using the same technique, we can show that a set with 4 elements has twice as many subsets as a set with 3 elements.

If S has 4 element, $P(S)$ has 16 elements.

Using the same technique in a more general setting, we can prove by mathematical induction that this pattern always holds.

Theorem Let S be a set. If S has n elements, then S has 2^n subsets.

Induction Proof Let $p(n)$: For all sets S, if S has n elements, S has 2^n subsets.

Part 1 Let n be a natural number. Assume that $p(n)$ is true.
Let S be an arbitrary set that has $n + 1$ elements:

$S = \{ a_1, a_2, a_3, \ldots, a_n, a_{n+1} \}$.

Divide the subsets of S into the following two sets:

Let $V = \{ X \mid X \subseteq S \text{ and } a_{n+1} \notin S \}$.
Let $W = \{ X \mid X \subseteq S \text{ and } a_{n+1} \in S \}$.

Let $S_n = \{a_1, a_2, a_3, \ldots, a_n\}$. Since S_n has n elements, by the induction hypothesis, S_n has 2^n subsets.
V is the set of all subsets of S_n. So, V has 2^n elements.

For each subset A of S_n, consider the following mapping:

$$A \xrightarrow{f} A \cup \{a_{n+1}\} \qquad f(A) = A \cup \{a_{n+1}\}$$

This mapping is a one-to-one function from V onto W.
So W has the same number of elements as V.
Thus, the total number of subsets of S is $2^n + 2^n$.
But $2^n + 2^n = 2\,(2^n) = 2^{n+1}$.

So S has 2^{n+1} subsets. Thus $p(n) \Rightarrow p(n+1)$.

Part 2 If S has 0 elements, $S = \emptyset$. The empty set has only 1 subset.
So, if S has 0 elements, S has 2^0 subsets. Thus, $p(0)$ is true.

Conclusion Thus, by mathematical induction, $p(n)$ is true for all nonnegative integers n.

As n gets larger, 2^n grows at an enormous rate. A set with 5 elements has 32 subsets; however, a set with 30 elements has 2^{30} subsets, which is 1,073,741,824. In a set S that contains 270 students, the number of different committees that we could form from these 270 students is the same as the number of subsets of S, which is 2^{270}. This number may not seem very large, but according to standard astronomy texts, it is more than 18 times larger than the number of atoms in our entire universe – which is less than 10^{80}. If S has only 270 elements, its power set $P(S)$ has more elements than our universe has atoms! In terms of quantity, the power set is indeed an extremely powerful set.

Partitions

A partition is a subdivision of a set into nonoverlapping subsets, similar to the way that we might partition a big room into smaller rooms. A *partition* P of a set S is a collection of nonempty subsets of S where each element in S is in one and only one of the subsets.

A set can be partitioned in many different ways. For example, consider the set, $S = \{1, 2, 3, 4, 5, 6\}$. We can group 1 and 2 together in a subset, group 3 and 4 together, and leave 5 and 6 in their own private sets, as illustrated in the adjacent sketch. This subdivision determines 4 sets which form a partition of S:

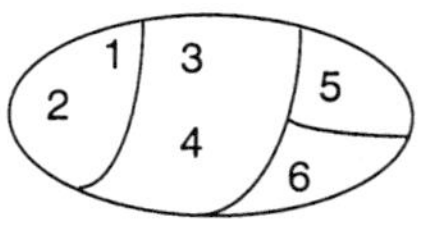

$$P = \{ \{1, 2\}, \{3, 4\}, \{5\}, \{6\} \}$$

We could also partition S into 2 nonoverlapping subsets by segregating 1 and 3, with the other elements in a separate set, as illustrated in the adjacent sketch. This subdivision determines 2 sets which form a partition of S:

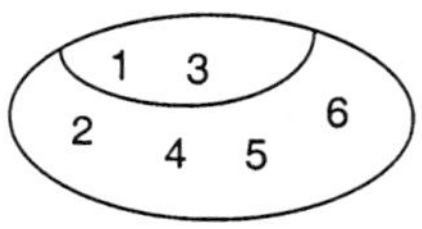

$$Q = \{ \{1, 3\}, \{2, 4, 5, 6\} \}$$

Let P be a collection of nonempty subsets of S. P is a *partition* of S if and only if each element in S is in one and only one element in P.

Now consider the following collection of subsets of S:

$$R = \{ \{1, 3\}, \{2, 3, 4\}, \{5, 6\} \}$$

Since two of the elements overlap, R is not a partition of S. Each element in S can be in only one of the elements in a partition. In the following collection of subsets of S, there is no overlap. However, 6 is not in any of the elements in T, so T is not a partition of S:

$$T = \{ \{1, 3\}, \{2, 4\}, \{5\} \}$$

Exercise Set 3.1

1. List the elements in each set. Translate $x \in S$.
   Also, translate $x \notin S$.
   a. $S = \{ x \mid x = 2n+1$ for some natural number $n \}$
   b. $S = \{ x \mid x = 2^n$ for some natural number $n \}$
   c. $S = \{ 3n \mid n$ is a natural number $\}$
   d. $S = \{ na \mid n$ is a natural number $\}$
   e. $S = \{ y^n \mid y$ is a natural number $\}$

2. Write the following sets in the form $\{x \mid p(x)\}$.
   a. $\{ 5, 7, 9, 11, 13, \ldots \}$
   b. $\{ 20, 25, 30, 35, \ldots, 100 \}$
   c. $\{ 2, 4, 8, 16, \ldots \}$
   d. $\{ 3, 5, 9, 17, \ldots \}$

3. Let $S = \{1,3\}$ and $T = \{\{1\},\{3\}\}$.
   a. Is $1 \in S$?
   b. Is $1 \in T$?
   c. Is $\{1\} \in S$?
   d. Is $\{1\} \in T$?
   e. Is $S \subseteq T$?
   f. Is $S = T$?

4. Let $S = \{5,1,3\}$ and $A = \{1,3,1,5,3\}$.
   a. How many elements does A have? b. Is $S = A$?

5. Let $A = \{ 3x+5y \mid x$ and y are integers $\}$.
   a. List 3 different elements in A. b. Is $1 \in A$?

6. Let $E = \{ \ldots, -6, -4, -2, 0, 2, 4, 6, \ldots \}$ and $E_n = \{ n+x \mid x \in E \}$.
   a. Write the definition of E_1 and E_2 by substituting for n.
   b. List the elements in E_1, E_2, E_3, and E_4.
   c. Let $P = \{ E_n \mid n$ is a natural number $\}$. List the elements in P. How many elements does P have?

7. Use definitions and the negation rules to translate the following.
   a. $C \neq B$ b. $D \not\subseteq C$

8. Let A, B, and C be arbitrary sets. Try to draw a Venn diagram where the hypothesis is true and the conclusion is false. If you cannot do it, try to prove the statement.
   a. If $A \not\subseteq B$ and $B \not\subseteq C$, then $A \not\subseteq C$.
   b. If $A \subseteq B$ and $B \not\subseteq C$, then $A \not\subseteq C$.
   c. If $A \subseteq B$ and $A \not\subseteq C$, then $B \not\subseteq C$.

9. Let A, B, and C be arbitrary sets. Prove or disprove each statement.
   Hint: Look at examples where you list the elements.
   a. $A \in A$.
   b. $A \subseteq A$.
   c. If $A \subseteq B$ and $x \notin B$, then $x \notin A$.

d. $\emptyset \in A$.

e. If $A \subseteq B$, then $A \in B$.

f. If $A \notin B$, then $A \subseteq B$. *Hint*: Construct A and B so that $A \in B$.

g. If $A \in B$ and $B \in C$, then $A \in C$.

10. Let S = $\{s_1, s_2, s_3, s_4, s_5, \ldots\}$, where s_n is the remainder of n divided by 3. List the elements in S. How many elements does S have?

11. Let $A = \{1,2,3\}$. Is $A \subseteq P(A)$? Is $A \in P(A)$?

12. How many different subcommittees can be formed from a class of 30 students? If you had a penny for each of these subcommittees, how much money would you have?

13. Let $S = \{1,2,3,4,5\}$. Is P a partition of S? If not, why not?
    a. $P = \{\{1,2\},\{3,4\},\{2,5\}\}$ b. $P = \{\{1,3\},\{2,4\}\}$

14. Let $S = \{1,2,3,4,5,6,7,8\}$. Form a partition of S.

15. A bit string is a finite sequence of 0's and 1's. 101 is a bit string of length 3. 11010 is a bit string of length 5. Let S_n denote the set of all bit strings of length n.
    a. How many elements are in S_2? S_3? S_4? How many elements are in S_n? Prove your answer using mathematical induction.
    b. Let $A = \{a, b, c\}$. Illustrate a one-to-one mapping from $P(A)$ onto S_3 that can be generalized for n elements. *Hint*: Let $f(\{a\}) = 100$, $f(\{a,c\}) = 101$, etc. How do you define it?
    c. Use part (a) to prove that if A has n elements, then $P(A)$ has 2^n elements. *Hint*: Generalize part (b).

Activity 3.2

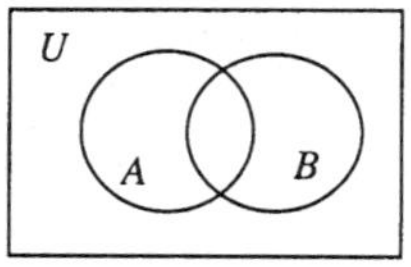

1. Shade each set on a Venn diagram.
    a. $\{x \mid x \in A \text{ or } x \in B\}$
    b. $\{x \mid x \in A \text{ and } x \in B\}$
    c. $\{x \mid x \notin A \text{ and } x \in B\}$
    d. $\{x \mid x \notin A \text{ or } x \in B\}$

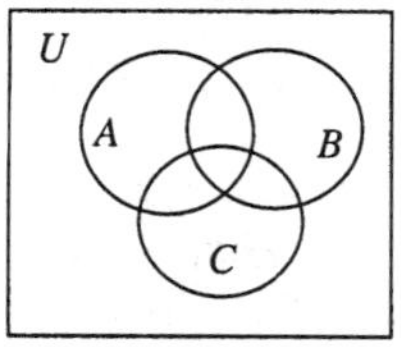

2. Shade each set on a Venn diagram. Are any of the 3 sets equal?
    a. $\{x \mid (x \in A \text{ or } x \in B) \text{ and } x \in C\}$
    b. $\{x \mid x \in A \text{ or } (x \in B \text{ and } x \in C)\}$
    c. $\{x \mid (x \in A \text{ and } x \in C) \text{ or } (x \in B \text{ and } x \in C)\}$

≡ 3.2 Operations on Sets ≡

One of the milestones in a child's education is mastering the operations of union, intersection, and set subtraction. A child must understand these simple operations before he or she can learn how to add and subtract. When we add, we form the union of two sets that do not intersect and then count the number of elements in the new set. When we subtract, we remove elements from a set and then count the number of elements left. Each set operation produces a new set from sets that we already have. Their definitions are based on the meaning of the logical operators. Consequently, they provide a way to visualize *or*, *and*, and *not*, as illustrated in the following sketches.

Union

To form the *union* of two sets, we combine their elements. The union of A and B, notated as $A \cup B$, is the set of elements that are in A or in B or in both sets.

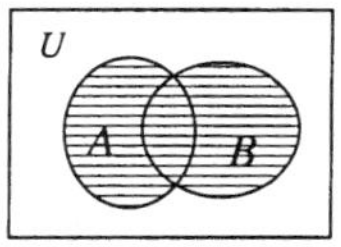

$A \cup B$

$$A \cup B = \{\, x \mid x \in A \ \text{ or } \ x \in B \,\}$$
$$x \in A \cup B \iff x \in A \ \text{ or } \ x \in B$$

Let $A = \{1, 3\}$ and $B = \{2, 3, 7\}$. Then $A \cup B = \{1, 2, 3, 7\}$.

Intersection

To form the *intersection* of two sets, we select those elements that are in both sets. The intersection of sets A and B, notated as $A \cap B$, is the set of elements that are in both A and B.

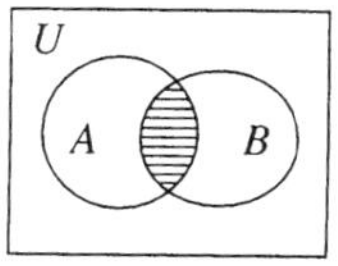

$A \cap B$

$$A \cap B = \{\, x \mid x \in A \ \text{ and } \ x \in B \,\}$$
$$x \in A \cap B \iff x \in A \ \text{ and } \ x \in B$$

Let $A = \{1, 3\}$ and $B = \{2, 3, 7\}$. Then $A \cap B = \{3\}$.

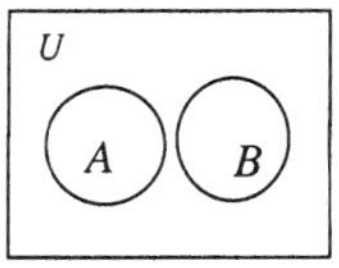

$A \cap B = \emptyset$

Visually, the intersection of two sets is where they overlap. If they do not overlap, we say that the sets are *disjoint*.

A and B are disjoint if and only if $A \cap B = \emptyset$.

Set Subtraction

With set subtraction, we remove elements in one set from another set. A *minus* B, denoted as $A - B$, is the set of all elements that are in A but not in B.

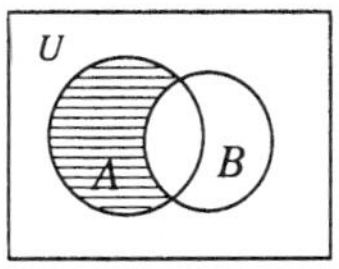

$A - B$

$$A - B = \{\, x \mid x \in A \ \text{ and } \ x \notin B \,\}$$
$$x \in A - B \iff x \in A \ \text{ and } \ x \notin B$$

Let $A = \{1, 3\}$ and $B = \{2, 3, 7\}$. $A - B = \{1\}$ and $B - A = \{2, 7\}$.

Complement

The *complement of A,* denoted as A', is the set of elements that are not in A.

$$A' = \{ x \mid x \notin A \}$$
$$x \in A' \iff x \notin A$$

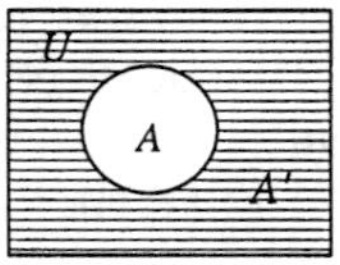

In the above definition, it is implicitly understood that x is in the universal set U. We can also express the complement in terms of set subtraction:

$$A' = U - A$$

Even though the set operations are simple concepts, mistakes are often made when more than one operation is used, which is not surprising, for these operations are manifestations in set form of *or*, *and*, and *not*. If we use the logical operators correctly, we will use the set operations correctly.

Size of a Set

The counting numbers were created to measure the different sizes of finite sets. We use the following notation for the size of a set:

$|A|$ represents the number of elements in A.

Let A and B be finite sets.
$|A \cup B| = |A| + |B| - |A \cap B|$

Let A and B be finite sets. If A and B have elements in common, those elements are counted twice in the sum $|A| + |B|$. So, the general formula for the number of elements in $A \cup B$ is:

$$|A \cup B| = |A| + |B| - |A \cap B|$$

Let A and B be finite sets with $B \subseteq A$.
$|A - B| = |A| - |B|$

For example, if A has 10 elements, B has 8 elements, and $A \cap B$ has 3 elements, then $A \cup B$ has 15 elements. If A and B have no elements in common, the number of elements in $A \cup B$ is the sum of the elements in the individual sets:

$$\text{If } A \cap B = \emptyset, \text{ then } |A \cup B| = |A| + |B|.$$

The number of elements in $A - B$ is determined by the number of elements in their intersection: $|A - B| = |A| - |A \cap B|$
If $B \subseteq A$, then $B = A \cap B$, which gives the adjacent formula.

Multiple Operations

To illustrate multiple set operations on a Venn diagram, shading with horizontal and vertical lines helps us locate the final set. For example, in the adjacent diagram, A' is marked with horizontal lines and B' with vertical lines.

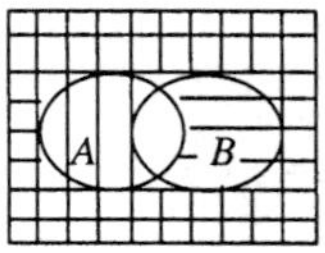

A': $=$
B': $||$

$\#$ $A' \cap B'$ is the grid region where both horizontal and vertical lines are present.

$= \ || \ \#$ $A' \cup B'$ is the region that has either a horizontal or a vertical line or both.

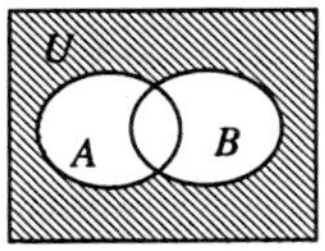

$(A \cup B)'$

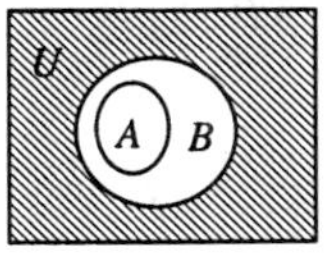

$(A \cup B)'$

In the previous diagram, $A' \cap B'$, the grid region, is all elements not in $A \cup B$, which is the same region as $(A \cup B)'$. This sketch suggests that $(A \cup B)' = A' \cap B'$ for all sets A and B. However, the way that A and B intersect in this example is only one of several possible cases. A could be a subset of B, or B could be a subset of A, or perhaps the sets do not intersect, or, at the other extreme, perhaps they are equal. However, if a statement about A and B is not always true, we will probably find a counterexample with the top sketch.

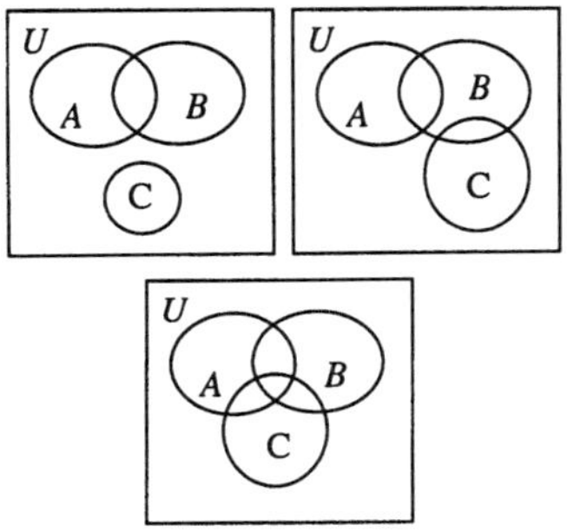

When checking examples of set operations that involve three sets, the number of cases increases dramatically. With each of the above cases for A and B, C could not intersect either A or B, C could intersect B and not intersect A or vice versa, or C could intersect both A and B. It is usually more fun to reason with pictures, but when the picture cases pile up, reasoning with words has a distinct advantage. A proof with words is simpler than checking all the picture cases.

When testing possible relations between operations on three sets, the best case to check is the adjacent case which has a region for each of the possible intersections. If a statement about operations on sets A, B and C is not always true, we will probably find a counterexample with this case, as illustrated in the next example.

⊕ *Example*

Is $A \cap (B \cup C) = (A \cap B) \cup C$ for all sets A, B, and C?

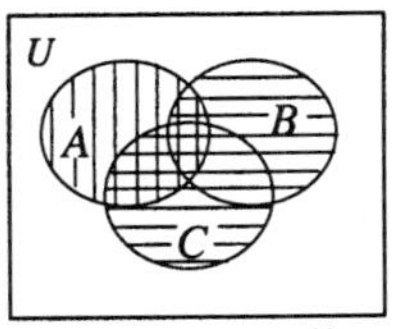

$A \cap (B \cup C)$: #

Let's compare these sets with Venn diagrams. In the first diagram, $B \cup C$ is shaded with horizontal lines and A is shaded with vertical lines. $A \cap (B \cup C)$ is the region shaded with both horizontal *and* vertical lines, the region with the grid.

In the second diagram, $A \cap B$ is shaded with horizontal lines and C is shaded with vertical lines. The union of these two sets, $(A \cap B) \cup C$, is the region that has horizontal lines or vertical lines or both horizontal and vertical lines.

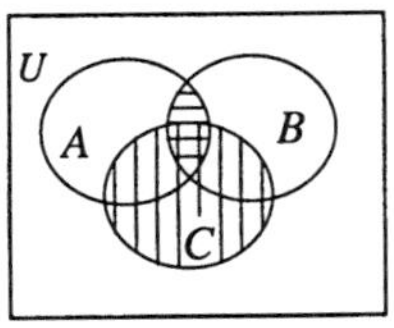

$(A \cap B) \cup C$: = || #

Note that the region shaded for $(A \cap B) \cup C$ is not the same as the region shaded for $A \cap (B \cup C)$. In this example:

$$A \cap (B \cup C) \neq (A \cap B) \cup C$$

Thus, the answer to the above question is no. We have a counterexample using the sets of points in the adjacent sketch. This example shows that the parentheses are essential in the expression $A \cap (B \cup C)$.

Distributive Property

The distributive property gives a relation between two operations. Multiplication distributes over addition, *and* distributes over *or,* and *or* distributes over *and* (page 57). Since intersection and union are defined in terms of these logical operators, we would suspect that intersection distributes over union and vice-versa.

$$p \wedge (q \vee r) \Leftrightarrow (p \wedge q) \vee (p \wedge r)$$

$$\text{Is } A \cap (B \cup C) = (A \cap B) \cup (A \cap C)?$$

Before we attempt a proof, let's compare these two sets with Venn diagrams.

Example

Is $A \cap (B \cup C) = (A \cap B) \cup (A \cap C)$ for all sets A, B, and C?

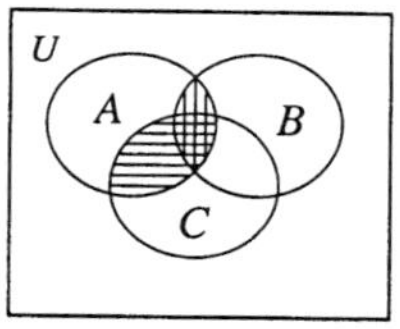

The set $(A \cap B) \cup (A \cap C)$ is illustrated in the adjacent diagram. $A \cap B$ is shaded with vertical lines and $A \cap C$ with horizontal lines. The union of these two sets is the region where we have vertical lines or horizontal lines or both:

$= \;||\; \#$ $\quad (A \cap B) \cup (A \cap C)$

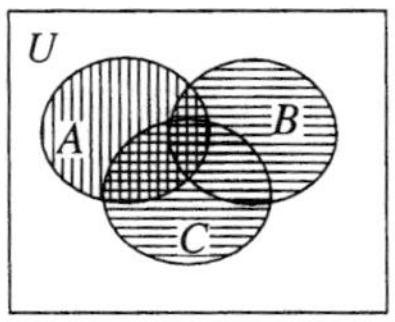

In the second diagram, $A \cap (B \cup C)$ is the grid region, which was explained in the previous example:

$\#$ $\quad A \cap (B \cup C)$

The grid in the second diagram is the same as the region shaded for $(A \cap B) \cup (A \cap C)$ in the first diagram. Hence, for the sets illustrated in the Venn diagram:

$$A \cap (B \cup C) = (A \cap B) \cup (A \cap C)$$

However, this one example does not prove that the above equation is true for all sets. A generalization that includes all possible sets requires a well-reasoned argument.

To prove the above statement for all sets A, B and C, we carefully substitute in the various definitions in the correct order. To translate the sentence, $x \in A \cap (B \cup C)$, we first view it as $A \cap Z$ and apply the definition of intersection. Whenever we deconstruct a set, we work from the outside to the inside, one step at a time.

The following outline proof is composed of two subproofs, so we have inserted claim statements to help the reader (and writer) keep focused on the immediate task at hand.

LEARNING NEVER ENDS

lib.nova.edu/school

lib.nova.edu/school

Explore the library's resources and services for grades K-5!

Book Lists

Databases

Events

NSU Florida

BROWARD COUNTY FLORIDA

ALVIN SHERMAN LIBRARY

Theorem For all sets A, B, and C, $A \cap (B \cup C) = (A \cap B) \cup (A \cap C)$.

Proof Let A, B, and C be arbitrary sets.

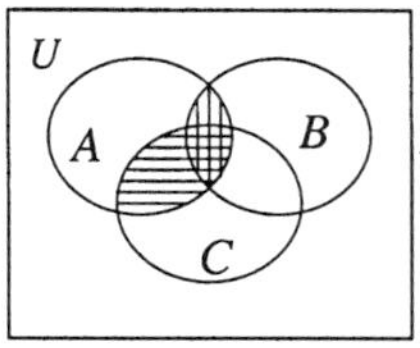

$(A \cap B) \cup (A \cap C)$

Claim: $A \cap (B \cup C) \subseteq (A \cap B) \cup (A \cap C)$

Assume that $x \in A \cap (B \cup C)$.

$x \in A$ and $x \in B \cup C$ . . . . *Definition of intersection*

$x \in A$ and ($x \in B$ or $x \in C$) . . . . *Definition of union*

($x \in A$ and $x \in B$) or ($x \in A$ and $x \in C$)

. . . . *Distributive property of "and" over "or"*

. . . . $p \wedge (q \vee r) \Leftrightarrow (p \wedge q) \vee (p \wedge r)$

$x \in A \cap B$ or $x \in A \cap C$ . . . . *Definition of intersection*

So, $x \in (A \cap B) \cup (A \cap C)$ . . . . *Definition of union*

Thus, $A \cap (B \cup C) \subseteq (A \cap B) \cup (A \cap C)$. . . . . . . *Def. of subset*

Claim: $(A \cap B) \cup (A \cap C) \subseteq A \cap (B \cup C)$

Assume that $x \in (A \cap B) \cup (A \cap C)$.

$x \in A \cap B$ or $x \in A \cap C$ . . . . *Definition of union*

($x \in A$ and $x \in B$) or ($x \in A$ and $x \in C$) . . . *Def. of intersection*

$x \in A$ and ($x \in B$ or $x \in C$)

. . . . *Distributive property of "and" over "or"*

. . . . $p \wedge (q \vee r) \Leftrightarrow (p \wedge q) \vee (p \wedge r)$

$x \in A$ and $x \in B \cup C$ . . . . *Definition of union*

So, $x \in A \cap (B \cup C)$. . . . . *Definition of intersection*

Thus $(A \cap B) \cup (A \cap C) \subseteq A \cap (B \cup C)$. . . *Definition of subset*

Therefore, $A \cap (B \cup C) = (A \cap B) \cup (A \cap C)$ . . . . *Def. of equality*

Note how the above proof is composed primarily of translations of definitions. Union also distributes over intersection:

Theorem For all sets A, B, and C, $A \cup (B \cap C) = (A \cup B) \cap (A \cup C)$.

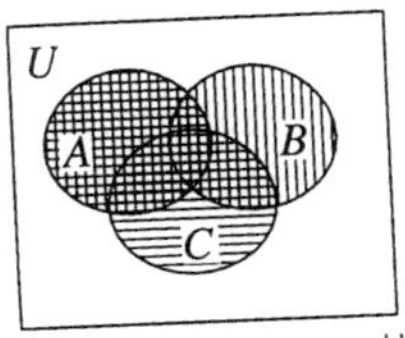

$(A \cup B) \cap (A \cup C)$: #

In the adjacent diagram, $A \cup B$ is shaded with vertical lines and $A \cup C$ is shaded with horizontal lines. The grid regions is the intersection of these two sets: $(A \cup B) \cap (A \cup C)$.

On the other hand, if we first focus on $B \cap C$, and then union this set with A, we see the same region. So, in this example, $A \cup (B \cap C) = (A \cup B) \cap (A \cup C)$. You are asked to prove that this relationship holds for all sets A, B, and C in (15) of the next exercise set.

Properties of Union and Intersection

Let A, B and C be sets.

Commutative Property

$A \cup B = B \cup A$

$A \cap B = B \cap A$

Associative Property

$A \cup (B \cup C) = (A \cup B) \cup C$

$A \cap (B \cap C) = (A \cap B) \cap C$

Distributive Property

$A \cap (B \cup C) = (A \cap B) \cup (A \cap C)$

$A \cup (B \cap C) = (A \cup B) \cap (A \cup C)$

The various properties of union and intersection are inherited from the corresponding properties of *or* and *and*. For example, *or* is commutative, which makes union commutative:

Claim: $A \cup B \subseteq B \cup A$

Assume that $x \in A \cup B$.

$x \in A$ or $x \in B$ .... *Definition of union*

$x \in B$ or $x \in A$ .... *Or is commutative*

So, $x \in B \cup A$. .... *Definition of union*

Therefore, $A \cup B \subseteq B \cup A$.

With an analogous argument, we can show that $B \cup A \subseteq A \cup B$. Thus, $A \cup B = B \cup A$.

Using similar arguments, we can prove the adjacent properties of union and intersection. Union and intersection are both commutative and associative, and they each distribute over the other in the same way that *and* distributes over *or*. Because of associativity, we may omit the parentheses when we form two unions or two intersections: $A \cup B \cup C$, $A \cap B \cap C$. However, we must use parentheses when we have both a union and an intersection because the position of the parentheses affects the meaning:

$$A \cap (B \cup C) \neq (A \cap B) \cup C$$

Complement Laws

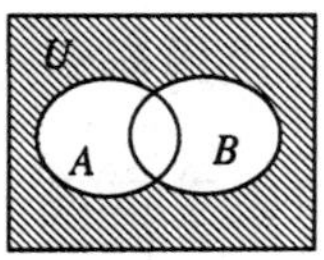

$(A \cup B)'$

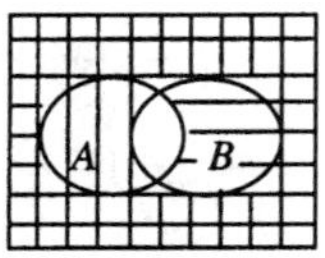

$A' \cap B'$
#

The complement laws give us another way to view complements of unions and intersections. For example, $(A \cup B)'$ is the shaded region in the adjacent diagram. In the second diagram, A' is shaded with horizontal lines and B' is shaded with vertical lines, so $A' \cap B'$ is the grid region, which is identical to the shaded region in the first diagram. Even though the process of constructing these two sets is very different, surprisingly, they end up as the same set. This property is inherited directly from the rule for negating an or-statement:

$$\sim(p \text{ or } q) \Leftrightarrow \sim p \text{ and } \sim q$$
$$(A \cup B)' = A' \cap B'$$

To prove that the above two sets are always equal, we deconstruct their meaning in the correct order. For example, to interpret the meaning of $x \in (A \cup B)'$, we first apply the definition of complement:

$$x \notin A \cup B$$

We must be careful not to slip into the casual way that we use negations in everyday language. We may be tempted to say the following:

So, $x \notin A$ or $x \notin B$. . . . . . . *Fatal Error!*

To avoid the above error, we should get in the habit of first translating the slash:

$x \notin A \cup B$

So, $\sim(x \in A \cup B)$. . . . . . *Translating* $\notin$

In the above form, we can substitute in the definition of union, as illustrated in the following proof.

Theorem For all sets A and B, $(A \cup B)' = A' \cap B'$.

Proof Let A and B be arbitrary sets.

Claim: $(A \cup B)' \subseteq A' \cap B'$

Assume that $x \in (A \cup B)'$.

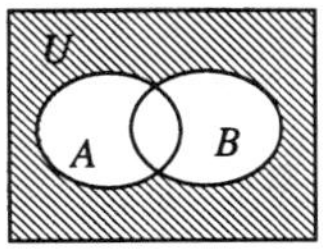

$\sim(x \in A \cup B)$ . . . . *Definition of complement*

$\sim(x \in A$ or $x \in B)$ . . . . *Definition of union*

$x \notin A$ and $x \notin B$ . . . . $\sim(p \vee q) \Leftrightarrow \sim p \wedge \sim q$

$x \in A'$ and $x \in B'$ . . . *Definition of complement*

So, $x \in A' \cap B'$. . . . . *Definition of intersection*

Therefore, $(A \cup B)' \subseteq A' \cap B'$. . . . . *Definition of subset*

Claim: $A' \cap B' \subseteq (A \cup B)'$

Assume that $x \in A' \cap B'$.

$x \in A'$ and $x \in B$ . . . . *Definition of intersection*

$x \notin A$ and $x \notin B$ . . . . *Definition of complement*

$\sim(x \in A$ or $x \in B)$ . . . . $\sim(p \vee q) \Leftrightarrow \sim p \wedge \sim q$

$\sim(x \in A \cup B)$ . . . . *Definition of union*

So, $x \in (A \cup B)'$. . . . . *Definition of complement*

Therefore, $A' \cap B' \subseteq (A \cup B)'$. . . . . *Definition of subset*

So, $(A \cup B)' = A' \cap B'$. . . . . *Definition of equality*

We have an analogous theorem for the complement of an intersection, which comes directly from the rule for negating an and-statement. You are asked to prove the following theorem in (13) of the next exercise set.

Theorem For all sets A, B, and C, $(A \cap B)' = A' \cup B'$.

To develop our reasoning powers beyond the mechanistic circuits of a computer, we must understand the meaning of symbolic equations like those in the previous two theorems. If we understand the meaning, we will be able to translate the symbols into verbal form:

> The complement of a union of two sets is the intersection of the individual complements.
>
> The complement of an intersection of two sets is the union of the individual complements.

Generalized Complement Laws

An obvious way to expand our knowledge is to try to generalize what we already know. When we generalize, we try to find a broader statement that includes the original statement and other statements as well. For example, can we generalize the following complement law?

$$(A \cup B)' = A' \cap B'$$

In the above form, we may see only a blank screen as we try to think of possible generalizations. However, if we rephrase the equation in terms of set subtraction, some ideas may surface:

$$U - (A \cup B) = (U - A) \cap (U - B)$$

What if we replace the universal set U with an arbitrary set C? Will the new equation be true?

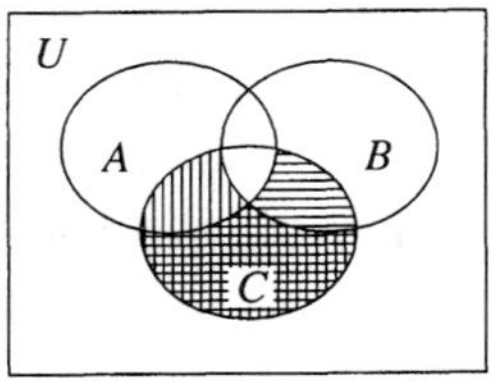

$$\text{Is } C - (A \cup B) = (C - A) \cap (C - B)?$$

Consider this question for the sets in the adjacent sketch.

> $C - A$ is shaded with horizontal lines.
>
> $C - B$ is shaded with vertical lines.
>
> $(C - A) \cap (C - B)$ is the region shaded with a grid.

If we visualize $C - (A \cup B)$ by removing $A \cup B$ from C, we are left with the grid region. So, $C - (A \cup B) = (C - A) \cap (C - B)$ in this example. This statement is also true for all sets A, B, and C, as stated in the following theorem which you are asked to prove in (15) of the next exercise set.

Theorem For all sets A, B, and C, $C - (A \cup B) = (C - A) \cap (C - B)$.

We have an analogous theorem for removing the intersection of two sets from a third set, C:

Theorem For all sets A, B, and C, $C - (A \cap B) = (C - A) \cup (C - B)$.

Operations & Subsets

Using visual reasoning, it is easy to see various subset relations that occur when we form unions and intersections of sets. For example, when we form the union of two sets, each of the original sets must be a subset of the union:

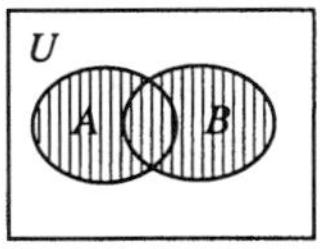

1. $A \subseteq A \cup B$

On the other hand, when we form the intersection of two sets, the new set must be a subset of each of the original sets:

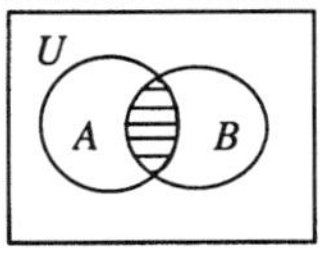

2. $A \cap B \subseteq A$

With more information, we can make further deductions about the union and intersection. For example, the adjacent illustration of $A \subseteq B$ indicates that the following are true.

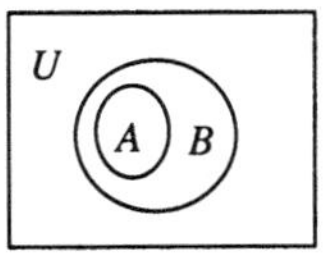

3. If $A \subseteq B$, then $A \cup B = B$.
4. If $A \subseteq B$, then $A \cap B = A$.
5. If $A \subseteq B$, then $B' \subseteq A'$.

If A and B are both subsets of C, the union of A and B will have to stay inside of C.

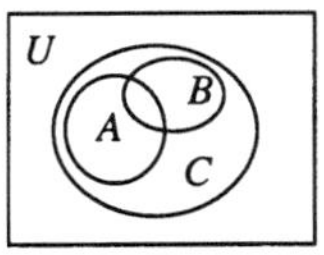

6. If $A \subseteq C$ and $B \subseteq C$, then $A \cup B \subseteq C$.

On the other hand, if C is contained in both A and B, C must be contained in their intersection.

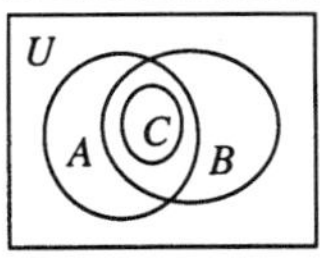

7. If $C \subseteq A$ and $C \subseteq B$, then $C \subseteq A \cap B$.

Even though the previous theorems are visually obvious, students sometimes have trouble writing proofs for them because they do not focus on what they want to derive. The structure of a proof is determined by the sentence that we want to derive, not by the assumptions we make in the beginning. When there are layers of assumptions, we must understand how each assumption fits into the overall structure of the proof.

Theorem: For all sets A and B, if $A \subseteq B$, then $A \cup B = B$.

Proof: Assume that $A \subseteq B$.
Claim: $A \cup B \subseteq B$
...
Claim: $B \subseteq A \cup B$
...
So, $A \cup B = B$.

For example, the adjacent theorem has four implications imbedded in it. Its outside structure is a implication. Imbedded in its hypothesis is the implication used in the definition of subset. Imbedded in the conclusion of the outside implication are two more implications that are used in the definition of equal sets. If we do not properly organize our thoughts and our writing, we can easily get confused by the layers of assumptions. However, if we carefully build the structure of our proof, as illustrated on the left, the confusion will evaporate and we will be able to see the inherent simplicity in it all.

In the following outline proof, we assume $A \subseteq B$ and then we immediately focus on what we want to derive. We keep the definition of $A \subseteq B$ on the back burner until we see a place to use it. We invoke it in Case 1, but we do not write out its definition. We expect the reader to know what it means.

Theorem Let A and B be arbitrary sets. If $A \subseteq B$, then $A \cup B = B$.

Proof Assume that $A \subseteq B$.

Claim: $A \cup B \subseteq B$

Assume that $x \in A \cup B$.

$x \in A$ or $x \in B$. .... *Definition of union*

Case 1: Assume that $x \in A$. Then $x \in B$. ..... $A \subseteq B$

Case 2: Assume that $x \in B$. Then $x \in B$.

In both cases, $x \in B$. So, $A \cup B \subseteq B$. .... *Def. of subset*

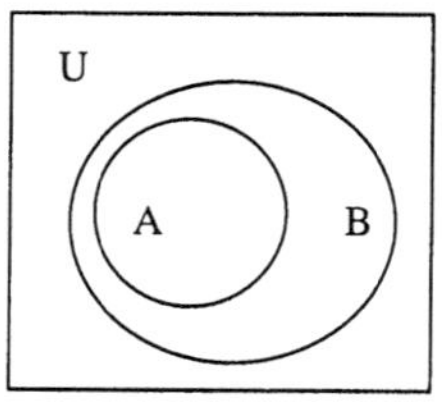

Claim: $B \subseteq A \cup B$

Assume that $x \in B$.

$x \in A$ or $x \in B$. .... *Valid argument*

So, $x \in A \cup B$. .... *Definition of union*

Therefore, $B \subseteq A \cup B$. .... *Definition of subset*

So, $A \cup B = B$. .... *Definition of equality*

Therefore, if $A \subseteq B$, then $A \cup B = B$.

Focused Thinking

To be a logical thinker, we have to train our mind to focus on what we're trying to do. When we structure a proof, we must focus on what we want to derive and not get sidetracked by what we have assumed.

In the following exercises, you are asked to write various proofs. If you first set up the outside structure by focusing on what you want to derive and then set up the inside structure by focusing on your simpler goals, these proofs will be quite easy, requiring only substitutions in definitions and the basic laws of logic. However, if you do not focus your thinking by setting up the appropriate structure, you will probably moan and groan and curse your fate for being assigned such a difficult task. Before you get to this state, though, please think about the visual picture of the statement, see how visually obvious it is, and then tell yourself how easy it will be to verbally validate the picture if you only focus your thinking and apply your knowledge of the little words *not*, *and*, *or*, and *implies*. We should be able to verbalize what we see, at least in terms of logical matters.

Exercise Set 3.2

1. $U = \{1,2,3,4,5,6\}$, $A = \{1,2,4\}$, $B = \{2,4,6\}$. Write each element of U in the appropriate region on the adjacent diagram. Then list the elements in the following sets:

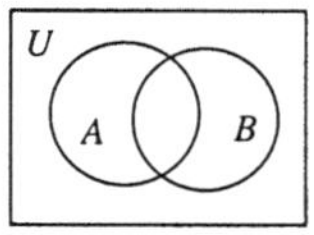

   a. $A \cup B$ b. $A \cap B$ c. $A - B$ d. $(A \cap B)'$ e. $A' \cup B'$

2. Illustrate each set on a Venn diagram. Are any of the sets equal?

   a. $(A \cup B)'$ b. $(A \cap B)'$ c. $A' \cup B'$ d. $A' \cap B'$

3. Write the following sets in interval notation.

   a. $[1,3] \cup [2,5]$ d. $[6,\infty) \cup (7,\infty)$ g. $(-\infty,2) \cup [1,\infty)$
   b. $[1,3] \cap [2,5]$ e. $[6,\infty) \cap (7,\infty)$ h. $(-\infty,2) \cap [1,\infty)$
   c. $[1,3] - [2,5]$ f. $[6,\infty) - (7,\infty)$ i. $(-\infty,\infty) - [1,2)$

4. Let A be a set. Write the following sets in a simplified form.

   a. $A \cup A$ b. $A \cap A$ c. $A \cup \emptyset$ d. $A \cap \emptyset$ e. $(A')'$

5. Illustrate each set on a Venn diagram, then write it in a simplified form.

   a. $(A \cap B) \cup (A \cap B')$ c. $A \cup (B \cap A)$ e. $A \cap (B \cup A)$
   b. $(A \cup B) \cap (A \cup B')$ d. $A \cup (B \cap A')$ f. $A \cap (B \cup B')$

6. Let $A \subseteq B$. Illustrate the given set on a Venn diagram, then write it in a simplified form.

   a. $A \cup B$ c. $A \cap B'$ e. $(A \cup B)'$ g. $A' \cup B'$

   b. $A \cap B$ d. $A - B$ f. $(A \cap B)'$ h. $A' \cap B'$

7. In the adjacent diagram, A, B, and C divide U into 8 nonoverlapping regions. Write each region in terms of operations on A, B, C and U. For example:

   Region 7: $A - (B \cup C)$

   Region 4: $(A \cap C) - B$

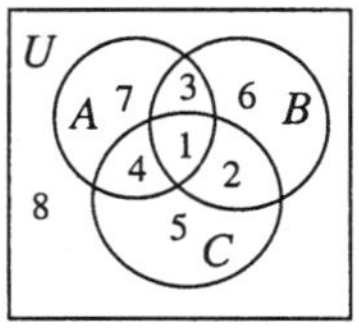

8. In the previous exercise, shade in Region 3 and Region 5. Then express the shaded area in terms of set operations.

9. Determine if the statement is true for all sets A, B, and C. If false, draw a counterexample using a Venn diagram.

   a. $A \cap A' = \emptyset$

   b. $A - B = A \cap B'$

   c. $A - B = B - A$

   d. $A - (B - C) = (A - B) - C$

   e. If $A \subseteq B$, then $A \cap B' = \emptyset$.

   f. $A \cap (B \cap C) = (A \cap B) \cap C$

   g. $A \cap (B \cup C) = (A \cap B) \cup C$

   h. $A \cup (B \cup C) = (A \cup B) \cup C$

10. With the given operations, are parentheses needed?

    a. $A \cup B \cup C$ b. $A - B - C$ c. $A \cap B \cup C$ d. $A - B \cup C$

11. Is the given expression a sentence?

    a. $A \subseteq B$ c. $x \in A \cup B$ e. Assume $A \cup B$.

    b. $A \cup B$ d. $A \subseteq A \cup B$ f. Assume $x \in A \cup B$.

12. Translate the following.

    a. $x \in (A \cap B)'$ b. $x \in (A \cup B \cup C)'$ c. $x \in A - (B \cup C)$

13. Prove the following for all sets A and B. Illustrate with a picture.

    a. $(A \cap B)' = A' \cup B'$

    b. If $A \subseteq B$, then $B' \subseteq A'$.

    c. $A \subseteq B$ if and only if $A \cap B = A$.

    d. $A \subseteq B$ if and only if $A \cup B = B$.

14. Use the theorems in (13) to make deductions about the following:

    a. If $B' \subseteq A'$, then ________

    b. If $A \subseteq B$, then $A' \cup B' =$___ and $A' \cap B' =$___.

15. Let A, B, and C be arbitrary sets. Prove the following. Illustrate each statement with a picture.

    a. $A \cup (B \cap C) = (A \cup B) \cap (A \cup C)$

    b. $C - (A \cup B) = (C - A) \cap (C - B)$

    c. If $A \subseteq C$ and $B \subseteq C$, then $A \cup B \subseteq C$.

    d. If $C \subseteq A$ and $C \subseteq B$, then $C \subseteq A \cap B$.

16. Explain why the syntax is wrong in the expression: $(A \subseteq B) \cup C$

17. U has 30 elements, A has 12, B has 9, and 4 elements are in both A and B. How many elements are in the given set?

a. $A \cup B$ c. $A - B$ e. $(A-B) \cup (B-A)$
b. $(A \cup B)'$ d. $B - A$ f. $(A-B) \cap (B-A)$

18. For security reasons, a spy operation is organized into 3 groups. The president is the only one who belongs to all 3 groups. There are a total of 6 people in the plumber group, 10 in the mole group, and 18 in the beaver group. A total of 2 people are both plumbers and moles, 3 people are both moles and beavers, and 4 are both plumbers and beavers. What is the total number of people involved in this operation? *Hint*: Use a Venn diagram to organize the information.

19. The symmetric difference of two sets, notated as $A \triangledown B$, is defined as follows: $A \triangledown B = (A-B) \cup (B-A)$

a. Represent $A \triangledown B$ on a Venn diagram. Is $A \triangledown B = B \triangledown A$?

b. Translate the following: $x \notin A \triangledown B$.
Does your answer agree with your sketch in part (a)?

c. Use your sketch in part (a) to draw a Venn diagram of $(A \triangledown B) \triangledown C$. Then draw a picture of $B \triangledown C$ and use it to draw $A \triangledown (B \triangledown C)$. Do you think that $\triangledown$ is associative?

d. If $A \in P(U)$ and $B \in P(U)$, is $A \triangledown B \in P(U)$?

20. Let $A_i = (i, \infty)$. Write the following sets in interval notation.
Hint: Visualize the sets on a number line.

a. A_{16} d. $A_1 \cup A_2 \cup A_3 \cup \ldots \cup A_{16}$
b. $A_1 \cup A_2$ e. $A_1 \cup A_2 \cup A_3 \cup \ldots \cup A_n$
c. $A_1 \cap A_2 \cap A_3$ f. $A_1 \cap A_2 \cap A_3 \cap \ldots \cap A_n$

21. Let $A_i = (\frac{1}{i}, \infty)$. Repeat the previous exercise.

Activity 3.3

$U = \{ x \mid x \in A_i \text{ for some } i \text{ in } I \}$ $V = \{ x \mid x \in A_i \text{ for all } i \text{ in } I \}$

1. Let $I = \{1, 2\}$, $A_1 = \{4, 6, 7\}$ and $A_2 = \{6, 8\}$.
List the elements in U. List the elements in V.

2. Let $A_i = (-\frac{1}{i}, \frac{1}{i})$ and I be the given set.
Write U and V in interval notation.

a. $I = \{1, 2\}$ c. $I = \{1, 2, 3, \ldots, n\}$
b. $I = \{1, 2, 3\}$ d. $I = \{1, 2, 3, \ldots\}$

3. Let $A_i = (4 - \frac{1}{i}, 4 + \frac{1}{i})$. Repeat the previous exercise.

≡ 3.3 Multiple Unions and Intersections ≡

The union of the three sets in the adjacent sketch is the shaded area. Its formal description comes from the union definition, which we apply twice:

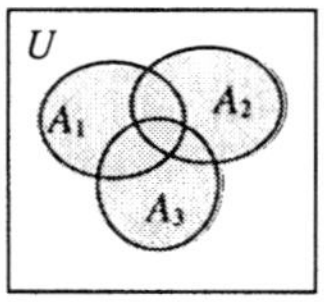

Assume that $x \in (A_1 \cup A_2) \cup A_3$.

Then $x \in (A_1 \cup A_2)$ or $x \in A_3$. ..... *Definition of union*

So, ($x \in A_1$ or $x \in A_2$) or $x \in A_3$. ... *Definition of union*

Since *or* is associative, we can omit the above parentheses. The sets A_1, A_2, and A_3 are indexed by the set $I = \{1, 2, 3\}$. Using the index set I, we can translate the above or-statement in terms of the existential quantifier:

$x \in A_1$ or $x \in A_2$ or $x \in A_3$
if and only if
for some i in I, $x \in A_i$.

We can use the above property to define the union of the three sets. Let $I = \{1,2,3\}$:

$$A_1 \cup A_2 \cup A_3 = \{ x \mid \text{for some } i \text{ in } I,\ x \in A_i \}$$

Multiple Unions

Using the above model, we will extend the definition of the union of two sets to include the union of any collection of sets. Let A_i be a set for each i in an index set I. The union of all A_i where i is in I is notated as $\bigcup_{i \in I} A_i$.

$x \in \bigcup_{i \in I} A_i$
if and only if
for some i in I, $x \in A_i$.

If $I = \{1,2,3\}$, then $\bigcup_{i \in I} A_i = A_1 \cup A_2 \cup A_3$.

If $I = \{1,2,3,\ldots\}$, then $\bigcup_{i \in I} A_i = A_1 \cup A_2 \cup A_3 \cup \ldots$

To form the set $\bigcup_{i \in I} A_i$, we combine the elements from each of the individual sets. As illustrated in the above example, the effect of combining elements from multiple sets can be verbalized with the existential quantifier. In fact, we use the same definition as the above set description for the union of three sets. Let A_i be a set for each i in an index set I:

$$\bigcup_{i \in I} A_i = \{ x \mid \text{for some } i \text{ in } I,\ x \in A_i \}$$

$$x \in \bigcup_{i \in I} A_i \iff \text{for some } i \text{ in } I,\ x \in A_i.$$

To be a logical thinker, we should read everything very carefully. However, with multiple unions we must take special care for the notation is a bit complex. Writing it out in expanded form helps us understand what it represents.

⊕ *Example*

Let $A_i = [i, i+1]$. Compute the multiple union.

1. Let $I = \{1,2,3\}$.

$$\bigcup_{i\in I} A_i = A_1 \cup A_2 \cup A_3 = [1,2] \cup [2,3] \cup [3,4] = [1,4]$$

2. Let $\mathbb{N}$ be the set of natural numbers.

$$\bigcup_{i\in \mathbb{N}} A_i = A_1 \cup A_2 \cup A_3 \cup \ldots$$
$$= [1,2] \cup [2,3] \cup [3,4] \cup \ldots = [1,\infty).$$

3. Let Z be the set of integers

$$\bigcup_{i\in Z} A_i = \ldots \cup A_{-2} \cup A_{-1} \cup A_0 \cup A_1 \cup A_2 \cup \ldots$$
$$= \ldots \cup [-2,-1] \cup [-1,0] \cup [0,1] \cup [1,2] \cup [2,3] \cup \ldots$$
$$= (-\infty,\infty)$$

In the above examples and the following example, we need to visualize the progression of the individual sets on a number line in order to see which elements will be in the multiple union.

⊕ *Example*

Let $\mathbb{N}$ be the natural numbers and $A_i = [\frac{1}{i}, 1]$.
Compute $\bigcup_{i\in\mathbb{N}} A_i$.

First, list the sets and observe the pattern:

$$\bigcup_{i\in \mathbb{N}} A_i = A_1 \cup A_2 \cup A_3 \cup \ldots A_n \cup \ldots$$
$$= [1,1] \cup [\tfrac{1}{2},1] \cup [\tfrac{1}{3},1] \cup \ldots [\tfrac{1}{n},1] \cup \ldots$$
$$= (0,1]$$

Since 0 is not in any of the A_i, 0 is not in the multiple union. If r is between 0 and 1, there exists a natural number n such that $\frac{1}{n} < r$. Therefore, $r \in A_n$. So, r is in the multiple union. Thus, the multiple union is the half-open interval $(0,1]$.

The Existential Quantifier

The existential quantifier seems to cause more confusion than the universal quantifier, especially when we write a proof that has several layers. If we say "for some i," several lines later – amid everything else that is going on – we may forget that i was existentially quantified. If we say "for some i_o," the subscript will remind us that i is existentially quantified:

$$x \in \bigcup_{i \in I} A_i \iff \text{there exists an } i_o \text{ in } I \text{ such that } x \in A_{i_o}.$$

Negation

We use the rule for negating an existential quantifier to translate the negation of $x \in \bigcup_{i \in I} A_i$:

$$x \notin \bigcup_{i \in I} A_i \iff \text{for all } i \text{ in } I,\ x \notin A_i.$$

Index Sets

Index sets give us a systematic procedure for naming sets. Usually the indices are involved in the definition of the individual sets, as in the previous examples. However, we can use any set that we wish as an index set.

Union over a Set

Instead of indexing sets with subscripts, we sometimes notate a union by using a letter to denote the collection of sets that we want to union. Let F be a collection of sets. The union of all the sets A that are in F is notated as $\bigcup_{A \in F} A$.

$x \in \bigcup_{A \in F} A$
if and only if
for some A in F, $x \in A$.

$$\bigcup_{A \in F} A = \{\, x \mid \text{for some } A \text{ in } F,\ x \in A \,\}$$

$$x \in \bigcup_{A \in F} A \iff \text{for some } A \text{ in } F,\ x \in A$$

If F is a collection of indexed sets, the above definition agrees with the definition for a collection of indexed sets:

$$\text{Let } F = \{A_1, A_2, A_3, \ldots\}. \text{ Then } \bigcup_{A \in F} A = \bigcup_{i \in N} A_i.$$

⊕ *Example*

1. Let $F = \{\{1,2\}, \{2,3,4\}, \{5\}\}$

   $\bigcup_{A \in F} A = \{1,2\} \cup \{2,3,4\} \cup \{5\} = \{1,2,3,4,5\}$

2. Let $F = \{\, [x, x+1] \mid x \text{ is a positive real number} \,\}$.

   $\bigcup_{A \in F} A = (0, \infty)$

3. Let $S = \{\, [x, x+1] \mid x \text{ is a real number} \,\}$.

   $\bigcup_{A \in S} A = (-\infty, \infty)$

Multiple Intersections

In a similar manner, we can generalize the definition of the intersection of two sets to a multiple intersection.

$$x \in A_1 \cap A_2 \cap A_3$$
$$\text{if and only if}$$
$$x \in A_1 \text{ and } x \in A_2 \text{ and } x \in A_3.$$

If we let $I = \{1, 2, 3\}$, we can translate the above and-statement in terms of the universal quantifier:

$$x \in A_1 \text{ and } x \in A_2 \text{ and } x \in A_3$$
$$\text{if and only if}$$
$$\text{for every } i \text{ in } I,\ x \in A_i.$$

Thus, the intersection of three sets can be described as follows. Let $I = \{1, 2, 3\}$.

$$A_1 \cap A_2 \cap A_3 = \{\, x \mid \text{for every } i \text{ in } I,\ x \in A_i \,\}$$

To generalize the above definition, we let A_i be a set for each i in an index set I. We notate the intersection of all the sets indexed by I as $\bigcap_{i \in I} A_i$.

$$\text{If } I = \{1, 2, 3, \ldots\}, \text{ then } \bigcap_{i \in I} A_i = A_1 \cap A_2 \cap A_3 \cap \ldots$$

To extend the definition of the intersection of two sets to the intersection of any collection of sets, we use the universal quantifier, which is a generalization of *and*.

$$\bigcap_{i \in I} A_i = \{\, x \mid \text{for every } i \text{ in } I,\ x \in A_i \,\}$$

$$x \in \bigcap_{i \in I} A_i \iff \text{for every } i \text{ in } I,\ x \in A_i.$$

Example

Compute the multiple intersection for the given index set.

1. Let $A_i = [i, \infty]$ and $I = \{1, 2, 3\}$.

$$\begin{aligned} \bigcap_{i \in I} A_i &= A_1 \cap A_2 \cap A_3 \\ &= [1, \infty] \cap [2, \infty] \cap [3, \infty] = [3, \infty] \end{aligned}$$

2. Let $A_i = [i, \infty]$ and $\mathbb{N} = \{1, 2, 3, \ldots\}$.

$$\begin{aligned} \bigcap_{i \in \mathbb{N}} A_i &= A_1 \cap A_2 \cap A_3 \cap \ldots \\ &= [1, \infty] \cap [2, \infty] \cap [3, \infty] \cap \ldots = \emptyset \end{aligned}$$

3. Let $A_n = [0,\frac{1}{n}]$ and $J = \{1,2,3,4\}$.

$$\bigcap_{n\in J} A_n = A_1 \cap A_2 \cap A_3 \cap A_4$$
$$= [0,1] \cap [0,\tfrac{1}{2}] \cap [0,\tfrac{1}{3}] \cap [0,\tfrac{1}{4}] = [0,\tfrac{1}{4}]$$

4. Let $A_n = [0,\frac{1}{n}]$ and $\mathbb{N} = \{1,2,3,\ldots\}$.

$$\bigcap_{n\in \mathbb{N}} A_n = A_1 \cap A_2 \cap A_3 \cap \ldots$$
$$= [0,1] \cap [0,\tfrac{1}{2}] \cap [0,\tfrac{1}{3}] \cap \ldots = \{0\}$$

5. Let $B_n = (0,\frac{1}{n}]$.

$$\bigcap_{n\in N} B_n = (0,1] \cap (0,\tfrac{1}{2}] \cap (0,\tfrac{1}{3}] \cap \ldots = \emptyset$$

Intersection over a Set

Instead of indexing sets with subscripts, we sometimes notate an intersection by using a letter to denote the collection of sets that we want to intersect. Let F be a collection of sets. The intersection of all the sets in F is notated as $\bigcap_{A\in F} A$.

$x \in \bigcap_{A\in F} A$ if and only if for every A in F, $x \in A$.

$$\bigcap_{A\in F} A = \{\, x \mid \text{for every } A \text{ in } F,\ x \in A\}$$

$$x \in \bigcap_{A\in F} A \iff \text{for every } A \text{ in } F,\ x \in A.$$

If F is a collection of indexed sets, the above definition agrees with the definition for indexed sets:

Let $F = \{A_1, A_2, A_3, \ldots\}$.

Then $\bigcap_{A\in F} A = \bigcap_{i\in N} A_i$.

⊕ *Example*

Compute the multiple intersection and the multiple union over the given set.

1. Let $F = \{\{1,2,3\}, \{2,3,5\}, \{2,4,6\}\}$.

$$\bigcap_{A\in F} A = \{1,2,3\} \cap \{2,3,5\} \cap \{2,4,6\} = \{2\}$$

$$\bigcup_{A\in F} A = \{1,2,3\} \cup \{2,3,5\} \cup \{2,4,6\} = \{1,2,3,4,5,6\}$$

2. Let $S = \{(0,x) \mid x \text{ is a real number and } x>1\}$.

$$\bigcap_{A\in S} A = (0,1], \quad \bigcup_{A\in S} A = (0,\infty)$$

Generalized Distributive Laws

Is it possible to generalize the following distributive law?

$$A \cup (B_1 \cap B_2) = (A \cup B_1) \cap (A \cup B_2)$$

What happens when we intersect 3 sets instead of 2 sets?

$$\begin{aligned} A \cup (B_1 \cap B_2 \cap B_3) &= A \cup (B_1 \cap (B_2 \cap B_3)) \quad \ldots\ldots \textit{Associative law} \\ &= (A \cup B_1) \cap (A \cup (B_2 \cap B_3) \quad \ldots\ldots\ldots \textit{Distributive law} \\ &= (A \cup B_1) \cap (A \cup B_2) \cap (A \cup B_3) \quad \ldots\ldots \textit{Distributive law} \end{aligned}$$

Thus, the distributive law can be generalized for the intersection of three sets. Let's translate this property in terms of multiple intersection notation. Let $I = \{1, 2, 3\}$.

$$\bigcap_{i \in I} C_i = C_1 \cap C_2 \cap C_3$$

Let $C_i = A \cup B_i$. If we substitute for C_1, C_2, and C_3, we get the following equation:

$$\bigcap_{i \in I} (A \cup B_i) = (A \cup B_1) \cap (A \cup B_2) \cap (A \cup B_3)$$

Thus, we can notate our original equation as follows:

$$A \cup (\bigcap_{i \in I} B_i) = \bigcap_{i \in I} (A \cup B_i)$$

$$A \cup (B_1 \cap B_2 \cap B_3) = (A \cup B_1) \cap (A \cup B_2) \cap (A \cup B_3)$$

$$A \cup (\bigcap_{i \in I} B_i) = \bigcap_{i \in I} (A \cup B_i)$$

The notation for multiple intersection enables us to express the generalized distributive law in a more concise form. The notation does look more complex; however, if we take the time to visualize or write it in expanded form, we can decipher its meaning. To see the real power of this notation, look how easy it is to generalize the above distributive law. We change only the index set. If $I = \{1, 2, 3, 4\}$, the above equation represents the following:

$$A \cup (B_1 \cap B_2 \cap B_3 \cap B_4) = (A \cup B_1) \cap (A \cup B_2) \cap (A \cup B_3) \cap (A \cup B_4)$$

For the ultimate generalization, let A be a set and let B_i be a set for each i in an index set I. Then the following is true:

$$A \cup (\bigcap_{i \in I} B_i) = \bigcap_{i \in I} (A \cup B_i)$$

To prove this theorem, we work with the definitions of union and multiple intersection. For example, to prove that the left side is a subset of the right side, we set up the outside structure as follows:

Claim: $A \cup (\bigcap_{i \in I} B_i) \subseteq \bigcap_{i \in I} (A \cup B_i)$

1. Assume that $x \in A \cup (\bigcap_{i \in I} B_i)$.

...

4. Therefore, $x \in \bigcap_{i \in I} (A \cup B_i)$.

The outside structure of Step 4 is $x \in \bigcap_{i \in I} Z_i$, where $Z_i = A \cup B_i$. In the following translation, we apply the definition of multiple intersection, treating $A \cup B_i$ as a single entity:

3. $x \in A \cup B_i$ for every i in I.

We then use Step 3 to structure the next layer of our proof:

Let i be an element in I.

...

Thus, $x \in A$ or $x \in B_i$.
Hence, $x \in A \cup B_i$.

Now that we understand what we need to derive, let's jump back to the beginning of the proof and work our way down.

1. Assume that $x \in A \cup (\bigcap_{i \in I} B_i)$.

Since the outside structure of the above sentence is the union of two sets, we first apply the definition of union, treating the multiple intersection set as a single entity:

2. Then $x \in A$ or $x \in \bigcap_{i \in I} B_i$.

From the above or-sentence, we branch into the following cases:

Case 1: Assume that $x \in A$.

Case 2: Assume that $x \in \bigcap_{i \in I} B_i$.

Within each case, we must stay focused on the goal, which is to end up with the sentence in Step 4. Having analyzed the various components of our proof, we are now ready to piece them together into a linearly ordered structure of interwoven valid arguments.

Theorem Let A be a set and let B_i be a set for each i in an index set I. Then $A \cup (\bigcap_{i\in I} B_i) = \bigcap_{i\in I} (A \cup B_i)$.

Claim $A \cup (\bigcap_{i\in I} B_i) \subseteq \bigcap_{i\in I} (A \cup B_i)$

Assume that $x \in A \cup (\bigcap_{i\in I} B_i)$.

$x \in A$ or $x \in \bigcap_{i\in I} B_i$. . . . . *Definition of union*

Case 1: Assume that $x \in A$.

Let i be an element in I.

$x \in A$ or $x \in B_i$ . . . . *Valid argument*

$x \in A \cup B_i$ . . . . *Definition of union*

So, $x \in \bigcap_{i\in I} (A \cup B_i)$. . . . . . *Def. of multiple intersection*

Case 2: Assume that $x \in \bigcap_{i\in I} B_i$.

Let i be an element in I.

$x \in B_i$ . . . . *Def. of multiple intersection*

$x \in A$ or $x \in B_i$ . . . . *Valid argument*

$x \in A \cup B_i$ . . . . *Definition of union*

So, $x \in \bigcap_{i\in I} (A \cup B_i)$. . . . . . *Def. of multiple intersection*

Since one of the two cases must happen, $x \in \bigcap_{i\in I} (A \cup B_i)$.

Therefore, $A \cup (\bigcap_{i\in I} B_i) \subseteq \bigcap_{i\in I} (A \cup B_i)$.

Claim $\bigcap_{i\in I} (A \cup B_i) \subseteq A \cup (\bigcap_{i\in I} B_i)$

See (9) in the next exercise set.

$A \cap (\bigcup_{i\in I} B_i) = \bigcup_{i\in I} (A \cap B_i)$

We have a similar generalization of the distributive law for intersection over union:

$$A \cap (B_1 \cup B_2 \cup B_3) = (A \cap B_1) \cup (A \cap B_2) \cup (A \cap B_3)$$

If $I = \{1, 2, 3\}$, the above equation can be written in terms of multiple unions as illustrated on the left. To prove that this equation holds for all sets B_i, we go through the same process, working from the outside to the inside, one step at a time, applying the definitions in the correct order. When we invoke the existential quantifier in the following proof, we use a subscript to remind us that i_o is existentially quantified. Notice how we work our way through the language by deconstructing the sentences and then putting them back together in a slightly altered fashion.

Theorem Let A be a set and let B_i be a set for each i in an index set I. Then $A \cap (\bigcup_{i \in I} B_i) = \bigcup_{i \in I} (A \cap B_i)$.

Claim $A \cap (\bigcup_{i \in I} B_i) \subseteq \bigcup_{i \in I} (A \cap B_i)$

Assume that $x \in A \cap (\bigcup_{i \in I} B_i)$.

Then $x \in A$ and $x \in \bigcup_{i \in I} B_i$. .... *Definition of intersection*

So, $x \in B_{i_0}$ for some i_o in I. .... *Definition of multiple union*

Hence, $x \in A$ and $x \in B_{i_0}$. .... *Valid argument*

Thus, $x \in A \cap B_{i_0}$. .... *Definition of intersection*

So, $x \in \bigcup_{i \in I} (A \cap B_i)$. .... *Definition of multiple union*

Therefore, $A \cap (\bigcup_{i \in I} B_i) \subseteq \bigcup_{i \in I} (A \cap B_i)$. .... *Def. of subset*

Claim $\bigcup_{i \in I} (A \cap B_i) \subseteq A \cap (\bigcup_{i \in I} B_i)$

See (9) in the next exercise set.

Generalized Complement Laws

Using a similar reasoning process, we can generalize the rules for complements of intersections and unions. For a given universal set, the complement of the intersection of two sets is the union of the complements of the two sets (page 240):

$$(A_1 \cap A_2)' = A_1' \cup A_2'$$

Applying the above rule twice, we can show that the complement of the intersection of three sets is the union of the complements of the three sets:

$(A_1 \cap A_2 \cap A_3)' = ((A_1 \cap A_2) \cap A_3)'$ .... *Associativity*

$= (A_1 \cap A_2)' \cup A_3'$ .... *Complement rule*

$= (A_1' \cup A_2') \cup A_3'$ .... *Complement rule*

Since union is associative, we can omit the union parentheses:

$$(A_1 \cap A_2 \cap A_3)' = A_1' \cup A_2' \cup A_3'$$

Let $I = \{1,2,3\}$. The above equation can be translated in terms of a multiple intersection and a multiple union:

$$(\bigcap_{i\in I} A_i)' = \bigcup_{i\in I} (A_i')$$

We can generalize further by letting I be any index set. The proof of the following theorem is straightforward if we work with the definitions in a step-by-step manner. When negations are involved, though, we must be careful for it is easy to make logical errors, especially when translating $\notin$. If we first translate the slash symbol as a negation at the beginning of the sentence, we can then make direct substitutions from the appropriate definitions, as illustrated in the following proof.

Theorem Let A_i be a set for each i in an index set I. Let U be a universal set that contains each A_i. Then $(\bigcap_{i\in I} A_i)' = \bigcup_{i\in I} A_i'$.

Claim $(\bigcap_{i\in I} A_i)' \subseteq \bigcup_{i\in I} A_i'$

Assume that $x \in (\bigcap_{i\in I} A_i)'$.

Then $x \notin \bigcap_{i\in I} A_i$. .... *Definition of complement*

$\sim (x \in \bigcap_{i\in I} A_i)$

$\sim$ (For every i, $x \in A_i$.) .... *Definition of multiple intersection*

There exists an i_0 such that $x \notin A_{i_0}$. .... *Negation law*

Hence, $x \in A_{i_0}'$. .... *Definition of complement*

So, $x \in \bigcup_{i\in I} A_i'$. .... *Definition of multiple union*

Therefore, $(\bigcap_{i\in I} A_i)' \subseteq \bigcup_{i\in I} A_i'$. .... *Definition of subset*

Claim $\bigcup_{i\in I} A_i' \subseteq (\bigcap_{i\in I} A_i)'$

See (9) in the next exercise set.

The analogous statement is true for the complement of a multiple union. You are asked to prove the following theorem in (9) of the next exercise set.

Theorem Let A_i be a set for each i in an index set I. Let U be a universal set that contains each A_i. Then $(\bigcup_{i\in I} A_i)' = \bigcap_{i\in I} (A_i')$.

Hopefully, the concepts of multiple intersections and unions seem simple to you now. If not, you may need to deepen your understanding of the two quantifiers by reviewing the material in Chapter 1. Or the difficulty may lie in the careful way in which one has to read subscript notation; take your time and write the compacted notation in an expanded form that will help you build an intuitive understanding of its meaning. Without this understanding, you are blinding yourself to the true meaning of these concepts, making them needlessly complex. If you understand the reasoning in the proofs in these last two sections, the index notation and subtle nuances of the parentheses, the order in which one applies the definitions, and how to structure the proofs, you can be assured that you have mastered the basic techniques of logical reasoning.

Exercise Set 3.3

1. Let $A_i = [-2, i]$, $I = \{1, 2, 3, 4\}$ and $\mathbb{N} = \{1, 2, 3, \ldots\}$. Compute the following sets. Write your answer in interval notation.

   a. $\bigcup_{i \in I} A_i$ b. $\bigcup_{i \in N} A_i$ c. $\bigcap_{i \in I} A_i$ d. $\bigcap_{i \in N} A_i$

2. Let $A_i = (-\infty, -i)$. Repeat (1).
3. Let $A_i = [5 + \frac{1}{i}, 6]$. Repeat (1).
4. Let $A_i = (5 - \frac{1}{i}, 5 + \frac{1}{i})$. Repeat (1).
5. Let $F = \{\{1, 3, 5\}, \{2, 3, 5, 7\}, \{1, 2, 9\}\}$.

   a. $\bigcup_{A \in F} A =$ __ b. $\bigcap_{A \in F} A =$ __

6. Let F be an arbitrary collection of sets. Is the statement always true? If not, give a counterexample.

   a. If there are sets A and B in F such that $A \cap B = \emptyset$, then $\bigcap_{A \in F} A = \emptyset$.

   b. If $\bigcap_{A \in F} A = \emptyset$, then there are sets A and B in F such that $A \cap B = \emptyset$.

7. Let A_i be a set for each i in the given index set. Let $I = \{1, 2, 3\}$, $J = \{1, 2, 3, 4\}$ and $\mathbb{N} = \{1, 2, 3, \ldots\}$. Is the statement always true? If not, give a counterexample.

   a. $A_3 \subseteq \bigcup_{i \in N} A_i$ c. $\bigcup_{i \in I} A_i \subseteq \bigcup_{i \in J} A_i$

   b. $A_3 \subseteq \bigcap_{i \in N} A_i$ d. $\bigcap_{i \in I} A_i \subseteq \bigcap_{i \in J} A_i$

8. To translate each statement, which definition do you use first?.

   a. $x \in \bigcup_{i \in I} (A \cap B_i)$ c. $x \in \bigcup_{i \in I} (A_i)'$

   b. $x \in A \cap (\bigcup_{i \in I} B_i)$ d. $x \in (\bigcup_{i \in I} A_i)'$

9. Let A be a set and let B_i be a set for each i in an index set I. Let U be a universal set that contains all the sets. Write easy-to-follow proofs of the following statements.

   Generalized Distributive Laws

   a. $A \cap (\bigcup_{i \in I} B_i) = \bigcup_{i \in I} (A \cap B_i)$

   b. $A \cup (\bigcap_{i \in I} B_i) = \bigcap_{i \in I} (A \cup B_i)$

   Generalized DeMorgan's Laws

   c. $(\bigcap_{i \in I} A_i)' = \bigcup_{i \in I} (A_i')$

   d. $(\bigcup_{i \in I} A_i)' = \bigcap_{i \in I} (A_i')$

Activity 3.4

A community of 2-dimensional beings, known as the Flatlanders, live in a plane. These poor creatures can see nothing outside of their plane. As you might suspect, they believe that their plane is the whole universe and that nothing could possibly exist outside of it. To the Flatlanders, the concept of a 3-dimensional universe is science fiction.

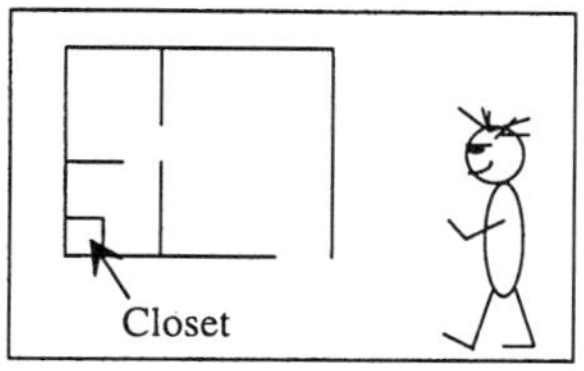

1. Can a Flatlander hide something from another Flatlander by putting it in a closet?
2. As a higher dimensional being with a 3-dimensional visual perception, can a Flatlander hide anything from you by putting it in a closet?
3. Could you hide any of your physical belongings from a higher being who has a 4-dimensional visual perception?
4. Do you think that perhaps we may be as limited in our visual perceptions as the 2-dimensional Flatlanders? Do you think that we could possibly extend our physical limitations through the power of reasoning? If so, do you have any ideas on how to tackle the dimension concept? What is a 4-dimensional space?

≡ 3.4 Cross Product ≡

The cross product is not as intuitively obvious as the other operations on sets. Given two circular regions in a plane, we can see their union, their intersection and their set difference, but we cannot see their cross product. Unlike the other set operations, the members of the new set formed under a cross product are a different kind of object. If we let the universal set U be the set of all subsets of a plane, for every A and B in U, the sets $A \cup B$, $A \cap B$, and $A - B$ are each in U. However, the cross product, $A \times B$, is not. It exists in a higher dimension, a 4-dimensional space that we cannot physically see. With the help of the mental construct of a cross product, though, we have the power to conceptualize beyond the limitations of our 3-dimensional vision. The cross product also provides us with a structure in which we can apply our knowledge of the real number line to figures in 2- and 3-dimensional space. The creation of this new set operation by René Descartes in the 17th century, like the invention of the wheel, opened the door to vast new universes for mental exploration.

Ordered Pairs

When we form the intersection of two sets, we do not change the individual elements. Instead, we select only those elements that are in both sets and put them in a new set. With a cross product, we make multiple copies of the elements in each set, then bond each element in the first set with each of the elements in the second set, thereby creating a new type of object – a bonded pair.

Let $A = \{1,2\}$ and $B = \{2,3\}$. The elements of $A \times B$ are the following bonded pairs:

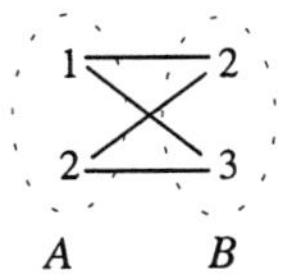

$$1 — 2 \quad 1 — 3 \quad 2 — 2 \quad 2 — 3$$

The bonding mechanism gives an ordered pair, which we represent with the notation (a, b). Using ordered pairs, we can represent $A \times B$ as follows:

$$A \times B = \{ (1,2), (1,3), (2,2), (2,3) \}$$

$(a,b) = (c,d)$ *if and only if* $a = c$ and $b = d$.

In formal set theory, an ordered pair is defined as the following set: $(a,b) = \{ \{a\}, \{a,b\} \}$. From this definition, it can be proved that two ordered pairs are *equal* if and only if their first terms are equal and their second terms are equal. With this characterization, which is stated on the left, we will not have to use the rather awkward formal definition.

The order in which elements are listed in a set has no effect on the set, but it does effect an ordered pair:

$$\{3,5\} = \{5,3\}, \text{ but } (3,5) \neq (5,3).$$

Unfortunately, the ordered pair notation has another meaning. We also use (a,b) to represent an open interval on a number line. Since (a,b) can represent a point in a plane or an open interval on a number line, we should make sure the context is clear when we use open parentheses in our writing.

$A \times B$

The *cross product* of sets A and B, denoted as $A \times B$, is the set of all ordered pairs whose first term is in A and whose second term is in B.

$$A \times B = \{ (a,b) \mid a \in A \text{ and } b \in B \}$$

The cross product is also called the *Cartesian product* in honor or René Descartes. Given two sets A and B, the cross product operation produces a third set, which is a set of ordered pairs.

Example

Let $A = \{1,2\}$ and $B = \{2,3,4\}$. List the elements in $A \times B$. How many elements are in $A \times B$?

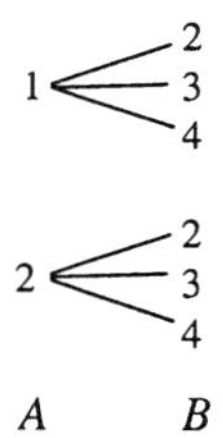

We first list the ordered pairs that have 1 as the first term and then list the ordered pairs that have 2 as the second term:

$$A \times B = \{(1,2), (1,3), (1,4), (2,2), (2,3), (2,4)\}$$

$A \times B$ has 6 elements: $|A \times B| = 6$

Note that A has 2 elements, B has 3 elements and $A \times B$ has $2 \cdot 3$ elements. Is this an interesting coincidence or the hint of an important relationship between the size of the individual sets and the size of their cross product?

Size of $A \times B$

If A has n elements and B has m elements, how many elements are in $A \times B$?

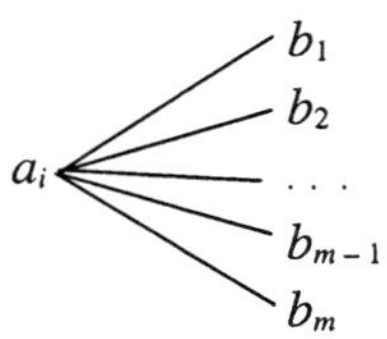

Let $A = \{a_1, a_2, a_3, \ldots, a_n\}$ and $B = \{b_1, b_2, b_3, \ldots, b_m\}$.

As illustrated on the left, for each a_i in A, there are m different ordered pairs in $A \times B$ that have a_i as the first term:

$$(a_i, b_1), (a_i, b_2), (a_i, b_3), \ldots, (a_i, b_m)$$

Thus, the total number of elements in $A \times B$ is the following sum:

$$\begin{array}{c} \quad\quad\quad\quad a_1 \quad a_2 \quad a_3 \quad \ldots \quad a_n \\ \quad\quad\quad\quad \downarrow \quad \downarrow \quad \downarrow \quad\quad\quad \downarrow \\ |A \times B| = \underbrace{m + m + m + \ldots + m}_{n \text{ terms}} \end{array}$$

Let A be a set with n elements and B a set with m elements. Then $|A \times B| = n \cdot m$.

Recall that multiplication of natural numbers is repeated addition: $5 + 5 + 5 = 3 \cdot 5$. The elementary school definition of $n \cdot m$ is to add m to itself n times. Hence, the above sum can be written as the product $n \cdot m$.

$$|A \times B| = n \cdot m$$

The above equation gives a simple formula for the number of elements in the cross product of two finite sets:

$$|A \times B| = |A| \cdot |B|$$

Multiplication

In the above formula, the product sign on the left side of the equation represents the product of two sets, whereas the product sign on the right side of the equation represents the product of two numbers. These two product operations are by no means equal, for they operate on very different types of objects; one operates on sets and the other operates on numbers. They do, though, have the intimate connection given by the above equation.

The cross product operation on sets gives a visualization of the multiplication operation on numbers. To illustrate the product $3 \cdot 4$, we could let $A = \{1,2,3\}$ and $B = \{a,b,c,d\}$, then arrange the elements in $A \times B$ in the following 3 rows and 4 columns:

$$\begin{array}{cccc} (1,a) & (1,b) & (1,c) & (1,d) \\ (2,a) & (2,b) & (2,c) & (2,d) \\ (3,a) & (3,b) & (3,c) & (3,d) \end{array}$$

If event A has n possible outcomes and event B has m possible outcomes, then event A followed by event B has $n \cdot m$ possible outcomes.

Many counting problems can be modeled with a cross product. If we need to count the number of possible outcomes for a sequence of two events that are independent of each other, we can list the possibilities in a cross product structure, which gives the adjacent counting principle. For example, suppose that one box contains the numbers 1, 2 and 3, and a second box contains the letters a, b, c, and d. The total number of possible outcomes for drawing one item from each box is the number of elements in the above listing of $A \times B$, which is $3 \cdot 4$.

⊕ *Example*

A die is thrown twice. How many possibilities are there for the outcome? What is the probability of throwing a 4 both times?

A die has 6 faces that contain the following numbers:

$$A = \{1, 2, 3, 4, 5, 6\}$$

To model two tosses of the die, we use $A \times A$. In the ordered pair (a, b), we let a represent the number on the first throw and b represent the number on the second throw. For example, (2,4) means that we threw a 2 on the first throw and a 4 on the second throw. So the total number of possibilities for throwing the die twice is the number of elements in $A \times A$. Since A has 6 elements, $A \times A$ has 36 elements.

Now that we know the total number of outcomes, we can compute the probability that two 4s will be thrown. The ordered pair (4, 4) is 1 out of 36 possibilities, so the probability that two 4s will be tossed is $\frac{1}{36}$.

$\mathbb{R} \times \mathbb{R}$

The set $\mathbb{R}$ of real numbers can be visualized as a number line. The set $\mathbb{R} \times \mathbb{R}$ can be visualized as a plane with two number lines perpendicular to each other. We usually denote the horizontal axis as the x-axis and the vertical axis as the y-axis.

$$\mathbb{R} \times \mathbb{R} = \{ (x, y) \mid x \text{ and } y \text{ are real numbers} \}$$

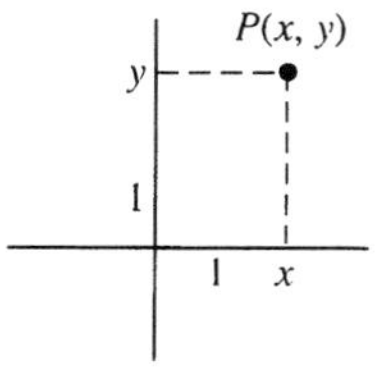

The ordered pair (x, y) is the coordinate of the point P in the plane whose projection on the x-axis is x and whose projection on the y-axis is y.

This ingenious method of assigning numbers to points in a plane was first published in 1637 in the famous text, *Discourse on Method*, by the French mathematician and philosopher, René Descartes. Descartes' new system, named in his honor as the Cartesian Coordinate System, revolutionized both mathematics and science. Since antiquity, number coordinates had been assigned to points on a ruler for measurement purposes. However, it was not until Descartes that the ruler concept was generalized to points in a plane. By providing a numerical notation for points in a plane, this new system made it possible to apply the methods of algebra to geometry and vice-versa. The new area of mathematics that emerged, called *analytic geometry*, provided the necessary tools on which calculus and modern science could be built. Descartes' revolutionary concept seems simple to us today, as does the invention of the wheel. We may even wonder whether or not we might have the ability to originate these concepts if we were placed back in time with no hindsight to guide us. There are, no doubt, many other mental constructs waiting to be discovered that no one has yet thought of that may, like Descartes' discovery, have a significant impact on the human race.

Pictures of $A \times B$

When A and B are sets of real numbers, we can visualize $A \times B$ as a subset of the plane. For example, let I represent the unit interval: $I = [0, 1]$.

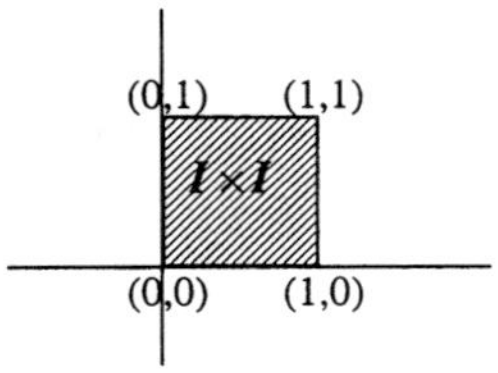

$$I \times I = \{ (x, y) \mid 0 \le x \le 1 \text{ and } 0 \le y \le 1 \}$$

The set of points in $I \times I$ form a square region in the plane with vertices at $(0, 0)$, $(1, 0)$, $(1, 1)$ and $(0, 1)$, as illustrated on the left. Each point in the shaded square has coordinates that fit the above description. Any point outside of the shaded square will have at least one of its coordinates greater than 1 or less than 0.

The cross product of two intervals will always form a rectangle in the plane. If $A = [2, 4]$ and $B = [1, 3]$, then $A \times B$ is the adjacent rectangle with vertices at the points (2,1), (2,3), (4,1), and (4,3).

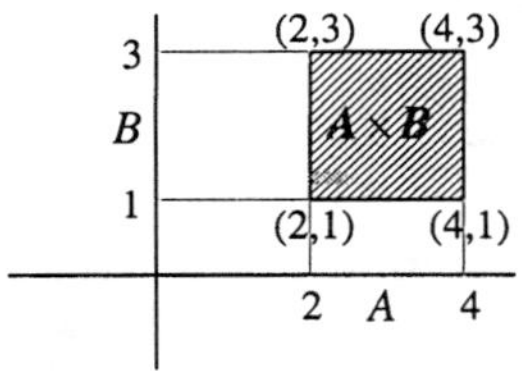

$$A \times B = \{ (x, y) \mid 2 \le x \le 4 \text{ and } 1 \le y \le 3 \}$$

Let's visualize $A \times (B \cup C)$ where A, B, and C are intervals. We imagine A on the horizontal axis and the other two sets on the vertical axis, for they will be the source for the second coordinates in the ordered pairs. We first union B and C on the vertical axis and then form the cross product, as illustrated on the left. On the other hand, if we individually construct $A \times B$ and $A \times C$ and then union these two sets, we end up with the same set. In this example, $A \times (B \cup C) = (A \times B) \cup (A \times C)$.

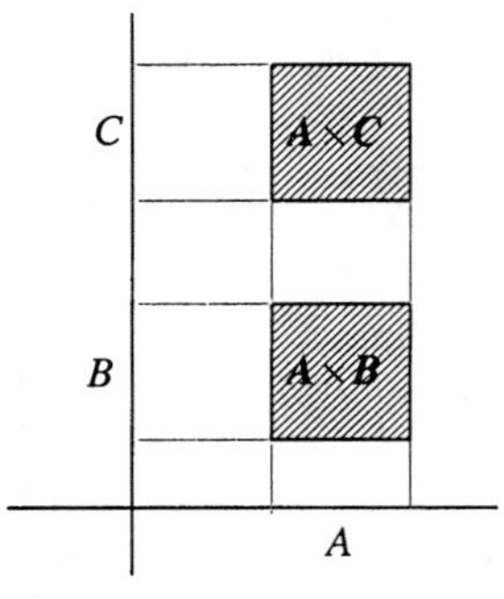

We will now prove that these two sets are always equal, using a standard subset argument. First, we prove that the left side is a subset of the right side and vice-versa. When we pick an arbitrary element z in the left side, we must translate what it means. In general, if z is an element in $A \times B$, then z must be an ordered pair whose first term comes from A and whose second term comes from B:

$$z \in A \times B$$

if and only if

$$z = (a, b) \text{ for some } a \text{ in } A \text{ and } b \text{ in } B.$$

Theorem For all sets A, B, and C, $A \times (B \cup C) = (A \times B) \cup (A \times C)$.

Claim $A \times (B \cup C) \subseteq (A \times B) \cup (A \times C)$

Assume that $z \in A \times (B \cup C)$.

Then $z = (x, y)$ for some x in A and y in $B \cup C$. .... *Def. of cross product*

So $y \in B$ or $y \in C$. .... *Definition of union*

Case 1: Assume that $y \in B$.

Since $x \in A$, $(x, y) \in A \times B$. ...... *Def. of cross product*

So $(x, y) \in A \times B$ or $(x, y) \in A \times C$. ... *Valid argument*

Thus, $(x, y) \in (A \times B) \cup (A \times C)$. .. *Definition of union*

Case 2: Assume that $y \in C$.

Since $x \in A$, $(x, y) \in A \times C$. ..... *Def. of cross product*

So $(x, y) \in A \times B$ or $(x, y) \in A \times C$. ... *Valid argument*

Thus, $(x, y) \in (A \times B) \cup (A \times C)$. .. *Definition of union*

In both cases, $(x, y) \in (A \times B) \cup (A \times C)$.

Since $z = (x, y)$, $z \in (A \times B) \cup (A \times C)$.

Therefore, $A \times (B \cup C) \subseteq (A \times B) \cup (A \times C)$. *Definition of subset*

Claim $(A \times B) \cup (A \times C) \subseteq A \times (B \cup C)$

Let $z \in (A \times B) \cup (A \times C)$.

Then $z \in A \times B$ or $z \in A \times C$. .... *Definition of union*

Case 1. Suppose that $z \in A \times B$.

$z = (x, y)$ for some x in A and y in B. .... *Def. of cross product*

Since $y \in B$, $y \in B$ or $y \in C$. .... *Valid argument*

So $y \in B \cup C$. .... *Definition of union*

So $(x, y) \in A \times (B \cup C)$. .... *Def. of cross product*

Case 2. Suppose $z \in A \times C$.

$z = (x, y)$ for some x in A and y in C. .... *Def. of cross product*

Since $y \in C$, $y \in B$ or $y \in C$. .... *Valid argument*

So $y \in B \cup C$ .... *Definition of union*

Thus $(x, y) \in A \times (B \cup C)$ .... *Def. of cross product*

In both cases, $(x, y) \in A \times (B \cup C)$.

Since $z = (x, y)$, $z \in A \times (B \cup C)$.

So $(A \times B) \cup (A \times C) \subseteq A \times (B \cup C)$.

Therefore, $A \times (B \cup C) = (A \times B) \cup (A \times C)$.

$A \times B \times C$

Unlike union and intersection, the cross product is not associative. For example, consider the following cross product:

Let $A = \{1\}$, $B = \{2\}$, and $C = \{3\}$.

$$(A \times B) \times C = \{ \; ((1,2),3) \; \}$$
$$A \times (B \times C) = \{ \; (1,(2,3)) \; \}$$

Each of the above sets has only one element, which is an ordered pair. $(1,2)$ is the first term in the ordered pair $((1,2),3)$, whereas 1 is the first term in the ordered pair $(1,(2,3))$. So the above two sets are not equal.

$$(A \times B) \times C \neq A \times (B \times C)$$

$(a_1, a_2, a_3) = (b_1, b_2, b_3)$
if and only if
$a_1 = b_1$, $a_2 = b_2$, and $a_3 = b_3$.

When we form the cross product of three sets, we visualize the elements as ordered triples instead of either of the above groupings. The notation (a, b, c) represents an ordered triple. Like ordered pairs, ordered triples are *equal* if and only if each of the corresponding terms are equal.

$(1,1,2) \neq (1,2,1)$ since their second terms are different.

We define the cross product of sets A and B and C as follows:

$$A \times B \times C = \{(a, b, c) \mid a \in A \text{ and } b \in B \text{ and } c \in C\}$$

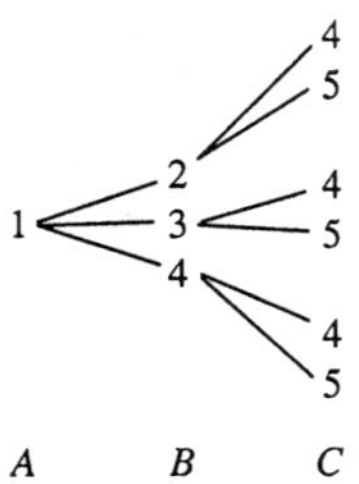

For example, let $A = \{1, 2, 6, 8\}$, $B = \{2, 3, 4\}$, and $C = \{4, 5\}$. The ordered triples in $A \times B \times C$ that have 1 as the first term are illustrated in the adjacent tree branches. We model this systematic branching procedure in the following listing of all elements in $A \times B \times C$:

(1,2,4) (2,2,4) (6,2,4) (8,2,4)
(1,2,5) (2,2,5) (6,2,5) (8,2,5)

(1,3,4) (2,3,4) (6,3,4) (8,3,4)
(1,3,5) (2,3,5) (6,3,5) (8,3,5)

(1,4,4) (2,4,4) (6,4,4) (8,4,4)
(1,4,5) (2,4,5) (6,4,5) (8,4,5)

Let A and B and C be finite sets.
Then $|A \times B \times C| = |A| \cdot |B| \cdot |C|$.

The above array has 4 columns and each column has $3 \cdot 2$ elements. So, the number of elements in $A \times B \times C$ is $4 \cdot 3 \cdot 2$. Using the definition of multiplication as repeated addition, it can be demonstrated that the number of elements in the cross product of three finite sets is the product of the number of elements in the individual sets.

$\mathbb{R}\times\mathbb{R}\times\mathbb{R}$

The set $\mathbb{R}\times\mathbb{R}\times\mathbb{R}$ can be visualized as 3-dimensional space with 3 number lines intersecting at right angles, as illustrated on the left. We usually notate this cross product as $\mathbb{R}^3$:

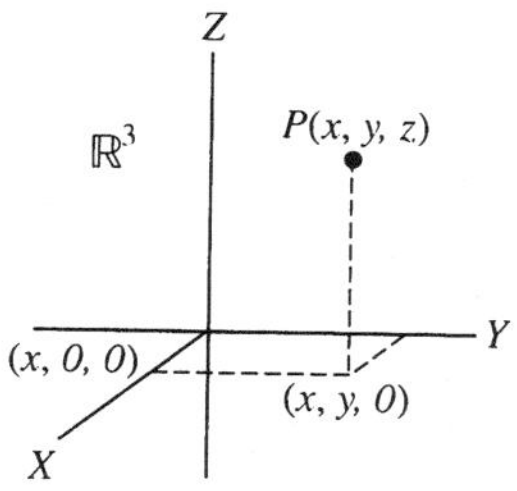

$$\mathbb{R}^3 = \{\ (x, y, z)\ |\ x, y \text{ and } z \text{ are real numbers }\}$$

To locate the point in 3-space represented by (x, y, z), we first find the point x on the x-axis, then we move in the y-direction, parallel to the y-axis, to the point $(x, y, 0)$. Next, we move z units in the z-direction to the point (x, y, z). As in 2-dimensional space, if we drop a perpendicular line from the point P to each of the 3 axes, the corresponding coordinates will be x on the x-axis, y on the y-axis, and z on the z-axis.

$I\times I\times I$

As we saw earlier, $I \times I$ is a unit square where I represents the unit interval: $I = [\ 0, 1\ ]$. Let's add another dimension to $I \times I$ and consider $I\times I\times I$:

$$I\times I\times I = \{\ (x, y, z)\ |\ 0 \le x \le 1 \text{ and } 0 \le y \le 1 \text{ and } 0 \le z \le 1\ \}$$

In 3-space, all of the points whose coordinates fit the above parameters must lie in a unit cube. The lower left back vertex is at the origin, as illustrated in the adjacent sketch. The coordinates of the 8 vertices of this cube are:

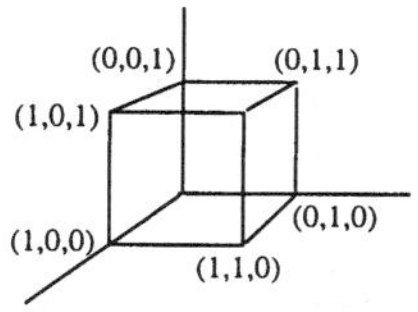

Lower level: (0,0,0), (1,0,0), (0,1,0), (1,1,0)

Upper level: (0,0,1), (1,0,1), (0,1,1), (1,1,1)

It seems perfectly natural to continue generalizing and consider a 4-dimensional cube, $I\times I\times I\times I$. Before we do this, though, let's talk about $\mathbb{R}\times\mathbb{R}\times\mathbb{R}\times\mathbb{R}$.

$\mathbb{R}\times\mathbb{R}\times\mathbb{R}\times\mathbb{R}$

In the 19th century, Arthur Cayley, Hermann Grassmann and Bernhard Riemann investigated generalizations of $\mathbb{R}^2$ and $\mathbb{R}^3$. To generalize to $\mathbb{R}^4$, all we have to do is add another coordinate. Since z is at the end of the alphabet, let's use subscripts to indicate the coordinates: x_1 for the first coordinate, x_2 for the second coordinate, etc. By indexing the coordinates, we have unlimited room for further generalizations:

$$\mathbb{R}^4 = \{\ (x_1, x_2, x_3, x_4)\ |\ x_1, x_2, x_3 \text{ and } x_4 \text{ are real numbers }\}$$

Even though we cannot visualize figures in $\mathbb{R}^4$, mathematicians soon found that they could talk about distances in $\mathbb{R}^4$ as easily as in spaces that we can see. Notice the striking similarity in the following distance formulas for $\mathbb{R}^2$ and $\mathbb{R}^3$, which both come from the Pythagorean Theorem (page 270, (11c)).

Distance on a line

Let d be the distance between $P(x_1)$ and $Q(y_1)$ in $\mathbb{R}$.

$$d = |x_1 - y_1| = \sqrt{(x_1 - y_1)^2}$$

Distance in a plane

Let d be the distance between $P(x_1, x_2)$ and $Q(y_1, y_2)$ in $\mathbb{R}^2$.

$$d = \sqrt{(x_1 - y_1)^2 + (x_2 - y_2)^2}$$

Distance in 3-space

Let d be the distance from $P(x_1, x_2, x_3)$ to $Q(y_1, y_2, y_3)$ in $\mathbb{R}^3$.

$$d = \sqrt{(x_1 - y_1)^2 + (x_2 - y_2)^2 + (x_3 - y_3)^2}$$

Distance in 4-space?

An obvious way to define distance in $\mathbb{R}^4$ is as follows. Let d be the distance from $P(x_1, x_2, x_3, x_4)$ to $Q(y_1, y_2, y_3, y_4)$.

$$d = \sqrt{(x_1 - y_1)^2 + (x_2 - y_2)^2 + (x_3 - y_3)^2 + (x_4 - y_4)^2}$$

It can be shown that the above equation gives a notion of distance that has the same fundamental properties that the distance concept has in $\mathbb{R}$, $\mathbb{R}^2$ and $\mathbb{R}^3$.

Hyperspheres

With a distance concept for $\mathbb{R}^4$, we can generalize familiar figures from 3-dimensional space. For example, we can generalize the notion of a circle and a sphere without even changing the definition:

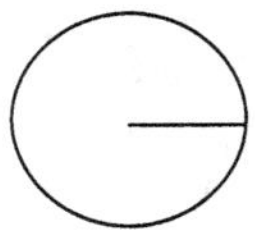

> A circle is the set of points in a plane at a fixed distance from a fixed point.
>
> A sphere is the set of points in 3-dimensional space at a fixed distance from a fixed point.

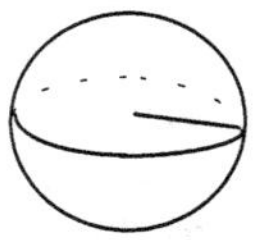

> A *hypersphere* is the set of points in 4-dimensional space at a fixed distance from a fixed point.

Let (c_1, c_2, c_3, c_4) represent the fixed point at the center of the hypersphere. Then the points (x_1, x_2, x_3, x_4) on the hypersphere will be determined by the following equation.

$$d = \sqrt{(x_1 - c_1)^2 + (x_2 - c_2)^2 + (x_3 - c_3)^2 + (x_4 - c_4)^2}$$

If we square both sides, we can eliminate the radical:

$$d^2 = (x_1 - c_1)^2 + (x_2 - c_2)^2 + (x_3 - c_3)^2 + (x_4 - c_4)^2$$

So, a hypersphere of radius 1 is the set of all points whose coordinates (x_1, x_2, x_3, x_4) satisfy the following equation:

$$(x_1 - c_1)^2 + (x_2 - c_2)^2 + (x_3 - c_3)^2 + (x_4 - c_4)^2 = 1$$

If the center of the sphere is (0,0,0,0), the equation becomes:

$$x_1^2 + x_2^2 + x_3^2 + x_4^2 = 1$$

With our algebraic skills, we can now work with a 4-dimensional sphere that we cannot fully visualize. Like the Braille system for the blind, the distance concept gives us a way to mentally visualize what we cannot physically see. However, if we stop and think about it, we cannot fully visualize 3-dimensional figures either. When we look at a box, we cannot see the complete box in one view. We must turn it around and look at it from various perspectives. We then imagine the complete totality of the box, even though we cannot see the front and back at the same time. In fact, we imagine a lot of things – "imagine" means to build an image in our mind.

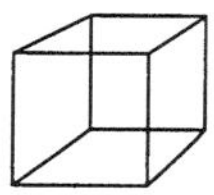

Cube

When we look at the adjacent sketch, we imagine it as a 3-dimensional cube, but it is really a 2-dimensional sketch. In a similar manner, all real objects that we see are constructed in our mind from 2-dimensional images sent by light rays through the portal of each eye, from which our mind miraculously builds a 3-dimensional understanding of the object. In the same way, we can learn to build our visual understanding of 4-dimensional space. The easiest way to enter into 4-dimensional viewing mode is through the dimension of time.

Hypercubes

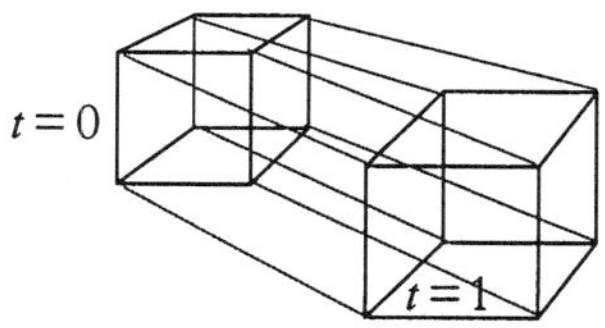

Suppose that the adjacent 3-dimensional solid cube is moving through time along a straight line. The trace of all points over a 1 unit time interval creates a 4-dimensional cube, which is called a *hypercube*. To build a mental construct of a hypercube, we first draw a picture of the cube when $t = 0$; then we draw another snapshot of the cube when $t = 1$ hour. To capture a fleeting essence of all the snapshots between $t = 0$ and $t = 1$, we connect the corresponding vertices with line segments, as illustrated on the left. A slice of this figure at any point along one of these line segments will be a cube, but let's clarify what we mean by a slice.

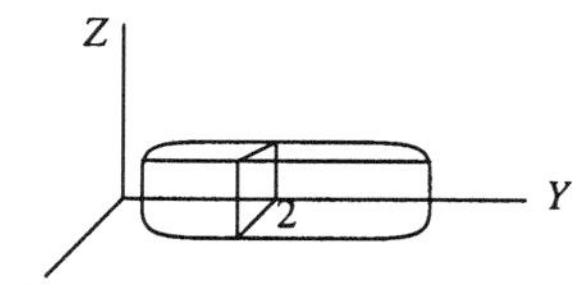

Slicing a Loaf

We can slice the 3-dimensional loaf of bread illustrated on the left by fixing the y-coordinate, say at $y = 2$. We then let the knife slice through all the other points where $y = 2$. This slice exists in a different plane, though, than the slice we get at $y = 3$. Similarly, when we slice our hypercube at $t = \frac{1}{2}$, we visualize all the points where $t = \frac{1}{2}$, which will be a cube, an identical twin of the cube pictured at $t = 0$. However, this slice exists in a different 3-dimensional universe than the 3-dimensional universe at $t = 0$. So, when we look at the above illustration of a hypercube, we must remind ourselves that the 3-dimensional universe that contains each slice is different.

Point Universe + Time

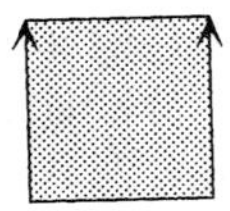

Line Universe + Time

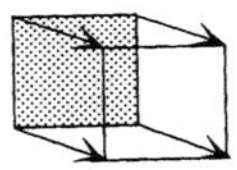

2-Dimensional Universe + Time

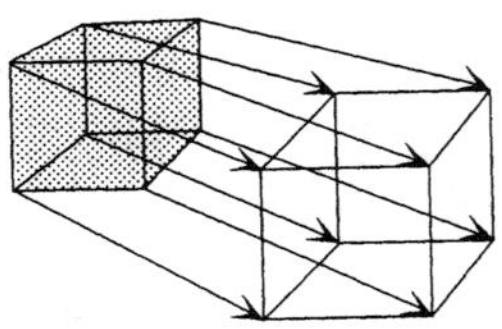

3-Dimensional Universe + Time

Reasoning by analogy is perhaps the best way to get a deeper understanding of what is happening here. Consider how a 0-dimensional point generates a 1-dimensional line; but first, suppose that you live on that point and cannot see nor move beyond it. The point is moving through space, which you experience as time, but you cannot see the extraterrestrial view of the 1-dimensional line in the adjacent illustration.

After eons of time, you evolve into a creature with 1-dimensional vision, having free reign now to move wherever you wish on a 1-dimensional line segment. Again, you can neither see nor move beyond your terrestrial line. But your line segment is moving through space, along a straight line segment of the same length, thereby spanning a 2-dimensional square region. Your vision is limited to only 1-dimension, so you cannot see the extraterrestrial view given on the left.

After another great span of time, you evolve into a creature with 2-dimensional vision, having free reign now to move wherever you wish in a 2-dimensional square region. As you move through time, your square region spans a cube, but to you the concept of a cube is pure science fiction. In fact, contemplation of such matters makes your head throb, for you think that your species is the most advanced in the universe. Perhaps it is your ego that prevents you from even considering that there may be higher-dimensional beings who can perceive things that you cannot.

Finally, you advance to your present place in the universe. You can move around in your 3-dimensional box and your brain has learned how to interpret visual information about this new space, but the next step up to a 4-dimensional cube still seems like science fiction, that is, until you use the power of the cross product to analyze the time space that you live in and conceptualize the mysterious hypercube.

It is rather surprising what a simple definition we can give for a hypercube H. H is the cross product of four unit intervals:

$$\begin{aligned} H &= I \times I \times I \times I \\ &= \{(x, y, z, w) \mid 0 \le x \le 1 \text{ and } 0 \le y \le 1 \text{ and } 0 \le z \le 1 \text{ and } 0 \le w \le 1\} \end{aligned}$$

If we interpret the 4th coordinate as representing time, (0,0,0,0) represents the position of the point (0,0,0) in the original cube when the time t is 0, whereas (0,0,0,1) represents the position of the same point when t is 1. The hypercube has twice as many vertices as the 3-dimensional cube. For each of the 8 vertices in the original cube, the 4th coordinate could be 0 or 1, so the hypercube has 16 vertices:

$t=0$: (0,0,0,0), (1,0,0,0), (0,1,0,0), (1,1,0,0)
(0,0,1,0), (1,0,1,0), (0,1,1,0), (1,1,1,0)

$t=1$: (0,0,0,1), (1,0,0,1), (0,1,0,1), (1,1,0,1)
(0,0,1,1), (1,0,1,1), (0,1,1,1), (1,1,1,1)

In a similar manner, we can mathematically generalize to $\mathbb{R}^5$, $\mathbb{R}^6$, $\mathbb{R}^n$, and even to $\mathbb{R}^\infty$. Even though we are visually impaired in these higher dimensions, the analytical tools of algebra retain their full power.

The ideas spawned by the mathematical generalization of $\mathbb{R}^3$ opened a new world for writers, scientists and artists, who no longer felt it necessary to depict the world from a 3-dimensional perspective. Cubist artists became preoccupied with developing an artistic language for depicting the spatial nature of 4 dimensions. Based on his reading of the new mathematics, H. G. Wells provided an interpretation of the 4th dimension as time in his novel, *The Time Machine*. Albert Einstein gave Wells' fantasy a scientific basis with his *Theory of Relativity*. The impact of the cross product on our perception of the universe has indeed been phenomenal.

Exercise Set 3.4

1. Let $A=[1,2]$, $B=[0,3]$, $C=[2,4]$. Sketch the following sets in $\mathbb{R}^2$.
   a. $\{3\}\times A$
   b. $A\times\{3\}$
   c. $A\times B$
   d. $A\times C$
   e. $(A\times B)\cap(A\times C)$
   f. $A\times(B\cap C)$
2. Prove or disprove:
   a. For all sets A and B, $A\times B = B\times A$.
   b. There exists sets A and B such that $A\times B = B\times A$.
   c. For all sets A, B, C, and D, if $A\subseteq C$ and $B\subseteq D$, then $A\times B\subseteq C\times D$.
   d. For all sets A, B, and C, $A\times(B\cup C) = (A\times B)\cup(A\times C)$.
   e. For all sets A, B, and C, $A\times(B\cap C) = (A\times B)\cap(A\times C)$.
3. Let $I=[0,1]$ and $J=[3,5]$. Let $F= I\times J\times I\times J$. List two different points in F.
4. Let A be a set with 6 elements and B be a set with 8 elements. How many elements does $A\times B$ have? Explain why.
5. At the race track, there are 6 horses in the first race, 8 horses in the second race, and 5 horses in the third race. Suppose that in each of

these three races you bet on one horse to win.

a. Model the possible ways that you can place these three bets in terms of a cross product. How many possibilities are there?

b. If you know nothing about the horses, what is the probability that you will win in all 3 races?

6. A square whose sides each measure 2 units is positioned so that each of its sides is parallel to one of the coordinate axes and its lower left vertex is located at the given point. Express the square as a cross product.

a. $(0,0)$ b. $(1,2)$ c. $(-2,3)$

7. A cube whose sides each measure 2 units is positioned with each face parallel to one of the coordinate planes. Express the cube as a cross product. Its lower left back vertex is at the given point.

a. $(0,0,0)$ b. $(1,2,0)$ c. $(1,2,5)$

8. Expand each cube in the previous exercise to a hypercube by adding a 4th coordinate that has a span of 2 units.

9. Suppose that you live in a 4-story dormitory.

a. Devise a coordinate system for representing all points within the dormitory. How many coordinates does it take?

b. Suppose that someone is using coordinates to track your location in the dorm throughout the evening. How many coordinates will it take?

c. Do you live in a 3-dimensional space?

10. A unit cube is placed on a table, but it suddenly disappears after 2 hours. Use a cross product to represent the life span of the cube on the table. Decide how you want to set up the coordinate axes.

11. Let a and b be real numbers with $a<b$.

a. Derive a formula for the distance from a to b.
Hint: Use 3 cases and a visual argument on the number line.

b. Use your work in part (a) to derive a formula for the distance between (a,c) and (b,c) in a plane.

c. Use the Pythagorean theorem to derive a formula for the distance between the points (a,c) and (b,d) in the adjacent sketch.

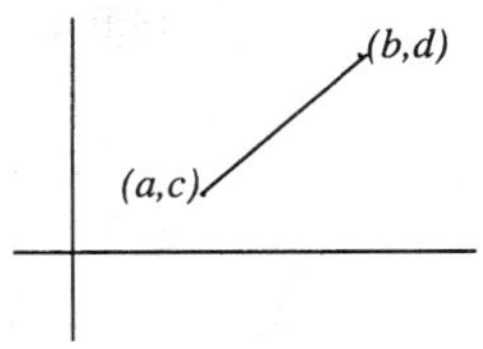

d. Generalize the distance formula to $\mathbb{R}^n$.

12. Let S_n denote the set of all bit strings of length n, as defined in (15) on page 232. Model S_n with a cross product.

Activity 3.5

1. Imagine that you live in a cave and have a flock of sheep. During the day the sheep wander off in search for food, and each evening before sunset, you must make sure that you get them all back in the cave. You have never heard of a number, most of your communications are visual rather than spoken, and, of course, you don't have paper or pencils. Using available supplies, devise several plans for making sure that you have gathered all your sheep in the evening.
2. Compare your cave system with the way we count today.
   a. How are they similar?
   b. What is the basic principle that is essential to any counting technique?
3. What does it mean to say that "two sets have the same size?"
4. Is the size of the set of natural numbers bigger than the size of the set of even natural numbers?

≡ 3.5 Finite Sets ≡

The term "infinite" is often used in the media to indicate a very large set, but this is not what it means in mathematics. The set of atoms in our universe is one of the largest physical sets that one could imagine; however, this set has less than 10^{80} elements, so it is a finite set. To understand the meaning of infinite, we must first understand the meaning of finite.

$$\begin{array}{cccc} & 1 & 2 & 3 \\ & \downarrow & \downarrow & \downarrow \\ S = & \{a, & d, & e\} \end{array}$$

The concept of finite is based on the counting process. If a set S is finite, it would be possible – given enough time – to count all the elements in the set. Counting, in turn, is based on the concept of a one-to-one correspondence. When we count the elements in a set S, we construct a one-to-one correspondence with the first n natural numbers, as illustrated on the left. We will now investigate the conditions that are essential for a one-to-one correspondence. After that, we will construct the natural numbers and then use the concept of a natural number and the concept of a one-to-one correspondence to give a simple definition of finite.

One-to-One Correspondences

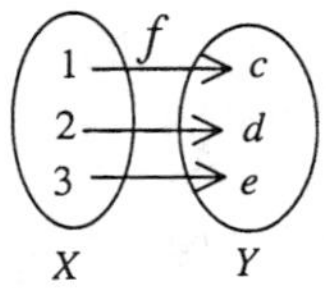

f is a *one-to-one correspondence* from *X* to *Y* if and only if *f* is a one-to-one function that maps *X* onto *Y*.

A *one-to-one correspondence* between the sets *X* and *Y* is a mapping in which each element in *X* is mapped to exactly one element in *Y*, and each element in *Y* has exactly one element in *X* mapped to it.

The adjacent mapping is a one-to-one correspondence between *X* and *Y*. Using function notation, we can describe this mapping as follows:

$$f(1) = c \quad f(2) = d \quad f(3) = e$$

A one-to-one correspondence can be defined as a function that is one-to-one and onto. These three concepts, function, one-to-one, and onto, are involved in any counting process. We will briefly examine the meaning of these concepts. In Chapter 4, we will investigate them in more detail.

Functions

f is a *function* from *X* into *Y* if and only if *f* maps each element in *X* to a unique element in *Y*.

A Function? No

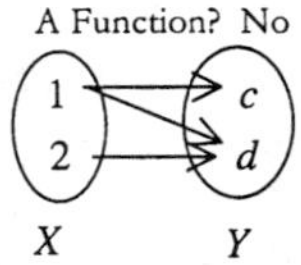

A Function? Yes

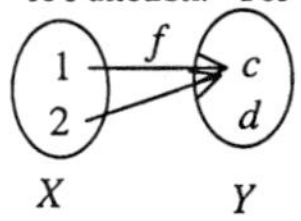

Functions, one of the most powerful concepts in mathematics, give a special kind of relation between two sets. A function is a one-way mapping between two sets where each element in the first set is mapped to a unique element in the second set.

The unique element to which x is mapped under a function f is notated as $f(x)$. The notation, $f(x) = y$, means that x is mapped to y under the function f. The great power of the function concept comes from the $f(x)$ notation, which allows us to manipulate our thoughts about functions in a very efficient manner.

To be a function, a mapping must send each element in the first set to only one location. The adjacent mapping is not a function because it maps 1 to two different places. We cannot use function notation with this mapping for we would have $f(1) = c$ and $f(1) = d$, but $c \neq d$.

In the adjacent mapping 1 is mapped to c and 2 is also mapped to c. Since each element in X is mapped to a unique element in Y, this mapping is a function from X into Y.

One-to-One Function

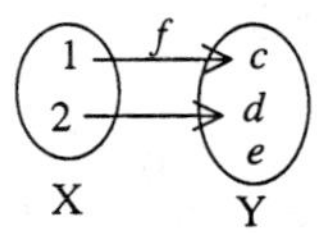

The previous function is not one-to-one because it maps different elements to the same location. The adjacent mapping, though, is a one-to-one function because it maps different elements to different places. A function is *one-to-one* if different elements in the domain always map to different elements in the range. Using function notation, this condition can be phrased as follows. Let a and b be elements in the domain of f:

f is *one-to-one*
if and only if
for all a and b in the domain,
if $a \neq b$, then $f(a) \neq f(b)$.

$$\text{If } a \neq b, \text{ then } f(a) \neq f(b).$$

We can state the property of being one-to-one in a more positive perspective by rephrasing the above implication in terms of its contrapositive:

$$\text{If } f(a) = f(b), \text{ then } a = b.$$

Onto Functions

f maps X *onto* Y
if and only if
for every y in Y, there exists
an x in X such that $f(x) = y$.

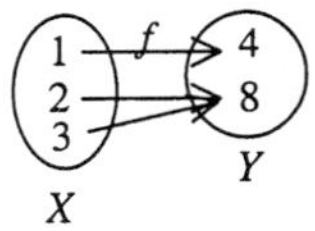

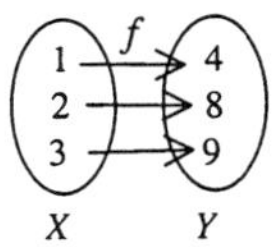

To be a function from X into Y, f must map each element in X to some element in Y, but each element in Y does not have to have someone map to it. In the last example on the previous page, the function f maps X into Y, but it does not map X onto Y because e is left out in the cold.

A function maps X *onto* Y if each element in Y has someone mapped to it. In other words, for every y in Y, there must exist an x in X such that $f(x) = y$.

The concept of onto is completely independent of the one-to-one concept. A function can be one-to-one, but not onto as was the case with the last example. Conversely, a function can be onto but not one-to-one, as illustrated by the following function:

$$X = \{1, 2, 3\},\ Y = \{4, 8\},\ f(1) = 4,\ f(2) = 8, \text{ and } f(3) = 8.$$

This function maps X onto Y, but it is not one-to-one since 2 and 3 both map to the same element. Some functions, though, are both one-to-one and onto. When this happens, the two sets have the same size.

Size of a Set

Let A and B be sets.
A has the same size as B
if and only if
there exists a one-to-one
function that maps A onto B.

Two sets that have the same size are said to have the same *cardinality*. We usually determine whether or not two sets have the same size by counting their elements; however, numbers are not necessary for this type of comparison. If we can find a one-to-one function that maps A onto B, then A and B have the same size. For example, let $X = \{a, b, c\}$ and $Y = \{r, s, \text{t}\}$. Define f as follows:

$$f(a) = r \quad f(b) = s \quad f(c) = t.$$

f is a one-to-one function that maps X onto Y, so X and Y have the same size.

The concept of "having the same size" is at the foundation of quantitative reasoning, laying the groundwork for the number concept. We do not need numbers to determine if two sets have the same size. If everyone is sitting in a chair at a meeting and all the chairs are taken, we know that there are the same number of chairs as people. However, if we need a written record of how many people were at the meeting, we must advance to a higher level of abstract thinking.

Natural Numbers

As early as 8500 B.C.E., around the same time as the first evidence of farming and herding, people in the Near East had developed ingenious ways to track sizes of different objects. Before numbers were conceived, long before writing was developed, prehistorical people tracked their supplies using physical objects, such as fingers, sticks, notches on sticks, strokes on a wall, knots on a string, and pebbles. Our word "calculate," comes from the Latin word "calculus," which was a small pebble used for counting. In the cradle of civilization, the calculi used by merchants in ancient Sumeria were clay pebbles stored in clay containers. The sale of a herd of sheep would be accompanied by a clay pot whose pebbles represented the size of the herd.

Pebble arithmetic is quite simple, even for the shepherd far removed from the bustling commerce of the emerging cities. To track his flock, a shepherd could place a pebble in a sack for each one of his sheep. When his flock returned in the evening, he would remove a pebble for each sheep. Leftover pebbles meant missing sheep, while a shortage of pebbles meant that he had picked up some extras.

To add the number of sheep in two different flocks, the shepherd could combine (*union*) the pebbles from the sacks that represent each flock. To subtract, he would remove pebbles (*set subtraction*). This simple, but effective, system has the tremendous advantage of being purely physical with nothing to memorize and no special schooling needed.

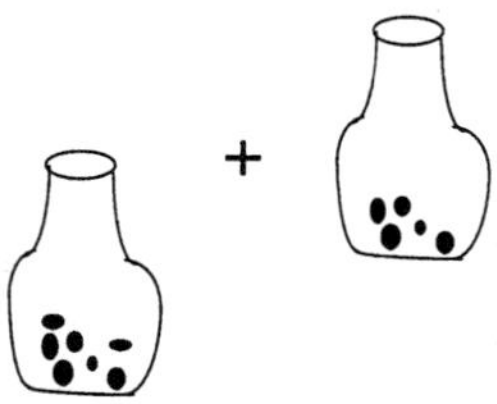

Our system of counting works essentially the same way, except that we carry the pebbles in our head, identifying them with number names, then we send our children to school to learn algorithms for adding and subtracting our abstract pebbles. In terms of the learning curve, the pebble method is far more efficient than our number system. However, as society became more complex with larger and larger quantities to be tracked, the pebbles, unlike our weightless numbers, were far too heavy and cumbersome to carry.

The clever merchants in ancient Sumeria cut down on the weight by drawing marks on empty clay pots. Later on they flattened the pots to clay tablets that could be stacked more efficiently. This simple act, like the invention of the wheel, changed the course of human history, for it was the birth of

$0 = \{\ \}$
$1 = \{*\}$
$2 = \{*\ *\}$
$3 = \{*\ *\ *\}$
$4 = \{*\ *\ *\ *\}$
$5 = \{*\ *\ *\ *\ *\}$
$6 = \{*\ *\ *\ *\ *\ *\}$
$7 = \{*\ *\ *\ *\ *\ *\ *\}$
$\ldots$

writing, centuries before words were written. Mathematics was not the product of an advanced civilization; mathematics was the stimuli, the intellectual catalyst that helped civilizations become advanced.

Numbers are not needed to determine if two sets have the same size; however, it is more convenient for commerce to have a standardized reference set for naming the various sizes. Like sacks of pebbles, the natural numbers serve as a collection of reference sets for all possible sizes of finite sets, as illustrated on the left.

Since those early days when commerce was transacted with pebble arithmetic, business deals can now be transacted on a Palm Pilot via satellite connection with someone in outer Mongolia. The amazing technological feats of the 20th century had their humble origins in the intellectual quest of the ancient Greeks to provide a logical foundation for geometry. Over two thousand years later, mathematicians at the turn of the 20th century focused their attention on developing a similar logical foundation for the natural numbers. Since the natural numbers are the foundation of quantitative reasoning, it may seem strange that the task was not tackled earlier. Perhaps the more physical aspect of the natural numbers with their pebble twins made them more concrete and less suspicious than the invisible points of geometry. Or perhaps the deep thinkers saw the difficulty in building a logical foundation for natural numbers because it would require that they take on the great challenge of infinity.

My theory stands as firm as a rock; every arrow directed against it will return quickly to its archer. How do I know this? Because I have studied it from all sides for many years; because I have examined all objections which have ever been made against the infinite numbers; and above all because I have followed its roots, so to speak, to the first infallible cause of all created things.

Georg Cantor
1845–1918

Since antiquity, the great nemesis of logic was the vast concept of infinity, which produced disturbing paradoxes. Like the Sumerian clay pots which could never contain an infinite number of pebbles, it was widely believed that a set could not contain an infinite number of elements.

Georg Cantor, though, considered a set as "a collection of definite, distinguishable objects of perception or thought conceived as a whole," and he saw no reason not to conceive of the collection of all natural numbers as a single entity labeled as a set. As he probed and explored the logical ramifications of his idea, he developed an intriguing theory of sets that included both infinite sets and infinite numbers. His ideas met with great resistance for many years. However, his theory provided answers to very deep questions in analysis, which led to its general acceptance. Today, we speak of infinite sets and infinite numbers with the same logical confidence as with their more worldly finite counterparts.

We will briefly examine the basic axioms that support Cantor's Theory of the Infinite. These axioms are designed to produce a sufficiently rich collection of sets without granting set status to the super large collections, like the one that produced Russell's paradox (page 206). Naturally, the axioms must be rich enough so that we can construct the natural numbers from sets instead of pebbles.

The wording in the following four axioms may seem a little awkward, but keep in mind that we are constructing representative sets for counting. What we build the sets from is not important; it doesn't matter if we use a sack of pebbles, strokes on a tablet, or abstract sets. The only thing that matters is that:

- We construct a representative set for each possible size that a finite set may have.
- We establish a pattern in our construction that enables us to continue forever. . . .

The latter requirement is where the great challenge lies, for it is where infinity is conceived. Using pebbles, we could never construct an infinite set of numbers, which is the advantage of using abstract sets. Some primitive cultures have a conception of only three numbers – one, two, and "many." Those with more developed systems of trading have more numbers, but most primitive cultures have a cutoff, an upper bound, beyond which they have no desire to distinguish between the sizes; and so they classify sets beyond the range of their numbers with a generic name such as "many," giving a one-size-fits-all solution to the naming problem. In our culture, most people use "infinity" in the same sense, as the size of a set whose size is beyond comprehension; however, as we will see in the next section, the well-reasoned mind can distinguish between different sizes of infinity.

The following method for creating the natural numbers may seem excessively laborious, but great precision is needed to logically escape the finiteness of mortal existence. After all, no human can count all the natural numbers. Even the days of existence for the sun that fuels our solar system is a finite number. The belief in an infinite set of natural numbers requires logical articles of faith, which can be boiled down to the following axioms. These axioms were the intellectual product of Georg Cantor's journey in following to its roots "the first infallible cause of all created things."

1. Axiom of Existence

There exists a set that has no elements.

$\exists S\ \forall x,\ x \notin S$

The first axiom of set theory postulates the existence of the empty set. Like the Big Bang, this axiom gives us all the material that we need to start building our universe. In the beginning was the empty set, and from this set we will build all our sets and numbers. In the sack and pebble method, we would represent the number 0 with an empty sack, which is analogous to an empty set. In a similar manner, we define the number 0 to be the empty set:

$$0 = \emptyset$$

2. Axiom of Equality

Let A and B be sets.

$A = B$

if and only if

$\forall x,\ x \in A \iff x \in B$

The second axiom gives the rule for determining when two sets are equal: two sets are equal if and only if they have the same elements. This axiom gives information on how we can manipulate the undefined terms, "set" and "is an element of."

As in any living organism, we need a reproductive system for producing new sets. The next two axioms allow us to build new sets from sets that we already have. Given two sets A and B, we can use Axiom 3 to build a new set: $\{A, B\}$. We can also use this axiom to build singleton sets by letting $B = A$:

$$\text{There exists a set } S \text{ such that } S = \{A, A\}.$$

3. Pair Axiom

Let A and B be sets.

There exists a set S such that $S = \{A, B\}$.

By the Axiom of Equality, $\{A, A\} = \{A\}$. Thus, from a set A, we can build a new set, $\{A\}$. This building technique will be used at each step in our construction of the natural numbers. Since we defined 0 as a set, we can form the set $\{0\}$, which is analogous to a sack with one pebble. Consequently, we define the number 1 to be this set:

$$1 = \{0\}$$

Since we defined both 0 and 1 as sets, we can use Axiom 3 to form the set $\{0, 1\}$, which is analogous to a sack with two pebbles. We define the number 2 to be this set:

$$2 = \{0, 1\}$$

4. Union Axiom

Let F be a set whose elements are sets.

There exists a set S such that

$S = \{\, x \mid x \in A \text{ for some } A \text{ in } F\}$

Alas, Axiom 3 will not get us to the number 3. We need another article of faith that allows us to produce larger sets, which is provided by the adjacent axiom. In Axiom 4, the set S is the multiple union of all sets in F (page 246):

$$S = \bigcup_{A \in F} A$$

The casual eye might miss the powerful control exerted at the beginning of Axiom 4 – the requirement that F be a set – which restricts the sets that we construct from getting too large.

Given a collection F of sets, we cannot arbitrarily form their union and call it a set. However, if we also know that F is a set, the Union Axiom allows us to do this. With this axiom, we can continue our construction of the natural numbers.

To construct a set with 3 elements, we first form the set $\{2\}$. Then we define the number 3 to be the union of the set 2 with the set $\{2\}$:

$$3 = 2 \cup \{2\} = \{0,1\} \cup \{2\} = \{0,1,2\}$$

Continuing in a recursive pattern, we define each new number in terms of the previous number:

$$4 = 3 \cup \{3\} = \{0,1,2\} \cup \{3\} = \{0,1,2,3\}$$
$$5 = 4 \cup \{4\} = \{0,1,2,3\} \cup \{4\} = \{0,1,2,3,4\}$$

At each step, the new natural number n is the following set:

$$n = \{0,1,2,3, \ldots, n-1\}$$

We notate its successor as $n+1$ and define it as follows:

$$n+1 = n \cup \{n\}$$

By Axiom 3, $\{n\}$ is a set and $\{\, n, \{n\} \,\}$ is a set. Using Axiom 4, we take the union over the latter set, which gives the desired result – a set with one more element:

$$\begin{aligned} n+1 &= \{0, 1, 2, 3, \ldots, n-1\} \cup \{n\} \\ &= \{0,1,2,3, \ldots, n-1, n\} \end{aligned}$$

Since the above set has $n + 1$ elements, it serves the same function as a sack with $n+1$ pebbles.

Let's run through the creation of the natural numbers in the style of Genesis, as portrayed on the left. If we believe that for any day, there is a next day, then this process can be continued forever. Through these verbal acrobatics, we now have a collection of sets with each a different size and with all possible sizes of finite sets represented. This step-by-step construction of the natural numbers gives them an inherent order which can be described by the subset relation:

$$n \le m \text{ if and only if } n \subseteq m.$$

Having justified the logical creation of the natural numbers, we can now forget about the rather complicated birthing process and view them as separate mature entities ordered by $\le$.

In the beginning was the empty set. Let us name this set 0.

$$0 = \{\ \}$$

On the first day, we put 0 in a set. Let us name this new set 1.

$$1 = \{0\}$$

On the next day, we take the two sets that we have created and put them in a new set. Let us name this set 2.

$$2 = \{0,1\}$$

On the next day, we form a set whose members are the previous sets. Let us name this set 3.

$$3 = \{0,1,2\}$$

Continuing in the same pattern, on the next day after the nth day, we form a set whose members are the previous sets. Let us name this set $n+1$.

$$n+1 = \{0,1,2,\ldots,n\}$$

Defining Sets

Axiom 5 empowers us to construct sets via the property method. For example, from the previous axioms we can form the sets, $A = \{0, 1, 2, 3\}$ and $C = \{0, 1, 2, 3, \ldots, 100\}$. Using Axiom 5, we can form the following subset of C.

$$B = \{ x \mid x \in C \text{ and } x \notin A \}$$

---

5. The Property Axiom

Let $p(x)$ be an open statement and let C be a set.

There exists a set S such that

$S = \{ x \mid x \in C \text{ and } p(x) \}$

The Property Axiom contains the restrictive clause "$x \in C$" in order to keep logical control of the sets that can be constructed via the property method. When we define a set S in terms of a property, each element in S must be a member of a set C that we already know exists. Otherwise, the property method cannot be used to grant the status of "set" to a collection. This subtle technicality eliminates Russell's Paradox (page 206), which was based on the following "set:"

$$V = \{ x \mid x \text{ is a set and } x \notin x \}$$

Under the new rules for set theory, V cannot be classified as a set, which defuses the paradox.

The axioms of set theory give a hierarchical technique for constructing sets. From sets that we already have, we can build more sets. Any property can be used to define a set, but it can only be applied to a set that we already have, which is why we have the notion of a universal set implicit in each property definition of a set. Given a set A, we cannot form the set of all x such that $x \notin A$, but given a universal set U, we can form the set of all x such that $x \in U$ and $x \notin A$.

Using these five axioms, we can construct a rich collection of finite sets. However, to construct infinite sets, we need another axiom, which we will discuss in the next section.

Cardinal Numbers

The natural numbers were constructed as standardized reference sets for all possible sizes of finite sets. The size of a set is called its *cardinal number*. If a set S has the same size as $\{1, 2, 3, \ldots, n\}$, then n is the cardinal number of S, which we notate as $|S|$. The empty set has 0 elements: $|\emptyset| = 0$.

$|S| = n$

if and only if

there exists a one-to-one function that maps $\{1, 2, 3, \ldots, n\}$ onto S.

To determine if S has n elements, we try to find a one-to-one function that maps $\{1, 2, 3, \ldots, n\}$ onto S. This definition uses the same counting technique we learned in first grade. When we count the elements in a set S, we are constructing a one-to-one function that maps $\{1, 2, 3, \ldots, n\}$ onto S. Actually, there will be quite a few functions that fit the one-to-one and onto requirement, for it doesn't matter where we start counting as long as we count all the elements in S.

Finite Sets

We can now give a simple definition of a finite set. Since 0 is not a natural number, we introduce the term *whole number* to indicate a number that is either a natural number or 0.

To say that a set is finite means that

$|S| = n$ for some whole number n.

S is *finite* if and only if $|S| = n$ for some whole number n.

A nonempty set is finite if and only if we can find a natural number n and a one-to-one function f that maps $\{1, 2, 3, \ldots, n\}$ onto S. We have traced the notion of finite back to the following four sources: natural numbers, the function concept, the one-to-one concept, and the onto concept.

From a mathematical perspective, an extremely large set, such as a set that has 10^{80} elements, is a finite set. When we say finite, we do not necessarily mean "small."

One-to-One Functions

A one-to-one function between two sets gives us information on the relative sizes of the sets. In the adjacent sketch, f is a one-to-one function that maps A into B. Note that $|A| \leq |B|$.

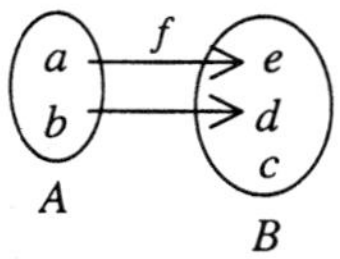

Suppose that f is a one-to-one function that maps A into B and A has n elements. When we count the elements in A, we can use the one-to-one mapping to simultaneously count the corresponding elements in B. So, B must have at least n elements. In general, whenever we have a one-to-one function that maps A into B, we can deduce that the size of A is less than or equal to the size of B.

On the other hand, if we are given sets with $|A| \leq |B|$, we can construct a one-to-one function that maps A into B. For example, suppose that A has n elements and B has m elements. Since A has n elements, there exists a one-to-one function g that maps $\{1, 2, 3, \ldots, n\}$ onto A. We can use g to index the elements in A. Label the elements in A as follows:

$$a_1 = g(1),\ a_2 = g(2), \ldots, \text{and } a_n = g(n).$$

$$A = \{a_1, a_2, \ldots, a_n\}$$

Let A and B be finite sets. $|A| \leq |B|$ if and only if there exists a one-to-one function that maps A into B.

We can index the elements in B in an analogous manner:

$$B = \{b_1, b_2, \ldots, b_n, \ldots, b_m\}$$

Define the function f on A as follows:

$$f(a_i) = b_i$$

f is a one-to-one mapping from A into B. Thus, to say that the size of A is less than or equal to the size of B guarantees the existence of a one-to-one function that maps A into B.

Onto Functions

Onto functions also give us information on the relative sizes of two sets. In the adjacent sketch, g is a function that maps A onto B. Note also that $|A| \geq |B|$.

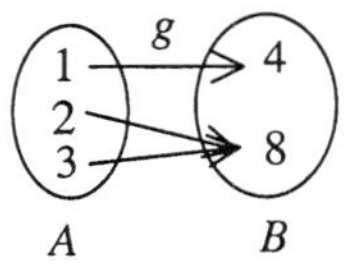

In an onto function, each element in the second set must have at least one element mapped to it. If we reverse the arrows, we may not have a function, but we can throw away extra arrows that emanate from the same source and construct a one-to-one mapping from B into A. Using our results for one-to-one mappings, we can deduce that $|B| \leq |A|$, which can be written as $|A| \geq |B|$. In general, whenever we have a function that maps A onto B, we can deduce that the size of A is greater than or equal to the size of B.

The converse is also true. Suppose that $|A| = n$ and $|B| = m$ and $n \geq m$. Using counting functions, we can index these sets as follows:

> Let A be a finite set and B a nonempty set.
> $|A| \geq |B|$
> if and only if
> there exists a function that maps A onto B.

$$
\begin{array}{l}
A = \{\, a_1,\ a_2,\ \ldots,\ a_m,\ \ldots,\ a_n \,\} \\
\qquad\ \downarrow\quad \downarrow \qquad\quad \downarrow \swarrow \\
B = \{\, b_1,\ b_2,\ \ldots,\ b_m \}
\end{array}
$$

We can construct a function from A onto B using the above arrows. We map each element in A to the corresponding element in B until we get to a_m and run out of corresponding elements. After that point, we map each element in A to b_m. We can formally define this function as follows:

$$
\begin{aligned}
f(a_i) &= b_i, && \text{if } i \leq m \\
f(a_i) &= b_m, && \text{if } i > m
\end{aligned}
$$

The function f maps A onto B. Thus, to say that the size of a set A is greater than or equal to the size of a set B guarantees the existence of a function that maps A onto B.

One-to-One and Onto

If f is a function from A into B is both one-to-one and onto, we can make the following deductions:

> $|A| = |B|$
> if and only if
> there exists a one-to-one function that maps A onto B.

$$|A| \leq |B| \text{ and } |A| \geq |B|$$

Thus, $|A| = |B|$. The condition under which $|A| = |B|$ is given in the adjacent box (page 273). The one-to-one property makes $|A| \leq |B|$. The onto property makes $|A| \geq |B|$.

Knowing that a function between two sets is one-to-one does not guarantee that the function is onto. However, if we know in addition that the sets are finite and have the same size, we can deduce that a one-to-one function must also be onto.

> Let A and B be finite sets with $|A| = |B|$ and f be a function from A into B.
>
> f is one-to-one if and only if f is onto.

Conversely, knowing that a function is onto does not guarantee that the function is one-to-one unless we also know that both sets are finite and have the same size. For functions between finite sets of the same size, the property of being one-to-one is equivalent to the property of being onto.

The adjacent statement is not true for infinite sets. For example, let $f(n) = n+1$. f is one-to-one and maps the set $\mathbb{N}$ of natural numbers into $\mathbb{N}$. However, f does not map onto $\mathbb{N}$, for there is no n in $\mathbb{N}$ such that $f(n) = 1$.

Sizes of Subsets

If A is a subset of B, the number of elements in A must be less than or equal to B: $|A| \leq |B|$. For finite sets, we have an analogous statement for "strictly less than." If we remove an element from a finite set, we have made its size smaller:

> Let A and B be finite sets.
>
> If $A \subset B$, then $|A| < |B|$.

$$\text{If } A \subset B, \text{ then } |A| < |B|.$$

The above property is not true for infinite sets. A proper subset of an infinite set can have the same size (page 287).

Exercise Set 3.5

1. Let $X = \{1, 2, 3, 4\}$ and $Y = \{a, b, c, d, e, f\}$.
   a. Does there exist a one-to-one function that maps X into Y?
   b. Does there exist a function that maps X onto Y?
2. Let $X = \{1, 2, 3, 4\}$ and $Y = \{a, b, c\}$.
   a. Does there exist a one-to-one function that maps X into Y?
   b. Does there exist a function that maps X onto Y?
3. Let X and Y be finite sets with $|X| < |Y|$.
   a. Does there exist a one-to-one function that maps X into Y?
   b. Does there exist a function that maps X onto Y?
4. Let X and Y be finite sets with $|X| > |Y|$.
   a. Does there exist a one-to-one function that maps X into Y?
   b. Does there exist a function that maps X onto Y?
5. Let X and Y be finite sets with $|X| = |Y|$.
   a. Does there exist a one-to-one function that maps X into Y?
   b. Does there exist a function that maps X onto Y?

6. Let $X = \{1, 2, 3, 4\}$ and $Y = \{a, b, c, d\}$. Does there exist a function f that maps X into Y with the following properties?
   a. f is one-to-one, but not onto.
   b. f is onto, but not one-to-one.
7. Let X and Y be finite sets. Let f be a function that maps X into Y.
   a. If f is one-to-one, does f have to be onto?
   If not, give a counterexample.
   b. If f maps X onto Y, does f have to be one-to-one?
   If not, give a counterexample.
   c. What if X and Y have the same number of elements?
   Will this change your answers in (a) and (b)?
8. Let X and Y be finite sets. What does the given information tell you about the relation between $|X|$ and $|Y|$?
   a. There exists a function that maps X onto Y.
   b. There exists a one-to-one function that maps X into Y.
   c. There exists a one-to-one function that maps X onto Y.
   d. X is a proper subset of Y.
9. *Pigeonhole Principle*: Suppose that p pigeons fly into h pigeonholes. Let f map each pigeon to the pigeonhole where it lands.
   a. Is f a function?
   b. If $p > h$, what information does this give you about f?
   c If $p < h$, what information does this give you about f?
10. Let f be a function that maps $\{1, 2, 3, \ldots, n\}$ into S.
   What statement must you verify in order to prove the following?
   a. f maps $\{1, 2, 3, \ldots, n\}$ onto S.
   b. f does not map $\{1, 2, 3, \ldots, n\}$ onto S.
11. Let S be a set and n a natural number.
   What statement must you verify in order to prove the following?
   a. S has n elements.
   b. S does not have n elements.
12. Let $\mathbb{N} = \{1, 2, 3, \ldots.\}$. Do you think that the given set has the same size as $\mathbb{N}$, a smaller size, or a larger size? Explain your rationale.
   a. $W = \{0, 1, 2, 3, \ldots\}$
   b. $A = \{2, 3, 4, 5, \ldots\}$
   c. $O = \{2, 4, 6, \ldots\}$
   d. $Z = \{\ldots, -3, -2, -1, 0, 1, 2, 3, \ldots\}$
13. Using the definition of "same size" on page 273, rethink your answers in the previous exercise.

Activity 3.6 – Hilbert's Infinite Hotel

1. Hilbert constructed a grand hotel with an infinite number of rooms on a beautiful mountain overlooking the Mediterranean sea. Feeling confident that there would always be a spare room, Hilbert advertised his hotel with the slogan, "Always Room for One More." The hotel became quite popular with tourists, and one day the manager was elated to find that all the rooms were full. When another tourist walked in and asked for a room, the manager was about to turn the tourist away, but Hilbert interrupted and said, "There is always room for one more." Is this true? If so, explain how the current residents can be reassigned to accommodate the new arrival. The rooms are numbered with the natural numbers. A guest may be moved to a different room, but the guests are not allowed to double up.
2. The next day 100 new guests arrive. Can the manager accommodate them? If so, explain how to do it.
3. Several months later, the hotel is still full, but the manager now knows how to accommodate any new finite group of tourists that arrive. On a particularly beautiful spring day, the manager glances at the infinite dispenser roll on the counter and gasps. All the waiting numbers, N1, N2, N3, . . . , have been taken, which means there are an infinite number of new people waiting to get a room. Can the manager accommodate them? If so, explain how to do it.

≡ 3.6 Infinite Sets ≡

From antiquity until the latter part of the 19th century, mathematicians, philosophers, and theologians struggled with the concept of infinity with little success, for paradoxical statements seemed to surface from every corner. It seemed as though the infinite were beyond the grasp of logical inquiry. What was missing was the language that would give a precise description of this very massive concept. Thanks to Georg Cantor and the other mathematical pioneers who developed axiomatic foundations for set theory, we can now work with infinite sets in a logical manner and even construct infinite numbers that make as much sense as finite numbers.

6. Axiom of Infinity

Define $n+1$ as follows:

$n+1 = n \cup \{n\}$

There exists a set W that has the following properties:

- $\emptyset \in W$
- If $n \in W$, then $n+1 \in W$.

In the previous section, we created an infinite list of natural numbers, but we have not yet created an infinite set. We need more than a giant step to get from the finite to the infinite; we need another axiom – similar to the existence axiom for the empty set – that guarantees the existence of an infinite set. After we have one infinite set, we can then construct many others.

The infinite set chosen to be the mother of all infinite sets was the simplest set possible, the set $\{0, 1, 2, 3, \ldots\}$. We can construct as many natural numbers as we like, but we cannot construct the set of all natural numbers without the *Axiom of Infinity*, which is stated on the left. The wording of this axiom parallels the construction method of the natural numbers on page 278:

$$n+1 = n \cup \{n\} = \{0,1,2,3, \ldots, n\}$$

The set W whose existence is postulated in the Axiom of Infinity is the following set:

$$W = \{0,1,2,3, \ldots, n, n+1, \ldots\}$$

Using the property method, we can define the set $\mathbb{N}$ as follows:

$$\mathbb{N} = \{\, x \mid x \in W \text{ and } x \neq 0 \,\}$$

Let S be a set.
S is *infinite*
if and only if
S is not finite.

Now we have captured infinity within the confines of a set. In 1850, the prevalent feeling amongst the deep thinkers was that this should not be allowed. Today, it is perfectly acceptable because we have a logical foundation to support it. The axioms of set theory provide us with tools to logically construct all of the standard infinite sets that we work with in mathematics: the set Z of integers, the set Q of rational numbers, the set $\mathbb{R}$ of real numbers, and the set $\mathbb{C}$ of complex numbers, as well as higher dimensional spaces.

In the previous section, we logically constructed the natural numbers and then used these numbers to define a finite set. Using the definition of finite, we can easily define "infinite:"

An *infinite* set is a set that is not finite.

Let's work backwards through the definition of finite to see exactly what this definition means:

A set S is infinite
if and only if
$\sim (|S| = n$ for some natural number.)

Thus, for a set S to be infinite means the following:

For every natural number n, $|S| \neq n$.

We can translate this latter statement as follows:

For every natural number n, ~(there exists a one-to-one function that maps $\{1, 2, 3, \ldots, n\}$ onto S.)

For every natural number n and for every one-to-one function f, f does not map $\{1, 2, 3, \ldots, n\}$ onto S.

To prove that a set is infinite, we would need to verify that the latter statement is true. Like the counting process, notice how our logical understanding of infinite sets comes from the three basic concepts of function, one-to-one, and onto.

Now that we've opened the door to a strange, new universe where infinite sets exist, let's investigate some of the inhabitants. Since mathematics originated from the analysis of the various sizes of finite sets, let's do the same for infinite sets.

Sizes of Infinite Sets

Like primitive tribes who use the general size of "many" for sets whose sizes are beyond two, most people use a generic label for the size of any infinite set. As with clothing, the one-size-fits-all label tends to hide rather than reveal the structure of the object to which it is applied. When we apply a critical eye to the various sizes of infinite sets, we will see some rather shocking demographics on the inhabitants of the familiar number line.

For example, we only see rational numbers as we divide the unit intervals into tenths and then divide each subinterval into tenths. We are mentally comfortable with the exact location of the rational number 3.47569 even though we may not be able to distinguish it with a pencil point. It is easy to understand how one might believe that the rational numbers are the most populous group on a number line, for they are indeed the most visible. The theory of infinite sets, though, tells us otherwise. The rational numbers are a very small minority.

Let A and B be sets.
A has the same size as B
if and only if
there exists a one-to-one
function that maps A onto B.

To pursue this fascinating topic, we must first decide how to compare sizes of infinite sets. As we saw earlier, we do not need natural numbers to determine if two finite sets have the same size; we only need to find a one-to-one function that maps one set onto the other (page 273). We will use the use definition for infinite sets. To say that two infinite sets have the same size means that there exists a one-to-one function that maps one set onto the other set.

Cardinality of a Set

The size of a set A is called its *cardinality*. As with finite sets, we use the notation $|A|$ to denote the size of A, or the *cardinal number* of A. The natural numbers were constructed to provide reference sets for the various sizes of finite sets. We will do the same for infinite sets, which is way we call $|A|$ a "number."

Let A and B be sets.

$|A| = |B|$

if and only if

there exists a one-to-one

function that maps A onto B.

Using the $|A|$ notation, we can translate the sentence "A has the same size as B" as $|A| = |B|$. So, we can phrase the definition of "having the same size" as stated on the left. Does the use of the equals relation between infinite numbers corrupt the basic properties of equality? If it does, we cannot hope to form reference sets that will serve as infinite numbers. For finite numbers, equality has the following 3 basic properties, which we first state in terms of the $|A|$ notation, and then translate in terms of the adjacent equivalence:

Let A, B and C be sets.

Reflexive Property

$|A| = |A|$
There exists a one-to-one function that maps A onto A.

Transitive Property

If $|A| = |B|$ and $|B| = |C|$, then $|A| = |C|$.
If there exists a one-to-one function that maps A onto B and there exists a one-to-one function that maps B onto C, then there exists a one-to-one function that maps A onto C.

Symmetric Property

If $|A| = |B|$, then $|B| = |A|$.
If there exists a one-to-one function that maps A onto B, then there exists a one-to-one function that maps B onto A.

It is not difficult to prove the function version of each of the above properties for infinite as well as finite sets:

> For the reflexive property, we use the identity function.
>
> For the transitive property, if f is a one-to-one function that maps A onto B and g is a one-to-one function that maps B onto C, we can prove that the composition of these two functions is a one-to-one function that maps A onto C.
>
> For the symmetric property, we can prove that if f is a one-to-one function that maps A onto B, then we can reverse the arrows and obtain a one-to-one function that maps B onto A.

You are asked to prove each of these properties in Section 4.3.

Sizes of Subsets

Before we create our first infinite number, we must comment on a paradoxical sounding aspect of infinite sets:

A proper subset of an infinite set can have the same size as the original set.

$$\mathbb{N} = \{1,\ 2,\ 3, \ldots n, \ldots\}$$
$$\downarrow\ \downarrow\ \downarrow\quad\ \downarrow$$
$$A = \{2,\ 3,\ 4, \ldots n+1, \ldots\}$$

For example, let's remove one element from the set $\mathbb{N}$ of natural numbers: $A = \{2, 3, 4, \ldots\}$. A appears to have a smaller size than $\mathbb{N}$ since it has one less element. However, we can set up a one-to-one mapping from $\mathbb{N}$ onto A: $f(n) = n+1$. So, A has the same size as $\mathbb{N}$, even though it is a proper subset of $\mathbb{N}$.

$$A \subset \mathbb{N} \text{ and } |A| = |\mathbb{N}|.$$

A set is *infinite* if and only if it has a proper subset of the same size.

B. Bolzano (1781–1848) noticed that this strange behavior happens with all infinite sets. In fact, it gives a property that distinguishes infinite sets from finite sets. By removing one element from an infinite set, we will always produce a proper subset that has the same size as the original set. If we take the viewpoint that infinite sets do not have to behave the same way as finite sets, the paradoxical nature of the following property disappears.

S is an infinite set
if and only if
there exists a set A such that $A \subset S$ and $|A| = |S|$.

The Size of $\mathbb{N}$

$\aleph_0$ repesents the number of elements in $\mathbb{N}$.

The ancient Sumerians represented the property shared by sets having the same size as $\{1,2,3\}$ with 3 pebbles in a clay pot. Later, a name was created for this size. We will do an analogous representation for the property shared by sets who have the same size as the set $\mathbb{N}$. In fact, we will use $\mathbb{N}$ as the reference set, but we will use another name to indicate when we are using it as a cardinal number. Georg Cantor selected $\aleph_0$ (aleph-null) to represent the size of the set $\mathbb{N}$ of natural numbers. Its name may appear strange, but "3" would also appear strange if we were not already familiar with it. Since $\aleph$ is the first letter of the Hebrew alphabet, $\aleph_0$ is an appropriate name for the beginning of a sequence of infinite numbers.

A set S has $\aleph_0$ elements
if and only if
there exists a one-to-one function that maps $\mathbb{N}$ onto S.

If we remove a finite number of elements from $\mathbb{N}$, we still have $\aleph_0$ elements. For example, if we remove the first 4 elements from $\mathbb{N}$, we get the following set:

$$\mathbb{N} = \{1, 2, 3, \ldots n, \ldots\}$$
$$\downarrow\ \downarrow\ \downarrow\ \quad \downarrow$$
$$B = \{5, 6, 7, \ldots, n+4, \ldots\}$$

$$B = \{5, 6, 7, \ldots\}$$

As illustrated on the left, we can construct a one-to-one function from $\mathbb{N}$ onto B: $f(n) = n + 4$. So, B has the same size as $\mathbb{N}$, which means that $|B| = \aleph_0$.

We can even remove an infinite number of elements from $\mathbb{N}$ and still have $\aleph_0$ elements. For example, if we remove the odd numbers from $\mathbb{N}$, we get the following set:

$$\mathbb{N} = \{1, 2, 3, \ldots n, \ldots\}$$
$$\downarrow\ \downarrow\ \downarrow\ \quad \downarrow$$
$$E = \{2, 4, 6, \ldots 2n, \ldots\}$$

$$E = \{2, 4, 6, 8, \ldots\}$$

As illustrated on the left, the function, $f(n) = 2n$, is a one-to-one mapping from $\mathbb{N}$ onto E. So, $|E| = \aleph_0$.

An infinite set S whose elements can be listed in sequence form has $\aleph_0$ elements. Suppose that we can list the elements of S as follows where no term is repeated:

$$\mathbb{N} = \{1, 2, 3, \ldots, n, \ldots\}$$
$$\downarrow\ \downarrow\ \downarrow\ \quad \downarrow$$
$$S = \{a_1, a_2, a_3, \ldots, a_n, \ldots\}$$

$$S = \{a_1, a_2, a_3, \ldots\}$$

To construct a one-to-one function that maps $\mathbb{N}$ onto S, we map 1 to a_1, 2 to a_2, 3 to a_3, and in general, n to a_n: $f(n) = a_n$. f is a one-to-one function from $\mathbb{N}$ onto S, so $|S| = \aleph_0$.

Conversely, any set S with $\aleph_0$ elements can be listed in sequence form by using a one-to-one function f from that maps $\mathbb{N}$ onto S:

$$S = \{f(1), f(2), f(3), \ldots\}$$

Countable Sets

A set S that has has the same size as the set $\mathbb{N}$ of natural numbers is called *countably infinite*. We use the term *countable* to cover both finite sets and countably infinite sets.

S is *countable* if and only if
S is finite or
has the same size as $\mathbb{N}$.

The set $\{5,8,7\}$ is a countable set that is finite, whereas the set $\{3, 6, 9, \ldots\}$ is a countable set that is infinite. A set that is not countable is called *uncountable*. The term "uncountable" represents a higher level of infinity, a level beyond the size of the natural numbers.

Do Uncountable Sets Exist?

Do there exist sets whose size is larger than $\mathbb{N}$? Or is there only one level of infinity? Consider the set Z of integers, which appears to be "twice" as large as $\mathbb{N}$.

$$Z = \{\ \ldots, -3, -2, -1, 0, 1, 2, 3, \ldots\ \}$$

$$\begin{array}{l} \mathbb{N} = \{1,\ 2,\ 3,\ 4,\ 5,\ 6,\ 7\ldots\} \\ \quad\ \ \downarrow\ \downarrow\ \downarrow\ \downarrow\ \downarrow\ \downarrow\ \downarrow \\ Z = \{0,\ 1, -1,\ 2, -2,\ 3, -3, \ldots\} \end{array}$$

Is Z uncountable, or does there exist a one-to-one mapping of $\mathbb{N}$ onto Z? We cannot use the technique from the previous page of mapping 1 to the first element on the left because Z does not have a first element. However, we can cross-stitch the counting process by starting at 0, then count 1, next –1, then 2, then to –2, alternating from side to side, as illustrated on the left. To give a formula for this mapping, note that the even numbers map to the positive integers with a simple pattern. The even number $2n$ is mapped to n:

$$f(2n) = n.$$

The next odd number on the right side of $2n$ is mapped to the negative of $f(2n)$:

$$f(2n+1) = -n.$$

$$A = \{a_1, a_2, a_3, \ldots\}$$
$$B = \{b_1, b_2, b_3, \ldots\}$$
$$A \cup B = \{a_1, b_1, a_2, b_2, a_3, b_3, \ldots\}$$

f is a one-to-one function that maps $\mathbb{N}$ onto Z, so Z is countably infinite. Even though Z may appear to have twice as many elements as $\mathbb{N}$, they have the same size. Z is a countable set.

Using a similar technique, we can prove that the union of any two countable sets is countable by interweaving the elements as illustrated on the left.

Theorem If A and B are countable sets, then $A \cup B$ is a countable set.

Using mathematical induction, we can generalize the above theorem to any finite union of countable sets:

Theorem Let n be a natural number. If $A_1, A_2, A_3, \ldots, A_n$ are countable sets, then $A_1 \cup A_2 \cup A_3 \cup \ldots \cup A_n$ is a countable set.

Proof Let $p(n)$: If $A_1, A_2, A_3, \ldots, A_n$ are countable sets, then $A_1 \cup A_2 \cup A_3 \cup \ldots \cup A_n$ is a countable set.

Let n be a natural number. Assume that $p(n)$ is true.

Let $A_1, A_2, \ldots, A_{n+1}$ be countable sets.
Since $p(n)$ is true, $A_1 \cup A_2 \cup A_3 \cup \ldots \cup A_n$ is a countable set.

Since the union of two countable sets is countable, the following set is countable:

$$(A_1 \cup A_2 \cup A_3 \cup \ldots \cup A_n) \cup A_{n+1}$$

Thus, $p(n+1)$ is true.

Therefore, for every natural number n, $p(n) \Rightarrow p(n+1)$.

$p(1)$: If A_1 is countable, then A_1 is countable.
$p(1)$ is obviously true.

So, by math induction, $p(n)$ is true for all natural numbers n.

Countable Unions of Countable Sets

Let's push a little further and see what happens when we union a countably infinite collection of countable sets. If the sets, A_1, A_2, A_3, . . . , A_n, . . ., are each countable, is the following set countable?

$$A_1 \cup A_2 \cup A_3 \cup \ldots \cup A_n \cup \ldots$$

$A_1 = \{ \frac{k}{1} \mid k \text{ is an integer} \}$

$A_2 = \{ \frac{k}{2} \mid k \text{ is an integer} \}$

$A_3 = \{ \frac{k}{3} \mid k \text{ is an integer} \}$

. . .

$A_n = \{ \frac{k}{n} \mid k \text{ is an integer} \}$

. . .

For example, consider the set Q of all rational numbers. Since each rational number can be expressed in fraction form, we can divide the rational numbers into the adjacent sets, according to the denominator of the fraction. A_2 contains all the rational numbers that can be represented as a fraction that has 2 in the denominator.

$$A_2 = \{ \ldots, \tfrac{-3}{2}, \tfrac{-2}{2}, \tfrac{-1}{2}, \tfrac{0}{2}, \tfrac{1}{2}, \tfrac{2}{2}, \tfrac{3}{2} \ \ldots \}$$

The function, $f(n) = \frac{n}{2}$, is a one-to-one mapping from Z onto A_2. Since Z is countable, A_2 is countable. Similarly, each set in the adjacent list is countable. Furthermore,

$$Q = \bigcup_{n \in \mathbb{N}} A_n$$

So, Q is a countable union of countable sets. Does Q have the same size as $\mathbb{N}$? Think of how the rational numbers are situated on a number line. Between any two rational numbers, no matter how close together they are, there is another rational number. In fact, between any two rational numbers, there are an infinite number of rational numbers. We have much more to deal with here than the two infinite tails in the set of integers. How on earth can we start counting them so that we count each and every one of them in a step-by-step manner? The cross-stitch technique that we used with the integers will not suffice here. One might suspect that the task is impossible. However, using a diagonal stitch, Georg Cantor found a clever way to construct a one-to-one counting procedure that covers all the elements in a countable union of countable sets.

Theorem A countable union of countable sets is countable.

Proof Suppose that B_n is a countable set for each natural number n. First, we remove any overlap in these sets as indicated on the left. In B_2, we remove any overlap with B_1, then label the new set as A_2. In B_3, we remove any overlap with $B_1 \cup B_2$, then label the new set as A_3. Continuing in this pattern, we produce a collection of mutually disjoint sets $A_1, A_2, \ldots, A_n, \ldots$, whose union is equal to the original union: $\bigcup_{n \in \mathbb{N}} A_n = \bigcup_{n \in \mathbb{N}} B_n$. Some of the A_ns may be finite or empty. Let's suppose that they are each infinite, which is the worst case scenario. If we can show that the union is countable for this case, then we can deduce that it is also true if some of the sets are finite or empty. Label the elements in each A_n as follows:

$A_1 = B_1$

$A_2 = B_2 - B_1$

$A_3 = B_3 - (B_1 \cup B_2)$

$\ldots$

$A_n = B_n - (B_1 \cup B_2 \cup \ldots \cup B_n)$

$\ldots$

$$A_n = \{a_{n1}, a_{n2}, a_{n3}, a_{n4}, a_{n5}, \ \ldots\}$$

In the adjacent infinite array, the elements in A_1 are listed in the first row, the elements in A_2 are listed in the second row, etc. Now, we are ready to count all of the elements in the adjacent array using the following diagonal stitch.

$$\begin{array}{ccccccc}
a_{11} & a_{12} & a_{13} & a_{14} & a_{15} & a_{16} & \cdots \\
a_{21} & a_{22} & a_{23} & a_{24} & a_{25} & a_{26} & \cdots \\
a_{31} & a_{32} & a_{33} & a_{34} & a_{35} & a_{36} & \cdots \\
a_{41} & a_{42} & a_{43} & a_{44} & a_{45} & a_{46} & \cdots \\
a_{51} & a_{52} & a_{53} & a_{54} & a_{55} & a_{56} & \cdots \\
a_{61} & a_{62} & a_{63} & a_{64} & a_{65} & a_{66} & \cdots \\
\cdots & \cdots & \cdots & \cdots & \cdots & \cdots & \cdots
\end{array}$$

Start in the upper left corner: a_{11}

Move to the next element in the top row and count down its diagonal towards the left side: a_{12}, a_{21}

Move to the next element in the top row and count down its diagonal towards the left side: a_{13}, a_{22}, a_{31}

Continuing in the same diagonal pattern, we can specify a countable listing of all the elements in the union:

$$\bigcup_{n \in \mathbb{N}} A_n = \{\underline{a_{11}},\ \underline{a_{12}, a_{21}},\ \underline{a_{13}, a_{22}, a_{31}},\ \underline{a_{14}, a_{23}, a_{32}, a_{41}},\ \ldots\}$$

Note the pattern in each diagonal – the sum of the two subscripts is constant. Using the above listing, we map n to the nth element. This mapping is a one-to-one function from $\mathbb{N}$ onto $\bigcup_{n \in \mathbb{N}} A_n$. So, $\bigcup_{n \in \mathbb{N}} A_n$ is a countable set.

Hence, $\bigcup_{n \in \mathbb{N}} B_n$ is countable.

The set of rational numbers is countable.

Since the set Q of rational numbers can be expressed as the countable union of a collection of countable sets, we can deduce from the previous theorem that Q is countable. The set of rational numbers has the same size as the set of natural numbers: $|Q| = \aleph_0$.

Size of $\mathbb{R}$

After the last theorem, we may be inclined to think that there is only one size of infinity – that every infinite set can be put into a one-to-one correspondence with the set of natural numbers. However, before we jump to an unsupported conclusion, let's look a little closer at the set $\mathbb{R}$ of real numbers.

The set of real numbers can be visualized as all the points on a number line. Like the rational numbers, they are densely distributed. Between any two real numbers, there are an infinite number of real numbers. This dense distribution did not keep us from finding a one-to-one function that maps $\mathbb{N}$ onto Q, but can we do the same for $\mathbb{R}$?

Suppose that f is a one-to-one function that maps $\mathbb{N}$ onto $\mathbb{R}$. This function gives a sequence of real numbers:

$$f(1), f(2), f(3), \ldots, f(n), \ldots$$

Is it possible that every real number is contained in the above list? Consider the decimal form of the above real numbers. Let's make up an example and suppose that f has the following values:

$$\begin{aligned} f(1) &= 32.4572876\ldots \\ f(2) &= 5.3766237\ldots \\ f(3) &= 7.2418621\ldots \\ f(4) &= 102.9837436 \\ f(5) &= 45.4543582\ldots \\ f(6) &= 331.3021737\ldots \end{aligned}$$

$$\begin{array}{cccccccc} . & \boxed{4} & 5 & 7 & 2 & 8 & 7 & 6 & \ldots \\ . & 3 & \boxed{7} & 6 & 6 & 2 & 3 & 7 & \ldots \\ . & 2 & 4 & \boxed{1} & 8 & 6 & 2 & 1 & \ldots \\ . & 9 & 8 & 3 & \boxed{7} & 4 & 3 & 6 & \ldots \\ . & 4 & 5 & 4 & 3 & \boxed{5} & 8 & 2 & \ldots \\ . & 3 & 0 & 2 & 1 & 7 & \boxed{3} & 7 & \ldots \\ . & \ldots & \ldots & \ldots & \ldots & \ldots & \ldots & \ldots & \ldots \end{array}$$

In the adjacent array, the digits to the right of each decimal point are entered, with the digits for $f(1)$ on the first row, the digits for $f(2)$ on the second row, etc. By contemplating this array, in particular, the digits on the diagonal, Georg Cantor saw a way to construct a real number that could not be in the list. His technique was to construct a real number y by making each digit in y different from the corresponding digit in the diagonal. For example, we could define each digit in y to be 1 except in those positions where 1 appears in the diagonal. In those positions, we could use 2. In the adjacent example, the first 6 digits of y would be the following: $y = 112111\ldots$

Using this technique, we can prove that $\mathbb{R}$ is so large that it is not possible for a function to map $\mathbb{N}$ *onto* $\mathbb{R}$.

Theorem There does not exist a function that maps the set $\mathbb{N}$ of natural numbers onto the set $\mathbb{R}$ of real numbers.

Proof Suppose that f is a function that maps $\mathbb{N}$ into the set $\mathbb{R}$. Then $f(1)$ is a real number, $f(2)$ is a real number, etc. Consider the list of all these real numbers:

$$f(1), f(2), f(3), \ldots\ f(n), \ldots$$

We will now construct a real number y that is not in this list. Let y_n denote the digit in the nth decimal place of y:

Set $y_n = 1$ if the digit in the nth decimal place of $f(n)$ is not 1.

Set $y_n = 2$ if the digit in the nth decimal place of $f(n)$ is 1.

Let $y = .y_1 y_2 y_3 y_4 \ldots$

The set of real numbers is uncountable.

Since y is a decimal, y is a real number. Furthermore, for every natural number n, $y \neq f(n)$ since they have different digits in the nth decimal place. We can make this deduction because the decimal representation of $f(n)$ is unique, except for 9s and 0s:

$$1.00000\ldots = .99999\ldots.$$

If the nth digit in $f(n)$ is either 9 or 0, the corresponding digit in $f(n)$ is 1. So $y \neq f(n)$.

Thus, the function f does not map $\mathbb{N}$ *onto* $\mathbb{R}$.

Since it is not possible to have a function that maps $\mathbb{N}$ onto $\mathbb{R}$, $\mathbb{N}$ and $\mathbb{R}$ are not the same size. Therefore, the number of real numbers is not $\aleph_0$; it is a larger infinite number. At last, we have an uncountable set, a set whose size is greater than the size of the set of natural numbers. The set of real number is uncountable.

The Irrational Numbers

Which is bigger, the set of rational numbers or the set of irrational numbers? Since we do not run into the irrational numbers as often as the rational numbers, we might suspect that there are more rational numbers than irrational numbers, but, surprisingly, this is not the case.

Theorem The set of irrational numbers is uncountable.

Proof Let Q be the set of rational numbers and S the set of irrational numbers.

$$\mathbb{R} = Q \cup S$$

Suppose that the set of irrational numbers is countable.

The set of rational numbers is countable.
The union of two countable sets is countable.
Therefore, $\mathbb{R}$ is countable. Contradiction!

So, the set of irrational numbers is uncountable.

The larger magnitude of the set of real numbers comes from the set of irrational numbers. Thus, in the demographics of the number line, the rational numbers, like stars in the night sky, may be more visible, but they are nonetheless a minuscule minority. The visibility of the rational numbers comes from the accessibility of their decimal form, which must eventually have a cycle that repeats. By working through the long division of dividing 1 by 7, we can see why this has to happen. After we start bringing down 0s, sooner or later one of the remainders has to repeat, which sets up the cycle. Thus, an irrational number is a number whose decimal form does not have a repeating cycle. As the above theorem demonstrates, all the decimals that do not have a repeating cycle form an uncountable set.

Extending $\leq$ For finite sets, $|A| \leq |B|$ if and only if there exists a one-to-one function that maps A *into* B (page 281). We use this same property to extend the meaning of the $\leq$ relation to infinite numbers.

Let A and B be sets.

$$|A| \leq |B|$$

if and only if

there exists a one-to-one function that maps A into B.

Does this generalization of $\leq$ to infinite numbers preserve the following basic properties that $\leq$ has between finite numbers?

Let A, B and C be sets.

Reflexive Property

$|A| \leq |A|$
There exists a one-to-one function that maps A into A.

Transitive Property

If $|A| \leq |B|$ and $|B| \leq |C|$, then $|A| \leq |C|$.
If there exists a one-to-one function that maps A into B and there exists a one-to-one function that maps B into C, then there exists a one-to-one function that maps A into C.

Antisymmetric Property

If $|A| \leq |B|$ and $|B| \leq |A|$, then $|A| = |B|$.
If there exists a one-to-one function that maps A into B and there exists a one-to-one function that maps B into A, then there exists a one-to-one function that maps A onto B.

To prove that $\leq$ has the reflexive property, we can use the identity function: $f(x) = x$.

To prove that $\leq$ is transitive, we can use the composition of the two functions (page 361).

Schröder-Bernstein Theorem

Let A and B be sets.

If $|A| \leq |B|$ and $|B| \leq |A|$, then $|A| = |B|$.

Unlike the other two properties, the proof of the antisymmetric property is very challenging, eluding even the illustrious Georg Cantor, who conceived the proposition. The theorem was proved by Ernst Schröder in 1896 and independently proved by Felix Bernstein two years later. Theorems whose proofs are of an Olympian stature are often named in honor of the creator, and so this theorem is called the Schröder-Bernstein Theorem. Even though its statement is outwardly simple, the Schröder-Bernstein Theorem provides a very powerful tool for working with infinite quantities.

Extending <

We extend the notion of < to infinite numbers using the same definition as for finite sets:

$$|A| < |B|$$
if and only if
$$|A| \leq |B| \text{ and } |A| \neq |B|.$$

The $<$ relation on infinite sets does not have the same properties as on finite sets. If A and B are finite and $A \subset B$, we can deduce that $|A| < |B|$. We cannot make a similar deduction for infinite sets: $\mathbb{N} \subset Z$, but $|\mathbb{N}| = |Z|$. The analogous property is preserved, though, for the $\leq$ relation:

Theorem For all sets A and B, if $A \subseteq B$, then $|A| \leq |B|$.

Proof Let A and B be sets with $A \subseteq B$. Consider the identity function on A: $f(x) = x$. Since $A \subseteq B$, f maps A into B. Since f is a one-to-one function that maps A into B, $|A| \leq |B|$.

Higher Levels of Infinity

At this stage, we know of only two different sizes of infinite sets: one represented by the set $\mathbb{N}$ of natural numbers and the other represented by the set $\mathbb{R}$ of real numbers. However, there are many more sizes of infinity, in fact there are an infinite number of sizes.

7. Power Set Axiom

Let S be a set.

There exists a set $P(S)$ such that

$P(S) = \{ A \mid A \subseteq S \}$.

To make such a bold assertion, though, we need another axiom, the *Power Set Axiom*. This axiom allows us to form the set of all subsets of a set S, which we notate as $P(S)$. The Axiom of Infinity (page 285) allowed us to progress from finite sets to sets of a countably infinite magnitude. The Power Set Axiom takes us to even higher realms of infinity.

If S is a finite set with n elements, $P(S)$ has 2^n elements (page 229). So, for every finite set S, $|S| < |P(S)|$. We will now demonstrate that this statement is also true for infinite sets. First, we will prove that $|S| \leq |P(S)|$, which is considerably easier than showing the strict inequality.

Theorem For every set S, $|S| \leq |P(S)|$.

Proof Let a be an element in S. Then $\{a\} \in P(S)$.
Define the function f as follows: $f(a) = \{a\}$.
Note that f maps S into $P(S)$.
If $a \neq b$, then $\{a\} \neq \{b\}$. So, f is one-to-one.

Since f is a one-to-one function that maps S into $P(S)$, $|S| \leq |P(S)|$.

Now we will tackle the more difficult task of showing that S cannot have the same size as $P(S)$.

Theorem For every set S, $|S| < |P(S)|$.

Proof Suppose that f is a function that maps S into $P(S)$.
Let x be an element in S.
Since f maps into $P(S)$, $f(x) \in P(S)$.
Thus, $f(x)$ is a subset of S.
So, $x \in f(x)$ or $x \notin f(x)$.

Let $C = \{\, x \text{ in } S \mid x \notin f(x) \,\}$
Since C is a subset of S, $C \in P(S)$.

Assume that f maps S onto $P(S)$.
Then there exists an x in S such that $f(x) = C$.
Either $x \in C$ or $x \notin C$.

Case 1. Suppose that $x \in C$.
By the definition of C, $x \notin f(x)$.
So $x \notin C$. Contradiction!

Case 2. Suppose that $x \notin C$.
By the definition of C, $x \in f(x)$.
So $x \in C$. Contradiction!

Either way, we get a contradiction.
Therefore, f does not map S onto $P(S)$.

Hence, there does not exist a function that maps S onto $P(S)$.
So, $P(S)$ does not have the same size as S: $|S| \neq |P(S)|$.

By the previous theorem, $|S| \leq |P(S)|$.
Since $|S| \neq |P(S)|$, it follows that $|S| < |P(S)|$.

> Pure mathematics is, in its way, the poetry of logical ideas.
>
> *Albert Einstein*

Einstein described pure mathematics as "the poetry of logical ideas." His comment is eloquently illustrated by the above proof, whose conception required a great deal of creativity. This proof – an ingenious analogue of the diagonalization proof that $\mathbb{R}$ is uncountable – tells us that the set of all subsets of $\mathbb{N}$ is also an uncountable set. Using this theorem, we can now produce an infinite string of infinite numbers.

First, we start with the set of natural numbers. The size of its power set is a larger infinite number:

$$|\mathbb{N}| < |P(\mathbb{N})|$$

Next we take the power set of $P(\mathbb{N})$, and then we take its power set, and so on:

$$|\mathbb{N}| < |P(\mathbb{N})| < |P(P(\mathbb{N})| < \ldots$$

Let's name the size of $P(\mathbb{N})$ as $\aleph_1$, and then label the sizes of successive power sets as $\aleph_2$, $\aleph_3$, and so on. Now we have an infinite chain of infinite numbers which continue on and on like the natural numbers:

$$\aleph_0 < \aleph_1 < \aleph_2 < \aleph_3 < \ldots$$

Thus, based on the axioms of set theory, we can deduce that there are an infinite number of different sizes of infinite sets.

Exercise Set 3.6

1. Does the given set have the same size as $\mathbb{N}$? If so, give a formula for a one-to-one function that maps $\mathbb{N}$ onto the set. *Hint*: Some of the patterns are closely related.
   a. $A = \{3, 6, 9, 12, \ldots\}$
   b. $B = \{2, 5, 8, 11, \ldots\}$
   c. $C = \{1, 4, 7, 10, \ldots\}$
   d. $D = \{6, 9, 12, 15, \ldots\}$
   e. $E = \{2, 4, 8, 16, 32, 64, \ldots\}$
   f. $F = \{1, 3, 7, 15, 31, 63, \ldots\}$
2. Define the following.
   a. A finite set
   b. An infinite set
   c. $\aleph_0$
   d. A countable set
   e. A countably infinite set
   f. An uncountable set
3. True of false? If false, give a counterexample.
   a. Every finite set has a largest element.
   b. An infinite set cannot have a largest element.
   c. If A is a proper subset of B, then the size of A is smaller than the size of B.
4. Use a diagonalization method to explain why $\mathbb{N} \times \mathbb{N}$ is countable. *Hint*: Arrange the ordered pairs in an infinite row/column array.
5. Let $R_n = \{\frac{k}{n} \mid k \in \mathbb{N}\}$.
   a. List the elements in the following sets: R_1, R_2, R_3

b. Let Q^+ denote the set of positive rational numbers. Use a diagonalization method to construct a function f that maps $\mathbb{N}$ onto Q^+. *Hint*: Arrange the R_n sets in an infinite row/column array. Then describe a step-by-step process for defining $f(n)$.

c. For your function f from part (b), is f one-to-one? What is $f(1)$? $f(2)$? $f(3)$? $f(8)$?

6. a. Explain how to use a diagonalization method to prove that the set of real numbers is not countable. Make up an example to illustrate your method.

   b. Apply your argument in part (a) to the set of rational numbers. Where does it fall apart?

7. Let X and Y be sets with $|X| = |Y|$. Let f be a function that maps X into Y. True or false? If false, give a counterexample.

   a. If f is one-to-one, then f is onto.

   b. If f is onto, then f is one-to-one.

8. True of false? If false, give a counterexample.

   a. Every finite set is countable.

   b. Every countable set is finite.

   c. Every subset of a countable set is countable.

   d. The union of two countable sets is countable.

   e. If A_i is countable for each i in $\mathbb{N}$, then $\bigcup_{i \in \mathbb{N}} A_i$ is a countable set.

   f. If A_i is countable for each i in an index set I, then $\bigcup_{i \in I} A_i$ is a countable set.

   g. The set of rational numbers is countable.

   h. Every set is countable.

9. Let $S = \{1, 2, 3, 4\}$. The function f maps S into $P(S)$. Determine if f is one-to-one. Draw an arrow mapping for each element in S. For example, in part (a), $1 \longrightarrow \{1, 3\}$.

   a. $f(1) = \{1, 3\}$, $f(2) = \{1, 3, 4\}$, $f(3) = \{2\}$, $f(4) = \{1, 2, 4\}$

   b. $f(1) = \{1, 3\}$, $f(2) = \{1, 2, 4\}$, $f(3) = \{2, 3\}$, $f(4) = \{4\}$

   c. $f(1) = \{2, 3, 4\}$, $f(2) = \{1\}$, $f(3) = \{1\}$, $f(4) = \{1, 3\}$

10. Let $C = \{\, x \text{ in } S \mid x \notin f(x) \,\}$. For each function in the previous exercise, list the elements in C. Then determine if C is in the range of f. Compare your results with the proof on page 298.

11. Let S be an arbitrary set.

    a. Define a one-to-one function f that maps S into $P(S)$.

    b. Let f be an arbitrary function that maps S into $P(S)$. Define a subset of S to which f does not map any element.

12. a. Let S_n denote the set of all bit strings of length n.
    How many elements are in S_n? (See (15), page 232.)
    b. Let S be the set of all bit strings of finite length.
    How many elements are in S? Is S countable?
    c. Let T be the set of all bit strings of infinite length. Suppose that you have a countable listing of the elements in T. Can you construct a bit string that is not in the list? Is T countable?
13. Any language has a finite number of symbols in its alphabet. Define a "word" to be a finite sequence of alphabet symbols.
    Let S_n denote the set of all words of length n.
    Let S denote the set of all words of finite length.
    If the alphabet has the following number of symbols,
    how many elements are in S_n? In S?
    a. 2 symbols b. 3 symbols c. x symbols
14. Let S denote the set of all possible computer programs in a given language. Is S countable? *Hint*: Counting spaces and punctuation marks as part of the alphabet, each computer program is composed of a finite sequence of symbols, so we could consider it a "word."
15. *Cantor's Paradox*: In 1899, George Cantor created a paradox from the "set" of all sets. Define the set V as follows:
    $$V = \{\ S \mid S \text{ is a set}\ \}$$
    a. From the definition of V, can we deduce that $P(V) \subseteq V$?
    Explain your reasoning.
    b. Is $|P(V)| \leq |V|$? Justify your answer.
    c. Is $|V| < |P(V)|$? Justify your answer. (Cite previous theorems.)
    d. Use the Schröder-Bernstein Theorem to derive a contradiction.
    e. Has Cantor's Paradox been defused? If so, explain how.
16. Explain why the decimal form of a rational number must have a repeating cycle.
17. Explain why $1 = .9999...$ *Hint*: Set $n = .9999...$ Consider $10n$.

Activity 3.7 – Hilbert's Infinite Hotel

A Forever Continuing Saga . . .

Hilbert's Infinite Hotel again has a swarm of people clamoring for rooms, but this time the numbers are overwhelming. $\aleph_o$ buses just arrived, with each containing $\aleph_o$ people. Can the manager accommodate all of the new guests? If so, explain how to do it.

Review

Set A collection of objects. In formal set theory, a set is undefined since there are no simpler concepts with which to define it.

Is an element of A relation between the members of a set and the collective unit to which the members belong. In formal set theory, "is an element of" is an undefined term since there are no simpler concepts with which to define it.

Universal set A set that serves as the universe for a particular discussion. When defining a set, all members of the set must come from a universal set. Otherwise, contradictions arise in set theory.

Empty set A set that has no elements. The empty set is analogous to an empty box, which exists even though it has nothing in it.

Power set The set of all subsets of a given set. $P(S) = \{X \mid X \subseteq S\}$. If S has n elements, $P(S)$ has 2^n elements. For every set S, $|S| < |P(S)|$.

Partition A subdivision of a set into nonoverlapping subsets. A partition P of a set S is a collection of nonempty subsets of S where each element in S is in one and only one of the subsets.

Equals relation Two sets are equal if they contain the same elements. $A = B$ if and only if for every x, $(x \in A \Rightarrow x \in B)$ and $(x \in B \Rightarrow x \in A)$. The equals relation is reflexive, transitive and symmetric.

Subset relation A is a subset of B if and only if every element in A is also in B. $A \subseteq B$ if and only if for every x, $x \in A \Rightarrow x \in B$. The subset relation is reflexive, transitive and antisymmetric.

Proper subset relation A is a proper subset of B if and only if $A \subseteq B$ and $A \neq B$.

Size relation A has the same size as B if and only if there exists a one-to-one function that maps A onto B. Two sets that have the same size are said to have the same *cardinality*.

Union A binary operation on two sets that produces a new set by combining their elements: $A \cup B = \{x \mid x \in A \text{ or } x \in B\}$. If A and B are finite sets, $|A \cup B| = |A| + |B| - |A \cap B|$.

Intersection A binary operation on two sets that produces a new set from their common elements: $A \cap B = \{x \mid x \in A \text{ and } x \in B\}$.

Set subtraction A binary operation on two sets that produces a new set by removing the elements in one set from another set:

$$A-B = \{ x \mid x \in A \text{ and } x \notin B \}$$

If A and B are finite sets and $B \subseteq A$, then $|A-B| = |A| - |B|$.

Complement A unary operation on a set which produces a new set composed of all the elements in the universal set that are not in the original set: $A' = \{ x \mid x \notin A \}$. $A' = U - A$.

Multiple unions A set formed by combining the elements in a collection of sets.

$$\bigcup_{i \in I} A_i = \{ x \mid \text{for some } i \text{ in } I,\ x \in A_i \}$$

Multiple intersections A set formed from the elements that are in each member of a collection of sets. $\bigcap_{i \in I} A_i = \{ x \mid \text{for every } i \text{ in } I,\ x \in A_i \}$

Ordered pair A pairing of elements where the order affects the meaning. $(a,b) = (c,d)$ if and only if $a = c$ and $b = d$.

Cross product $A \times B$ is the set of all ordered pairs whose first term is in A and whose second term is in B:

$$A \times B = \{ (a,b) \mid a \in A \text{ and } b \in B \}$$

$$A \times B \times C = \{ (a,b,c) \mid a \in A \text{ and } b \in B \text{ and } c \in C \}$$

The number of elements in a cross product is the product of the number of elements in the individual sets.

Commutative property Let $*$ be a binary operation on a set S. $*$ is commutative if and only if for every a and b in S, $a * b = b * a$. Union and intersection are commutative.

Associative property Let $*$ be a binary operation on a set S. $*$ is associative if and only if for every a, b, and c in S, $a * (b * c) = (a * b) * c$. Union and intersection are associative.

Distributive property for sets Intersection distributes over union, and union distributes over intersection. Let A be a set and B_i be a set for each i in I:

$$A \cap (B_1 \cup B_2) = (A \cap B_1) \cup (A \cap B_2) \ \ldots \ A \cap (\bigcup_{i \in I} B_i) = \bigcup_{i \in I} (A \cap B_i)$$

$$A \cup (B_1 \cap B_2) = (A \cup B_1) \cap (A \cup B_2) \ \ldots \ A \cup (\bigcap_{i \in I} B_i) = \bigcap_{i \in I} (A \cup B_i)$$

Complement laws The complement of a union is the intersection of the complements. The complement of an intersection is the union of the complements. Let A_i be a set for each i in I:

$$(A_1 \cup A_2)' = A_1' \cap A_2' \ \ldots \ (\bigcup_{i \in I} A_i)' = \bigcap_{i \in I} (A_i')$$

$$(A_1 \cap A_2)' = A_1' \cup A_2' \ \ldots \ (\bigcap_{i \in I} A_i)' = \bigcup_{i \in I} (A_i')$$

One-to-one correspondence	A one-to-one and onto function between two sets.
Function	f is a function from X into Y if and only if f maps each element in X to a unique element in Y.
One-to-one function	A function that maps different elements in the domain to different elements in the range. For all a and b in the domain of f, if $a \neq b$, then $f(a) \neq f(b)$. If f is a one-to-one function that maps X into Y, then $\|X\| \leq \|Y\|$.
Onto function	f maps X *onto* Y if and only if for every y in Y, there exists an x in X such that $f(x) = y$. If f is a function that maps X onto Y, then $\|X\| \geq \|Y\|$. If $\|X\| = \|Y\|$ and both sets are finite, f being onto is equivalent to f being one-to-one.

Finite set	A set S is finite if and only if S is the empty set or $\|S\| = n$ for some natural number n.
Infinite set	A set that is not finite.
$\aleph_0$	The number of elements in the set $\mathbb{N}$ of natural numbers. $\|S\| = \aleph_0$ if and only if there exists a one-to-one function that maps $\mathbb{N}$ onto S.
Countable	A set that is either finite or can be placed in a one-to-one correspondence with the set $\mathbb{N}$ of natural numbers. The union of every countable collection of countable sets is countable.
Countably infinite	An infinite set that is countable. S is countably infinite if and only if $\|S\| = \aleph_0$. The set of natural numbers, the set of integers, and the set of rational numbers are countably infinite sets.
Uncountable	A set that is not countable. An uncountable set is a larger size of infinity than a countably infinite set. The set of irrational numbers is uncountable, which makes the set of real numbers uncountable.
Cardinal number of a set	The number of elements in a set A, notated as $\|A\|$. Let n be a natural number. $\|A\| = n$ if and only if there exists a one-to-one function f that maps $\{1, 2, 3, \ldots, n\}$ onto A. Two sets that have the same size have the same *cardinality*. $\|A\| = \|B\|$ if and only if there exists a one-to-one function that maps A onto B. $\|A\| \leq \|B\|$ if and only if there exists a one-to-one function that maps A into B. $\|A\| \geq \|B\|$ if and only if there exists a function that maps A onto B. Let A and B be finite sets. If $A \subset B$, then $\|A\| < \|B\|$.

Chapter Review

1. Define the following operations on sets and give examples of each: union, intersection, set subtraction, complement, multiple union, multiple intersection, cross product.

2. Define the following relations between sets and give examples of each: A is equal to B, A is a subset of B, A is a proper subset of B, A has the same size as B.

3. Use definitions and negation rules to translate each sentence.

   a. $A \neq B$ c. $A \not\subset B$ e. $x \notin A \cap B$ g. $x \notin \bigcap_{i \in I} A_i$

   b. $A \not\subseteq B$ d. $x \notin A \cup B$ f. $x \notin \bigcup_{n \in I} A_n$

4. Is the given expression a grammatically correct sentence?

   a. $A \cup B$ c. Assume $A \cup B$. e. $(A \subseteq B) \cup C$

   b. $A \subseteq B$ d. Assume $x \in A \cup B$. f. $A \subseteq (B \cup C)$

5. Use Venn diagrams to illustrate sets and properties of sets, such as the following.

   a. $A \cap (B \cup C)$ c. $(A \cup B)' = A' \cap B'$

   b. $(A - B) - C$ d. $A \cup (B \cap C) = (A \cup B) \cap (A \cup C)$

6. Write sets, such as the following, in a simplified form.

   a. $(C \cap B) \cup (C \cap B')$ c. $C \cup (B \cap C)$ e. $E \cap (B \cup E)$

   b. $(D \cap B) \cap D'$ d. $F \cap (B \cup B')$ f. $A \cup (B - A)$

7. Suppose that $C \subseteq B$. Simplify the following sets.

   a. $C \cup B$ c. $C \cap B$ e. $C - B$

   b. $C' \cap B'$ d. $C' \cup B'$ f. $C \cap B'$

8. Given the description of a set S, list the elements in the set and translate the expression, $y \in S$.

   a. $C_x = \{x^n \mid n \text{ is a natural number}\}$

   b. $S = \{a^n \mid a \text{ is a natural number}\}$

   c. $B_n = \{\frac{k}{n} \mid k \text{ is a natural number}\}$

   d. $F_x = \{x + k \mid k \text{ is a natural number}\}$

9. Is the given statement true for every set A? If not, give a counterexample.

   a. $\emptyset \subseteq A$ c. $A \subseteq A$ e. $A \in P(A)$

   b. $\emptyset \in A$ d. $A \in A$ f. $\emptyset \in P(A)$

10. Is the given statement true for all sets A and B? If not, give a counterexample.

    a. $A - B = B - A$

    b. $A \times B = B \times A$

c. If $A \subseteq B$, then $A \in B$.
d. If $A \in B$, then $A \subseteq B$.
e. If $|A_{,}| = |B|$ and f is a function from A onto B, then f must be one-to-one.

11. Is the given statement true for all finite sets A and B? If not, give a counterexample.
a. $|A \cup B| = |A| + |B|$
b. $|A - B| = |A| - |B|$
c. If $|A| = |B|$ and f is a function from A onto B, then f must be one-to-one.

12. Is the given statement true for all sets A, B and C? If not, give a counterexample.
a. If $A \subseteq B$ and $B \subseteq C$, then $A \subseteq C$.
b. If $A \not\subseteq B$ and $B \not\subseteq C$, then $A \not\subseteq C$.
c. If $A \in B$ and $B \in C$, then $A \in C$.
d. If $A \subseteq B$ and $B \not\subseteq C$, then $A \not\subseteq C$.
e. If $A \subseteq B$ and $A \not\subseteq C$, then $B \not\subseteq C$.

13. Is the given statement true for all sets A, B and C? If not, give a counterexample.
a. $A \cup (B \cup C) = (A \cup B) \cup C$
b. $A - (B - C) = (A - B) - C$
c. $A \cap (B \cap C) = (A \cap B) \cap C$
d. $A \cap (B \cup C) = (A \cap B) \cup C$

14. Let A_i be a set for each i in $\mathbb{N}$. Is the given statement true? If not, give a counterexample.
a. $A_1 \subseteq \bigcup_{i \in \mathbb{N}} A_i$
b. $A_1 \subseteq \bigcap_{i \in \mathbb{N}} A_i$

15. Write easy-to-follow proofs of the following theorems. State the reason each time you use a definition, valid argument, or previous theorem, having only one reason for each step in your proof. Let A, B, C, and D be sets.
a. $(A \cap B)' = A' \cup B'$ . . . . $(\bigcap_{i \in I} A_i)' = \bigcup_{i \in I} (A_i')$
b. $(A \cup B)' = A' \cap B'$ . . . . $(\bigcup_{i \in I} A_i)' = \bigcap_{i \in I} (A_i')$
c. $A \cap (B \cup C) = (A \cap B) \cup (A \cap C)$ . . . $A \cap (\bigcup_{i \in I} B_i) = \bigcup_{i \in I} (A \cap B_i)$
d. If $A \subseteq B$, then $A \cap B = A$ and $A \cup B = B$.
e. If $A \subseteq B$, then $B' \subseteq A'$.
f. If $A \subseteq C$ and $B \subseteq C$, then $A \cup B \subseteq C$.
g. If $C \subseteq A$ and $C \subseteq B$, then $C \subseteq A \cap B$.
h. $A \cup (B \cap C) = (A \cup B) \cap (A \cup C)$
i. $A \cup (\bigcap_{i \in I} B_i) = \bigcap_{i \in I} (A \cup B_i)$
j. If $A \subseteq C$ and $B \subseteq D$, then $A \times B \subseteq C \times D$.

k. $A \times (B \cup C) = (A \times B) \cup (A \times C)$

l. $A \times (B \cap C) = (A \times B) \cap (A \times C)$

16. There are 3 students at Level A, 6 students at Level B, and 10 students at Level C. The coach must select one student from each level.
    a. What is the total number of possibilities for the selection?
    b. What general property of sets did you use in your computation?

17. What is the power set of a set? If a set S has n elements, how many subsets does S have?

18. Let X and Y be finite sets. What does the given information tell you about the relation between $|X|$ and $|Y|$?
    a. There exists a function that maps X onto Y.
    b. There exists a one-to-one function that maps X into Y.
    c. There exists a one-to-one function that maps X onto Y.
    d. X is a proper subset of Y.

19. What do you have to demonstrate in order to prove the following?
    a. A set is not finite. b. A set does not have n elements.

20. True or false? If false, give a counterexample.
    a. There are an infinite number of infinite numbers.
    b. For all sets A and B, if A is a proper subset of B, then the size of A is smaller than the size of B.
    c. For all sets A and B, if $A \subset B$, then $|A| < |B|$.
    d. For all finite sets A and B, if $A \subset B$, then $|A| < |B|$.
    e. For every set A, $|A| < |P(A)|$.
    f. If $|A| < |B|$, there exists a one-to-one function from A into B.
    g. Every finite set has a largest element.
    h. If a set is infinite, it does not have a largest element.

21. Explain why the following are true:
    a. The set of integers has the same size as the set of natural numbers.
    b. If A and B have $\aleph_o$ elements, then $A \cup B$ has $\aleph_o$ elements.
    c. The set of rational numbers has the same size as the set of natural numbers.
    d. The set of real numbers does not have the same size as the set of natural numbers.

22. Are there more rational numbers than irrational numbers? Or vice-versa? Justify your answer.

◈ Chapter 4

Relations

– The Action

The fundamental through-line that runs through all of mathematics is the concept of a relation. Playwrights use a through-line as a dramatic device to hold a play together. The through-line used by Anton Chekhov in his famous play, *The Three Sisters*, was the great desire of the sisters to go to Moscow. There is obviously much more to a play than the through-line, but Chekhov knew that if he didn't have one, he would soon lose the audience. The same is true in mathematics. We need a through-line to help us work our way through the varied and sometimes dense areas of mathematics.

Mathematical activity has always focused on relations. In the reasoning process, we are usually trying to figure out how various objects may be related to each other. When we work with sets, we do not individually analyze the elements in a set; instead, we compare the set with other sets by looking for relations between them. Within the grand house of mathematics, there are many diverse areas of study, but within each area, the focus is on relations. Relations are where the action is in mathematics, which is why they are the verbs of mathematical language. In fact, relations provide a simple through-line for describing mathematics:

Mathematics is the study of relations.

The three sisters wanted to go to Moscow. What mathematicians want is to find interesting relations, as did the three sisters, and we will go wherever we have to, within the unbounded confines of our mind, to find them. We will create

> Mathematicians do not study objects, but relations among objects; they are indifferent to the replacement of objects by others as long as relations do not change.
> Matter is not important, only form interests them.
>
> *Henri Poincaré*
> 1854–1912

the most bizarre spaces imaginable, or unimaginable, and, before long, people in other disciplines are using them, which is not too surprising since their research has the same underlying structure as ours. They cannot analyze a physical element by itself, be it a rock, a chemical compound, or a person, so they, too, are looking for relations, which is the reason mathematics has so many applications in so many disciplines. Mathematicians do abstractions of relations, and scholars in other disciplines flesh it out.

A relation embeds a structure between sets which we can visualize in various ways. Some relations, such as $f(x) = 2x$, have a simple pattern. Binary operations, like addition, produce a more complex structure whose abstract form may be shared by different types of operations. One of the distinguishing features of modern mathematics is the focus on structures. This new direction started in 1801 when Carl Gauss, at the young age of 24, introduced the notion of a congruence in his famous text, *Disquisitiones arithmeticae.* With the seed planted by Gauss, the concept has blossomed until it now occupies a central focus in all areas of mathematics. The fundamental question is:

> When are two different types of objects essentially the same with respect to the structure embedded by a relation?

> Mathematics, rightly viewed, possesses not only truth, but supreme beauty – a beauty cold and austere, like that of sculpture, without appeal to any part of our weaker nature, without the gorgeous trappings of paintings or music, yet sublimely pure and capable of a stern perfection such as only the greatest art can show.
>
> *Bertrand Russell*
> 1872–1970

Similar questions are asked in all disciplines. In biology, plants and animals are classified into groups that have a similar structure. Even in music, art, and literature, pieces are classified through the lens of a structure style. So it is not surprising that we have a similar focus in mathematics.

The centuries-long quest to find the relation between Euclid's Fifth Postulate and his other four postulates might have seemed frivolous to a scientific mind focused on understanding tangible aspects of reality. In the end, though, it produced a completely new kind of structure that scientists are now using to build a mental picture of our universe. The obsession that drives the mind to explore such abstract relations is the same obsession that drives a true artist in any field. Even though scientific disciplines make the heaviest use of mathematics, the spirit that drives mathematicians is more closely aligned with art.

This chapter covers the basic terminology associated with relations and provides an opportunity for you to compare structures created by various types of relations. We will focus on the three most frequently used types of relations: functions, equivalence relations, and order relations. These special types of relations help us organize and classify information in meaningful structures. Even more important, they give us ideas for how to use creative reasoning to extend our knowledge by building other structures.

Activity 4.1

1. The equals relation has the following 3 properties on a set S. Let x, y and z be elements in S.

   Reflexive: $x = x$

   Transitive: If $x = y$ and $y = z$, then $x = z$.

   Symmetric: If $x = y$, then $y = x$.

   Generalize these 3 properties for an arbitrary relation R on a set S.

2. Determine if the given relation is reflexive, transitive, or symmetric. Let S be the set of students at your school.
   a. $x \mathsf{R} y$ if and only if x is shorter than y.
   b. $x \mathsf{R} y$ if and only if x is not taller than y.
   c. $x \mathsf{R} y$ if and only if x and y were born in the same year.
   d. $x \mathsf{R} y$ if and only if x and y are taking the same course.

3. The $\leq$ relation is not symmetric. In fact, it has the opposite property, which is called *antisymmetry*. The only time we can reverse the order for $\leq$ is when the two elements are equal. Let x and y be real numbers.

   Antisymmetric: If $x \leq y$ and $y \leq x$, then $y = x$.

   Generalize antisymmetry for an arbitrary relation R on a set S.

4. Determine if each relation in (2) is antisymmetric.

5. Let a and b be integers. Define the divides relation on the set Z of integers as follows:

   $a \mid b$ if and only if there exists an integer k such that $b = ak$.

   Determine if the divides relation has the given property on Z:

   a. Reflexive c. Symmetric
   b. Transitive d. Antisymmetric

≡ 4.1 Relations ≡

$\{3,4\} = \{4,3\}$

$\{1\} \subseteq \{1,2\}$

$x < y$

A relation embeds a structure between two sets by giving a connection between various elements. There are many different ways to view the connection. In terms of language, relations serve as verbs, providing connections between objects on either side of the relation, as illustrated in the adjacent sentences. To generalize these sentences, let R represent a relation. The notation $x \mathrm{R} y$ means that x *is related to* y *under the relation* R. For example, let R represent the $<$ relation. Then $x \mathrm{R} y$ means that $x < y$.

We sometimes represent the connection between related elements with an arrow:

$$x \to y \quad \text{represents} \quad x \mathrm{R} y.$$
$$1 \to 4 \quad \text{represents} \quad 1 \leq 4.$$

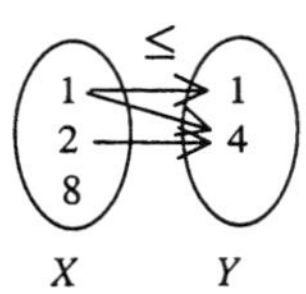

From this perspective, a relation can be viewed as a mapping where the arrows tell us which elements are related. For example, we can represent the $\leq$ relation from the set $\{1, 2, 8\}$ to the set $\{1, 4\}$ as the adjacent mapping. The one-way arrow conveys the message that the order of the elements matters.

Using a mapping picture, we can describe a relation R from a set X into a set Y as a mapping where some of the elements in X are mapped to some of the elements in Y. An element in the first set can be mapped to more than one element in the second set, as illustrated in the adjacent example. Furthermore, not everyone in the first set has to be mapped somewhere.

Let X and Y be sets.

A *relation* R from X into Y is a mapping where some of the elements in X are mapped to some of the elements in Y.

Since order is essential to the concept of a relation, ordered pairs provide a good language tool for describing a relation. Instead of the arrow picture, we use ordered pair notation:

$$(x, y) \quad \text{represents} \quad x \mathrm{R} y.$$
$$(1, 4) \quad \text{represents} \quad 1 \leq 4.$$

A *relation* R from X into Y is a set of ordered pairs whose first terms come from X and whose second terms come from Y.

Using ordered pairs, we can formally define a relation as stated on the left. For example, the above mapping picture of $\leq$ can be translated as follows:

$$\leq \; = \; \{ (1,1), (1,4), (2,4) \}$$

Ordered pairs provide a concise way to define a relation, which is the main advantage of this type of representation. In the ordered pair representation of a relation, we interpret each ordered pair to mean that the first term is related to the second term. The ordered pairs are merely a technical device to show who is related to whom:

$$x \text{ is related to } y \text{ if and only if } (x,y) \in \mathsf{R}.$$

⊕ Example

Let $X = \{ 1, 2, 3 \}$ and $Y = \{ 4, 5, 7, 8 \}$. Let R be the following relation between X and Y:

$$\mathsf{R} = \{ (2,5), (2,7), (3,4) \}$$

The ordered pairs tell us that 2 is related to 5, 2 is related to 7, and 3 is related to 4.

In the above example, note that $\mathsf{R} \subseteq X \times Y$. Using the ordered pair definition of a relation, a relation R from X into Y is a subset of $X \times Y$. Conversely, if R is a subset of $X \times Y$, then R is a set of ordered pairs, which means that R can be viewed as a relation from X into Y. Thus, we can rephrase the definition of a relation in terms of a cross product:

R is a *relation* from X into Y if and only if $\mathsf{R} \subseteq X \times Y$.

When using the ordered pair interpretation of a relation, we should keep in mind that the motivating idea is the relation between the first and second terms of each ordered pair.

Domain & Range

The *domain* of a relation R is the set of all first terms of the ordered pairs in R. The *range* of R is the set of all second terms of the ordered pairs in R.

If $(x,y) \in \mathsf{R}$, then $x \in$ Domain (R) and $y \in$ Range (R).

Let R be a relation from X into Y.

Domain (R) $= \{ x \mid (x,y) \in \mathsf{R} \}$

Range (R) $= \{ y \mid (x,y) \in \mathsf{R} \}$

For example, let $\mathsf{R} = \{ (2,1), (3,1), (3,2), (4,2) \}$.

Domain (R) $= \{ 2,3,4 \}$ Range (R) $= \{ 1,2 \}$

The translation of $x \in$ Domain (R) and $y \in$ Range (R) requires an existential quantifier:

$x \in$ Domain (R) if and only if
there exists an y in Y such that $(x,y) \in \mathsf{R}$.

$y \in$ Range (R) if and only if
there exists an x in X such that $(x,y) \in \mathsf{R}$.

Graphs of Relations

Graphs provide a powerful visual tool for analyzing relations between sets of real numbers. A *graph* is a visual picture of a relation where we use points in a coordinate plane to represent the ordered pairs in a function. Unlike arrow pictures, graphs provide visual information on magnitudes. As x gets larger, we can see whether or not y is getting larger.

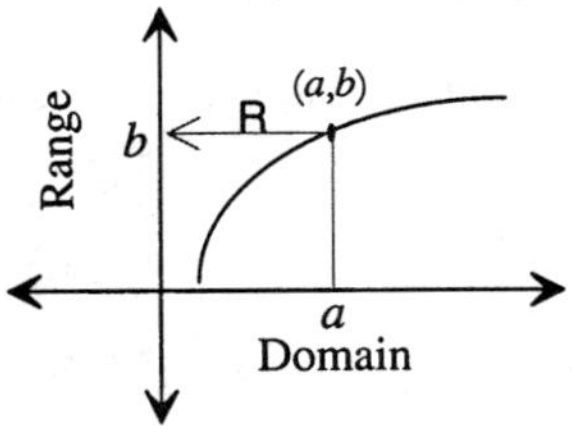

Each point on a graph indicates that the first coordinate is related to the second coordinate under the given relation. To see the arrow relation represented by a point (a, b) on the graph, draw an arrow from a to b with a right angle bend on the graph, as illustrated in the adjacent graph.

If (a, b) is on the graph of a relation, its perpendicular projection onto the horizontal axis is a, which is an element of the domain, and its projection onto the vertical axis is b, which is an element of the range. When a relation is represented on a graph, we can find its domain by projecting the graph onto the horizontal axis. Similarly, we can find its range by projecting the graph onto the vertical axis, as illustrated on the left.

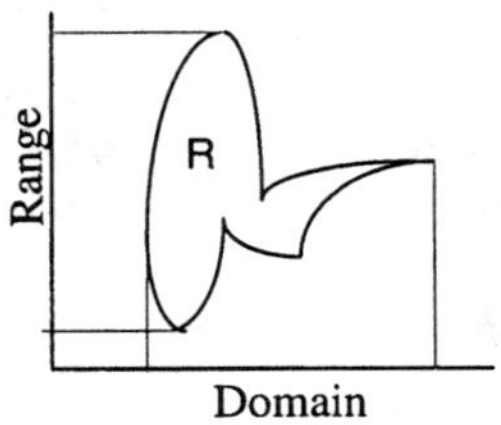

When R is a relation from a set S into S, we call R a relation *on S*. Any picture that can be drawn in a coordinate plane, such as the figure on the left, represents a relation on the set of real numbers. Each point in the region R indicates a relation between its first coordinate and its second coordinate.

Example

1. Let R be the > relation on S, where $S = \{1, 2, 3, 4\}$.

$$\mathsf{R} = \{ (x, y) \mid x > y,\ x \in S \text{ and } y \in S \}$$
$$\mathsf{R} = \{ (2,1), (3,1), (3,2), (4,1), (4,2), (4,3) \}$$

$$\text{Domain } (\mathsf{R}) = \{2, 3, 4\}$$
$$\text{Range } (\mathsf{R}) = \{1, 2, 3\}$$

2. Let R be the ≥ relation on S, where $S = \{ x \mid 0 \le x \le 4 \}$. The graph of R is the adjacent shaded region, which is bounded by the line $y = x$.

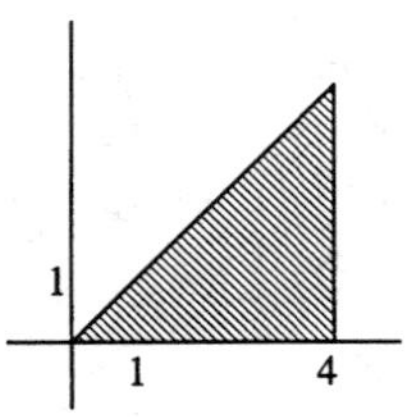

If (x, y) is on the line, then $x = y$.

If (x, y) is below the line, then $x \ge y$.

If (x, y) is above the line, then $x < y$.

Directed Graphs

A directed graph is similar to an arrow mapping, but we chain the arrows without segregating the domain from the range. If an element is in both the domain and range, we only list it once. For example, in the adjacent directed graph of the < relation on the set $\{1,2,3\}$, 2 is in both the domain and range, but we only list it once. In a directed graph, the domain is the set of starting points for the arrows, while the range is the set of terminal points for the arrows. A regular graph gives information on magnitudes, whereas a directed graph gives information on a different type of structure, as illustrated in the following example.

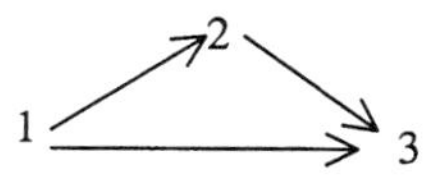

$<$ on $\{1, 2, 3\}$

⊕ *Example*

Let R be the following relation on S, where $S = \{1,2,3,4,5\}$:

$$a \mathrel{\mathsf{R}} b \text{ if and only if } 2 \text{ divides } a-b.$$

To determine if one element is related to another, we substitute in the above sentence. Is $1 \mathrel{\mathsf{R}} 3$?

$$1 \mathrel{\mathsf{R}} 3 \text{ if and only if } 2 \text{ divides } 1-3.$$

2 divides $1-3$, so 1 is related to 3. Since 2 also divides $3-1$, 3 is related to 1. In the adjacent directed graph of R, the arrow from 1 to 3 represents $1 \mathrel{\mathsf{R}} 3$, and the arrow from 3 to 1 represents $3 \mathrel{\mathsf{R}} 1$.

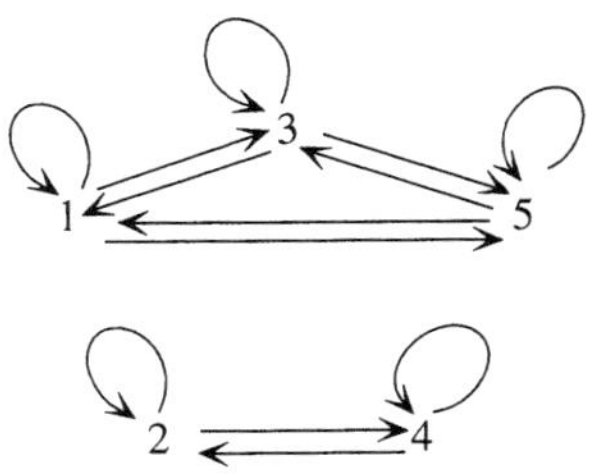

Is $3 \mathrel{\mathsf{R}} 3$? Does 2 divide $3-3$? $\frac{0}{2}$ is an integer, so $3 \mathrel{\mathsf{R}} 3$. In this relation, each element is related to itself, which is represented in the directed graph with loop arrows.

As illustrated in the directed graph, this relation partitions the elements into two nonoverlapping subsets: $\{1,3,5\}$ and $\{2,4\}$. Within each of these subsets, everyone is related to everyone else. This relation is an example of an equivalence relation, which we will study in the next section.

Tables

One of the earliest structures for recording relations were tables where the row/column position indicated some type of relation. Tables, like multiplication tables, often indicate a relation between the pair of elements formed by the entry at the head of the row and the head of the column with the entry in the intersection of the row and column. Some tables have no column heads and the row arrangement indicates that each element in a row is related to the first element in the row.

Matrix Representation

Another way to represent a relation between two sets is with a *matrix*, which is a row/column table structure. Instead of using position to indicate who is related to whom, we use 1s and 0s. In a matrix for a relation between two sets, we list each element in the first set down the first column and each element in the second set across the top row. In the intersection of the a-row and the b-column, we write 1 if a is related to b, and 0 if a is not related to b, as illustrated in the following example.

⊕ *Example*

Let R be the < relation from X to Y, where $X = \{ 1, 2, 8 \}$ and $Y = \{ 1, 4, 7 \}$. Represent R in a matrix.

<	1	4	7
1	0	1	1
2	0	1	1
8	0	0	0

We list the elements of X in the first column and the elements of Y in the top row. Then we make the appropriate entry in each cell, as illustrated on the left.

$1 < 1$ is false, so we enter 0 in the intersection of the 1-row and 1-column.

$1 < 4$ is true, so we enter 1 in the intersection of the 1-row with the 4-column.

We can visually represent a relation with an arrow mapping, a graph, a directed graph, a table, or a matrix. An arrow mapping helps our intuition when we are thinking in general terms about relations. Graphs are more popular in algebra and calculus because of the visual information they provide on relative sizes of related numbers. A directed graph provides an excellent visual tool for analyzing relations when we are interested in traveling from one related element to another. A matrix provides a convenient form for storing a relation in a computer.

Since we cannot draw an infinite set of arrows, complete pictures of arrow mappings and directed graphs are restricted to relations on finite sets. Matrix representations are also limited to relations between finite sets. A graph in a Cartesian plane is the best visual picture of a relation on the uncountable set of real numbers. However, we must be aware that no one, not even a computer, can plot an infinite number of points accurately. So, we must always do some extra analysis on a graph to make sure that the shape we see is preserved under a more microscopic or macroscopic view.

Functions

f is a *function* from X into Y
if and only if
f maps each element in X
to a unique element in Y.

A *function* is a special type of relation. In a relation, an element in the domain can be mapped to more than one element, but under a function, an element can be mapped to only one element (page 272). Also, a relation from X into Y does not have to map each element in X to an element in Y, but a function does. When we say that f is a function from X into Y, X must be the domain for f.

If f is a function from X into Y, then f must map each element in X to a unique element in Y. The unique element to which x is mapped is notated as $f(x)$. This notation cannot be used with an arbitrary relation because x could be mapped to more than one element.

In the mapping representation of a function, two arrows can end up at the same place, but they cannot emanate from the same source, as illustrated in the following example.

⊕ *Example*

Let $X = \{ 1, 2 \}$ and $Y = \{ 4, 8, 9 \}$. Define f and g as follows:

$$f = \{ (1,4), (2,4) \}$$
$$g = \{ (1,4), (1,8), (2,9) \}$$

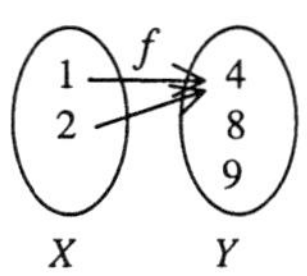

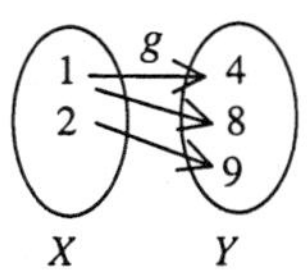

Both f and g are relations from X into Y. In the adjacent arrow mapping, note that f maps each element in X to exactly one element in Y. So f is a function. Since f maps each x to only one element, we can use $f(x)$ notation to represent the mapping:

$$f(1) = 4, \quad f(2) = 4$$

The relation g is not a function because it maps 1 to different elements. If we tried to use function notation with g, we would have $g(1) = 4$ and $g(1) = 8$, which implies that $4 = 8$. For this reason, we cannot use function notation with g.

Because of the uniqueness requirement, we must check the definition carefully when we define a function. For example, suppose that we define a "function" as follows on the set $\mathbb{N}$ of natural numbers:

$$f(n) = a \text{ if and only if } a \text{ is a factor of } n.$$

$f(6) = 2$ and $f(6) = 3$. With this definition, f is not a function, so we cannot use function notation with it. The definition does give a relation: $n \; \mathsf{R} \; a$ if and only if a is a factor of n.

Properties of Relations

In addition to the function property, we have other properties that a relation may or may not have. The most important of these are generalizations of the properties of the equals and subset relations. The equals relation has the following three properties (page 224):

- Reflexive Property
- Transitive Property
- Symmetric Property

Instead of the symmetric property, the subset relation has the following property (page 225):

- Antisymmetric Property

We will now generalize these properties for an arbitrary relation R on an arbitrary set S.

Reflexive Property

In a *reflexive* relation, each element is related to itself:

$$a \mathrel{\mathsf{R}} a$$

R is *reflexive*
if and only if
for every a in S, $a \mathrel{\mathsf{R}} a$.

$=$ and $\leq$ are reflexive relations on every set of real numbers.
$=$ and $\subseteq$ are reflexive relations on every set of sets.

$$a = a$$
$$a \leq a$$
$$A \subseteq A$$

The $\subset$ and $<$ relations are not reflexive. Neither is the following relation:

$\mathsf{R} = \{ (1,3), (3,5), (1,1), (3,3) \}$
Since $(5,5) \notin \mathsf{R}$, 5 is not related to 5.

Transitive Property

In a *transitive* relation, we can make inferences analogous to the transitivity rule for implications:

If $a \mathrel{\mathsf{R}} b$ and $b \mathrel{\mathsf{R}} c$, then $a \mathrel{\mathsf{R}} c$.

R is *transitive*
if and only if
for every a, b, and c in S,
if $a \mathrel{\mathsf{R}} b$ and $b \mathrel{\mathsf{R}} c$, then $a \mathrel{\mathsf{R}} c$.

$=$, $\leq$, and $<$ are transitive relations on every set of real numbers.
$=$, $\subseteq$, and $\subset$ are transitive relations on every set of sets.

If $a = b$ and $b = c$, then $a = c$.
If $a \leq b$ and $b \leq c$, then $a \leq c$.
If $A \subseteq B$ and $B \subseteq C$, then $A \subseteq C$.

The following relation is not transitive:

$$R = \{ (1,3), (3,5), (1,1), (3,3) \}$$
$$1 \text{ R } 3 \text{ and } 3 \text{ R } 5, \text{ but } \sim(1 \text{ R } 5).$$

Consider the set S of students at Hanover College. Define a relation on S as follows: a R b if and only if a and b are in the same class. This relation is not transitive, for A could be in a class with B, and B could be in another class with C, but A is not in any class with C.

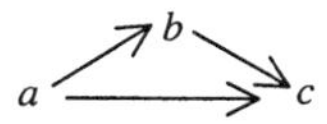

Transitive

In the directed graph of a transitive relation, arrows that are lined up head-to-tail must have the triangle completed, as illustrated on the left. In the second graph, the triangle formed by the arrow from a to b, and the arrow from b to c is not completed. So, a R b and b R c, but ~(a R c). Thus, this relation does not have the transitive property.

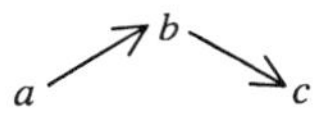

Not Transitive

The transitive property is a powerful tool for making deductions. If we're trying to prove that a is related to c, we can look for an intermediate stepping stone b to get us there. Consequently, transitive relations are frequently used in mathematics.

Symmetric Property

In a *symmetric* relation, the order of the elements does not affect the relation:

If a R b, then b R a.

> R is *symmetric*
> if and only if
> for every a and b in S,
> if a R b, then b R a.

If a is related to b, then b must also be related to a. It does not matter which element comes first. The subset relation does not have this property; neither does the $\leq$ relation. In fact, this particular property helps to distinguish those relations that order elements from those relations that arrange elements into nonoverlapping groups. For example, let a R b mean that a and b were born in the same month. This relation has the symmetric property:

If a and b were born in the same month,
then b and a were born in the same month.

In a room full of people, this particular relation would divide the people into nonoverlapping subgroups – the subgroups formed by people born in the same month.

In the directed graph of a symmetric relation, any two related elements must be connected with two-way arrows. If there is an arrow in one direction, there must be an arrow in the opposite direction. The adjacent directed graph is symmetric. However, it is neither reflexive nor transitive.

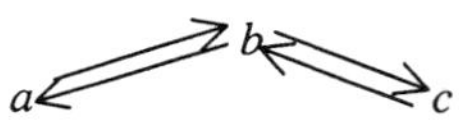

Symmetric

Antisymmetric Property

Some relations instill a sense of order on a set. For example, the $\leq$ relation orders the set of real numbers. In an order relation, it makes a difference as to which number we write first: $3 \leq 5$, but $\sim(5 \leq 3)$. The only time we can reverse the order with the $\leq$ relation is when the two elements are equal. This property is called *antisymmetric* because it is the opposite extreme of symmetric.

R is *antisymmetric*
if and only if
for every a and b in S,
if a R b and b R a, then $a = b$.

If a R b and b R a, then $a = b$.

For example, $\subseteq$, $\leq$, and $<$ are antisymmetric relations:

If $A \subseteq B$ and $B \subseteq A$, then $A = B$.

If $a \leq b$ and $b \leq a$, then $a = b$.

If $a < b$ and $b < a$, then $a = b$.

For the $<$ relation, the hypothesis of the above implication is always false, so the implication is true. Similarly, the proper subset relation is antisymmetric by default. With $<$ and $\subset$, we can never reverse the order:

If $a < b$, then $b \nless a$.

If $A \subset B$, then $\sim(B \subset A)$.

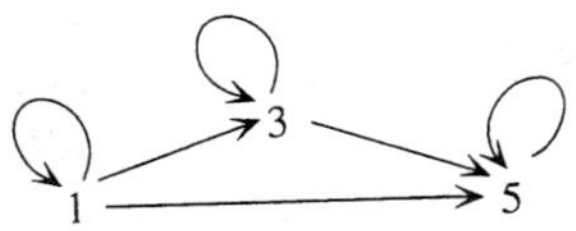

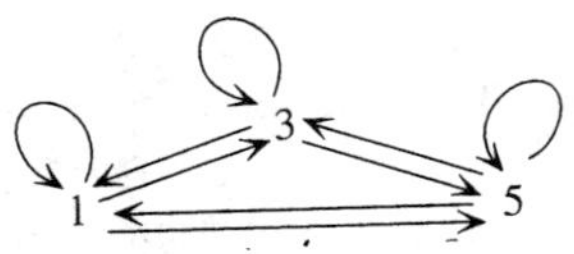

The directed graph of an antisymmetric relation will not have any two-way arrows between different elements. Conversely, if there are no two-way arrows between different elements, the relation is antisymmetric.

The first graph on the left is reflexive, transitive, and antisymmetric. A relation that has these three properties is called a *partial order*. We will examine order relations in the last section of this chapter.

The second graph is reflexive, transitive, and symmetric. A relation that has these three properties is called an *equivalence relation*. We will examine equivalence relations in the next section.

Inverse Relations

Each relation has an associated relation, called its *inverse*, which we obtain by reversing the order. The inverse of R is notated as R^{-1}.

$a\,R^{-1}b$
if and only if
b R a.

$a\,R^{-1}b$ if and only if b R a.

a is related to b under the inverse relation R^{-1} means that b is related to a under the relation R.

Using ordered pairs, the definition of R^{-1} can be phrased as follows:

$$R^{-1} = \{ (a,b) \mid (b,a) \in R \}$$
$$(a,b) \in R^{-1} \text{ if and only if } (b,a) \in R.$$

Domain (R^{-1}) = Range (R)
Range (R^{-1}) = Domain (R)

The inverse relation is obtained by switching the first and second terms in each ordered pair. Since the mapping is reversed in the inverse relation, the domains and ranges are switched. The domain of R^{-1} is the range of R, while the range of R^{-1} is the domain of R.

Example

1. Let $R = \{ (1,6), (1,8), (2,8), (3,10) \}$.

$R^{-1} = \{ (6,1), (8,1), (8,2), (10,3) \}$

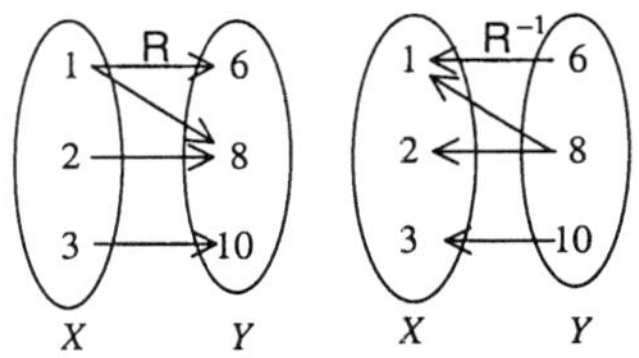

Since the ordered pairs in R are reversed in R^{-1}, their domains and ranges are switched:

Domain $(R) = \{ 1, 2, 3 \}$	Domain $(R^{-1}) = \{ 6, 8, 10 \}$
Range $(R) = \{ 6, 8, 10 \}$	Range of $(R^{-1}) = \{ 1, 2, 3 \}$

In the arrow picture of R^{-1}, we reverse the direction of each arrow in R, as illustrated on the left.

2. Let $R = \{ (x,y) \mid x^2 + 3y^2 = 1 \}$. Write R^{-1} in set notation.

$R^{-1} = \{ (y,x) \mid (x,y) \in R \}$ .... *Definition of* R^{-1}
$R^{-1} = \{ (y,x) \mid x^2 + 3y^2 = 1 \}$ .. *Substitute for* R
$R^{-1} = \{ (x,y) \mid y^2 + 3x^2 = 1 \}$ .. *Substitute x for y, y for x.*

Switching the x and y in the ordered pair is equivalent to switching the x and y in the equation, but we cannot switch in both places at the same time.

For every relation R, $(R^{-1})^{-1} = R$.

3. Let $R = \{ (1,2), (2,4), (2,5) \}$. Find $(R^{-1})^{-1}$.

$(R^{-1})^{-1}$ represents the inverse of R^{-1}.

$R^{-1} = \{ (2,1), (4,2), (5,2) \}$
$(R^{-1})^{-1} = \{ (1,2), (2,4), (2,5) \}$

When we switch the order twice in the above example, we are back to the original position of the ordered pairs. Thus, the inverse of R^{-1} is the relation R: $(R^{-1})^{-1} = R$

The R^{-1} notation may cause some confusion because the exponent does not have the same meaning as it does with a number. If x is a number, $x^{-1} = \frac{1}{x}$. However, when -1 is used as an exponent with a relation, it does not mean the reciprocal:

$$R^{-1} \neq \frac{1}{R}$$

n-ary Relations

A *3-ary relation* R between the sets X, Y and Z is a set of ordered triples whose first terms come from X, second terms from Y, and third terms from Z.

The relations we have examined so far are called *binary relations* because they provide a relation between two elements. Sometimes, though, we need relations between three or more elements. For example, a business may want a connection between a person's name, address, and account number, which is called a *3-ary relation.* To define a 3-ary relation, we use ordered triples instead of ordered pairs. With 3 sets involved in a relation, we no longer use the term range. Instead, we call X the first domain, Y the second domain, and Z the third domain. For example, let R be the following 3-ary relation:

$$R = \{ (1,2,5), (2,3,1), (3,3,4) \}$$

The first domain of R is $\{1, 2, 3\}$, its second domain is $\{2, 3\}$, and its third domain is $\{1,4,5\}$.

An *n-ary relation* R between the sets $X_1, X_2, X_3, \ldots, X_n$ is a set of ordered n-tuples where the ith term comes from X_i.

We can generalize the definition of a relation further to an *n-ary* relation. An n-ary relation is a set whose elements are ordered n-tuples. Each n-tuple is interpreted as giving a relation between each of the terms. An n-ary relation can also be described as a subset of a cross product of n sets.

Relational database theory is based on n-ary relations. For example, an airline may use a 5-ary relational database in which ticket information is stored in terms of (N, F, S, A, D) where N represents the name of the ticket-holder, F the flight number, S the departure site, A the destination site, and D the date. Each ordered 5-tuple in the database is called a *record.*

Exercise Set 4.1

1. Let x and y be real numbers. Sketch the graph of the relation R. Find the domain and range of R.

   a. $R = \{ (x,y) \mid x^2+y^2 = 1\}$
   b. $R = \{ (x,y) \mid x^2+y^2 \leq 1\}$
   c. $R = \{ (x,y) \mid x = y^2 \}$
   d. $R = \{ (x,y) \mid 0 \leq x \leq 1 \text{ and } 0 \leq y \leq 1 \}$
   e. $R = \{ (x,y) \mid 0 \leq x \leq 2 \text{ and } y = 2x \}$
   f. $R = \{ (x,y) \mid 1 \leq x \leq 3 \text{ and } y = 3x-5 \}$

2. Using arrow mappings, draw an example of the following:
   a. A relation that is not a function.
   b. A function whose inverse relation is not a function.
   c. A function whose inverse relation is a function.
3. Let f be a function. If f^{-1} is a function, what special property must f possess?
4. For the given R, represent R and R^{-1} as a set of ordered pairs, an arrow mapping, a graph, a directed graph, and a matrix.
   Find the domain and range of both R and R^{-1}.
   Determine if either R or R^{-1} is a function..
   a. R = { (0,2), (0,3), (0,4), (1,2) }.
   b. R = { (1,2), (2,2), (3,2), (4,3) }.
   c. R is the < relation on the set $S = \{3,7,8,9\}$.
   d.

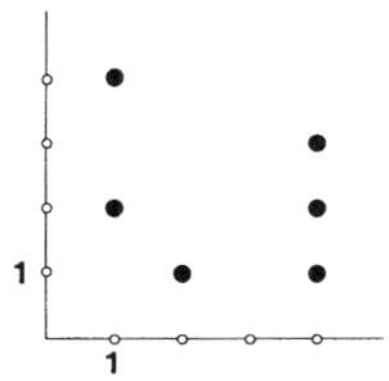

   e.

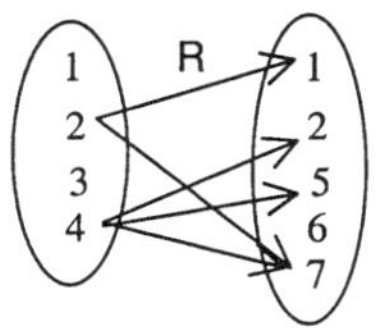

5. For each relation in the previous exercise, describe the visual relation between:
   a. The graphs for R and R^{-1}. b. The matrices for R and R^{-1}.
6. Compute R^{-1} in the indicated stages.
   a. Let R = $\{(x,y) \mid y = 2x+1$, where x and y are real numbers $\}$
      $R^{-1} = \{(y,x) \mid \_\_\_\_\_\_\_\_\}$
      $R^{-1} = \{(x,y) \mid \_\_\_\_\_\_\_\_\}$
   b. Let R = $\{(x,y) \mid y = x^2+1$, where x and y are real numbers $\}$
      $R^{-1} = \{(y,x) \mid \_\_\_\_\_\_\_\_\}$
      $R^{-1} = \{(x,y) \mid \_\_\_\_\_\_\_\_\}$
7. What is the inverse relation for the $\subseteq$ relation? The $\leq$ relation?
8. Explain why the following statements are true.
   a. Every relation has an inverse relation.
   b. The domain of the inverse relation is the range of the original relation.
   c. The range of the inverse relation is the domain of the original relation.
   d. For every relation R, $(R^{-1})^{-1} = R$.

9. Determine if the given relation is reflexive, transitive, symmetric, or antisymmetric.
   a. $R = \{(1,1), (1,3), (3,4), (1,4), (2,2), (3,3)\}$
   b. $R = \{(1,1), (1,3), (3,1), (3,4), (4,3), (2,2), (3,3), (4,4)\}$
   c. The $\leq$ relation on the set of real numbers.
   d. The $<$ relation on the set of real numbers.
   e. The subset relation on a collection of sets.
   f. The proper subset relation on a collection of sets.
10. Let R be the divides relation on the set $\mathbb{N}$ of natural numbers.
    $a \, R \, b$ if and only if a divides b.
    a. Prove that R is reflexive, transitive, and antisymmetric on $\mathbb{N}$.
    b. Is R antisymmetric on the set Z of integers?
11. On a graph, plot the points $A(2,7)$, $B(7,2)$ and $C(7,7)$. Draw ΔABC and the line $y = x$.
    a. Examine ΔABC. What kind of triangle is it?
    b. Find the coordinates of the midpoint of $\overline{AC}$ and $\overline{BC}$. Use these coordinates to find the midpoint of $\overline{AB}$. Label this point as D. Is D on the line $y = x$?
    c. Draw $\overline{DC}$. Closely examine ΔACD and ΔBCD. Do you notice any relationship between them?
    d. Prove that the line $y = x$ is the perpendicular bisector of $\overline{AB}$.
    e. Generalize your proof in part (d) and prove the following:
       For every point (a,b), the line $y = x$ is the perpendicular bisector of the line through (a,b) and (b,a).
12. Let R be a relation on the set $\mathbb{R}$ of real numbers.
    a. Given the location of a point (a, b), describe how you could locate (b,a) using the line $y = x$ instead of the coordinate axes.
    b. Given the graph of R in a coordinate plane, describe how to graph the relation R^{-1} using only the line $y = x$. Draw a graph and illustrate your technique.
13. Make up a random example of a relation R where R is a set of ordered pairs. Is $R = R^{-1}$? Is R symmetric? Are these two questions related?
14. Let R be a relation on a set of real numbers. How can you determine if $R = R^{-1}$ from the following visual information?
    a. Graph of R b. Matrix for R c. Directed graph of R
15. Given the graph of a relation R on a set of real numbers, how can you visually determine if R has the given property?
    a. Reflexive b. Symmetric c. Antisymmetric

Activity 4.2

Let R be a relation on a set S. If $a \in S$, $[a]$ is called the *equivalence class* of a. $[a]$ denotes all the elements in S that are related to a:

$$[a] = \{ x \mid a \mathrm{R} x \}$$

1. Let S be the set of students in your school.
   $a \mathrm{R} b$ if and only if a and b were born in the same month.
   a. What is your equivalence class for this relation?
   b. How many different equivalence classes are there?
      Do they form a partition of S?
2. Let S be the set of students in your school.
   $a \mathrm{R} b$ if and only if a and b have the same birthday.
   a. What is your equivalence class for this relation?
   b. How many different equivalence classes are there?
      Do they form a partition of S?
3. Let $S = \{0,1,2,3,4,5,6\}$. $a \mathrm{R} b$ if and only if 2 divides $a-b$.
   a. For each a in S, list the elements in $[a]$.
   b. Are there any elements in S for which $[a] = [b]$?
   c. Let $P = \{ [a] \mid a \in S. \}$ List the elements in P.
   d. How many elements are in P? Is P a partition of S?
4. Repeat the previous exercise for the given relation.
   a. $a \mathrm{R} b$ if and only if 3 divides $a-b$.
   b. $a \mathrm{R} b$ if and only if $a \le b$.

≡ 4.2 Equivalence Relations ≡

An equivalence relation identifies a property that makes elements essentially the same with respect to that property. For example, different dollar bills are different entities, but most people only care about the face value. Two different bills are essentially the same if they have the same face value. The relation of "having the same face value" is an equivalence relation on a set of money. To be an equivalence relation, a relation must possess the three basic properties of equality.

Let R be a relation on the set S.
R is an *equivalence relation*
if and only if
R has the following 3 properties:

- Reflexive
- Transitive
- Symmetric

Let R be a relation on a set S. R is an *equivalence relation* if and only if the following three statements are true for every a, b, and c in S.

Reflexive Property: $a \mathrel{R} a$

Transitive Property: If $a \mathrel{R} b$ and $b \mathrel{R} c$, then $a \mathrel{R} c$.

Symmetric Property: If $a \mathrel{R} b$, then $b \mathrel{R} a$.

The reflexive property guarantees that every element is related to itself. The transitive property states that any middle term in a relation lineup can be eliminated. The symmetric property states that the order in which the elements are listed does not affect the relation.

Example

On any set of money, the relation of "having the same face value" satisfies the definition of an equivalence relation:

Reflexive: a has the same face value as a.

Transitive: If a has the same face value as b and b has the same face value as c, then a has the same face value as c.

Symmetric: If a has the same face value as b, then b has the same face value as a.

We use symbols similar to the = sign to represent equivalence relations:

$$a \cong b, \quad a \equiv b, \quad a \approx b$$

Congruent Figures

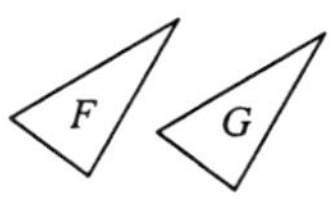

In Euclidean geometry, we focus on the size and shape of figures instead of their location. The relation of having the same shape and size, which is called *congruence*, is notated as $F \cong G$. We treat two different congruent triangles as essentially the same; however, they are not equal as sets since triangles in different locations represent different sets of points. The congruence relation has the three basic properties of equality, so it is an equivalence relation on every set of figures:

Reflexive: $F \cong F$

Transitive: If $F \cong G$ and $G \cong H$, then $F \cong H$.

Symmetric: If $F \cong G$, then $G \cong F$.

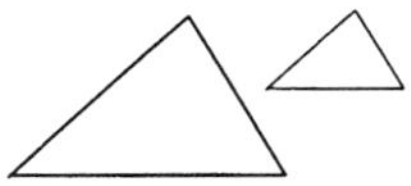

Another important relation on figures is the relation of being *similar*. *F is similar to G* means that F and G have the same shape, but not necessarily the same size. Similarity has the reflexive, transitive and symmetric properties, so it is an equivalence relation on every set of figures.

Congruent Numbers

One of the most important equivalence relations on the set of integers is *congruence mod n*. Even though its name is similar, it is not related to congruence on figures. Let n be a natural number and let a and b be integers.

$$a \equiv_n b \text{ if and only if } n \text{ divides } a - b.$$

The notation $a \equiv_n b$ is read as "a is congruent to b mod n." It is sometimes notated as $(a \equiv b)_{\text{mod } n}$. When the context is clear, we may omit the subscript and write $a \equiv b$. We will prove that congruence mod n is an equivalence relation on page 338.

Congruence mod 2

Let's first look at *congruence mod 2*:

$$a \equiv_2 b \text{ if and only if } 2 \text{ divides } a - b.$$

The difference of two even integers is always even, so each two even integers will be congruent. Likewise, the difference of two odd integers is always even, so each two odd integers will be congruent:

$$\begin{array}{ll} 2 \equiv_2 4 & 1 \equiv_2 3 \\ 14 \equiv_2 536 & 413 \equiv_2 75 \\ 96 \equiv_2 -8 & -15 \equiv_2 63 \end{array}$$

Moreover, the difference of an odd integer and an even integer is always odd, so an odd integer will never be congruent to an even integer under this relation. Congruence mod 2 divides the set of integers into two nonoverlapping subsets: the even integers and the odd integers.

Congruence mod 3

Let's take a brief look at *congruence mod* 3:

$$a \equiv_3 b \text{ if and only if } 3 \text{ divides } a - b.$$

If b is one of the following integers, 3 divides $0 - b$.

$$\ldots, -6, -3, 0, 3, 6, 9, 12, 15 \ldots$$

So, 0 is related to each of the above integers.

If b is one of the following integers, 3 divides $1 - b$.

$$\ldots, -8, -5, -2, 1, 4, 7, 10, 13, \ldots$$

So, 1 is related to each of the above integers.

2 is related to each of the following integers:

$$\ldots, -7, -4, -1, 2, 5, 8, 11, 14, \ldots$$

3 is related to the same integers that are related to 0.

4 is related to the same integers that are related to 1.

5 is related to the same integers that are related to 2.

And so on. . . .

Notice how congruence mod 3 divides the set of integers into 3 nonoverlapping subsets, which are called equivalence classes.

Equivalence Classes

$[a] = \{ x \mid a \mathsf{R} x \}$

An equivalence relation R on a set S partitions S into nonoverlapping subsets called *equivalence classes*. Let a be an element in the set S. The equivalence class of a is the set of all elements in S that are related to a; it is notated as $[a]$.

$$[a] = \{ x \mid a \mathsf{R} x \}$$

The equivalence class of a is a subset of the original set S. In the above definition, x must be in S, otherwise $a \mathsf{R} x$ would not be true. To be more explicit, we could write the equivalence class of a in either of the following forms:

$$[a] = \{ x \mid a \mathsf{R} x \text{ and } x \in S \}$$
$$[a] = \{ x \text{ in } S \mid a \mathsf{R} x \}$$

Example

Find the equivalence classes of the given equivalence relation.

1. Let R be the relation of congruence mod 2 on S, where $S = \{ 1, 2, 3, 4, 5 \}$

   To compute $[1]$, we make the appropriate substitutions:

   $[a] = \{ x \mid a \mathsf{R} x \}$
   $[1] = \{ x \mid 1 \equiv_2 x \}$. $1 \equiv_2 1,\ 1 \equiv_2 3,\ 1 \equiv_2 5$
   So, $[1] = \{ 1, 3, 5 \}$.

   $[3] = \{ x \mid 3 \equiv_2 x \} = \{ 1, 3, 5 \}$

Even though the notation is different for $[1]$ and $[3]$, they represent the same set: $[1] = [3]$. Congruence mod 2 has only two different equivalence classes on S:

$$[1] = [3] = [5] = \{1, 3, 5\}$$
$$[2] = [4] = \{2, 4\}$$

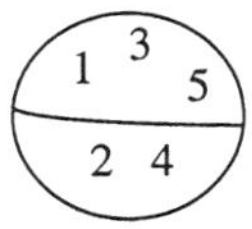

S

Let P denote the set of equivalence classes for this relation. Then P has two elements:

$$P = \{\{1, 3, 5\}, \{2, 4\}\}$$

P partitions the set S into 2 nonoverlapping subsets, as illustrated on the left.

2. Consider the congruence relation on the adjacent set S of figures. $[A]$ denotes the set of all figures congruent to A.

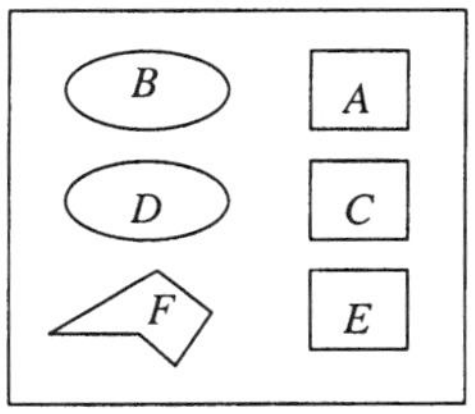

$$[A] = \{X \mid A \cong X\} = \{A, C, E\}$$
$$[C] = \{X \mid C \cong X\} = \{A, C, E\}$$
$$[E] = \{X \mid E \cong X\} = \{A, C, E\}$$
$$[B] = \{X \mid B \cong X\} = \{B, D\}$$
$$[D] = \{X \mid D \cong X\} = \{B, D\}$$
$$[F] = \{X \mid G \cong X\} = \{F\}$$

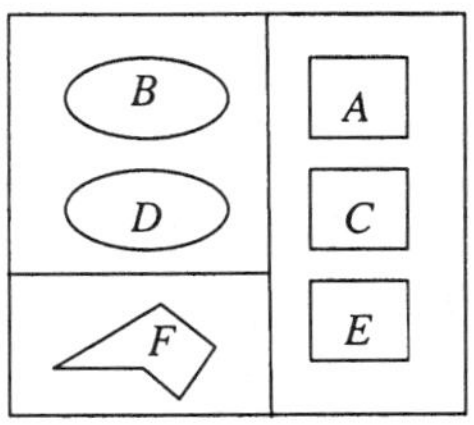

Let P denote the set of equivalence classes for this relation. P has 3 elements:

$$P = \{\{A, C, E\}, \{B, D\}, \{F\}\}$$

P partitions S into 3 nonoverlapping subsets, as illustrated on the left.

Partitions

The set of equivalence classes of an equivalence relation *partitions* the original set S into nonoverlapping subsets. Since the elements in a partition are sets, students sometimes get confused by a proof of this statement. However, if we first set up the outside structure and then pay close attention to the meaning of each term, the proof is fairly straightforward. Let's analyze how to structure the proof. Let P denote the set of all equivalence classes of an equivalence relation R on a set S:

$$P = \{[a] \mid a \in S\}$$

To prove that P is a partition of S, we must demonstrate that the adjacent definition of a partition is satisfied (page 230). Thus, we must prove the following three statements:

1. P is a collection of *nonempty subsets* of S.
2. Each element in S is in *at least one* element of P.
3. Each element in S is in *only one* element of P.

Let P be a collection of nonempty subsets of S.
P is a *partition* of S
if and only if
each element in S is in one and only one element in P.

These three statements give us the outside structure for the proof. Next we translate each statement in terms of variables so that we can see how to structure each subproof.

1. P is a collection of nonempty subsets of S.

   If $[a] \in P$, then $[a] \subseteq S$ and $[a] \neq \emptyset$.

   By the definition of $[a]$, $[a] \subseteq S$. In order to show that $[a]$ is not empty, we will use the fact that R is reflexive. Since $a \mathrel{R} a$, $a \in [a]$. Thus, $[a] \neq \emptyset$.

2. Each element a in S is in at least one element of P.

   There exists an X in P such that $a \in X$.

   Since we know that $a \in [a]$, we have the set that we need, namely, $X = [a]$.

3. Each element a in S is in only one equivalence class.

   If $a \in [b]$ and $a \in [c]$, then $[b] = [c]$.

   Since we already know that $a \in [a]$, we can simplify the above translation by eliminating c:

   If $a \in [b]$, then $[a] = [b]$.

   To prove that $[a] = [b]$, we use the standard strategy for proving two sets are equal:

   $[a] \subseteq [b]$ and $[b] \subseteq [a]$

   In this argument, which will be a little longer than the other two parts, we need to stay focused on our goal.

All of the above tasks can be accomplished using the three properties of an equivalence relation. In the first two parts, we need the reflexive property. In the third part, we need to use the both the symmetric and transitive properties. Before reading the following proof, try to construct it yourself using the above analysis to set up your structure.

Theorem Let R be an equivalence relation on a set S. The set of all equivalence classes of R is a partition of S.

Proof Let a and b be in S. Since R is an equivalence relation, R is reflexive, symmetric and transitive. Let P be the set of all equivalence classes of R:

$$P = \{ [a] \mid a \in S \}$$

Claim 1: P is a collection of nonempty subsets of S.
Let $[a]$ be an element in P.
$[a] = \{ x \text{ in } S \mid a \, \mathsf{R} \, x \}$. So, $[a] \subseteq S$.
Since R is reflexive, $a \, \mathsf{R} \, a$. So $a \in [a]$. Thus, $[a] \neq \emptyset$.
So P is a collection of nonempty subsets of S.

Claim 2: Each element in S is in at least one element of P.
In Claim 1, we demonstrated that $a \in [a]$.
Since $[a] \in P$, a is in at least one element of P.

Claim 3: Each element in S is in only one element of P.
Assume that $a \in [b]$.
$[b] = \{ x \mid b \, \mathsf{R} \, x \}$. Since $a \in [b]$, $b \, \mathsf{R} \, a$.
Since R is symmetric, $a \, \mathsf{R} \, b$.

Subclaim: $[a] = [b]$
Assume that $x \in [a]$.
By the definition of $[a]$, $a \, \mathsf{R} \, x$.
In the beginning of Claim 3, we established that $b \, \mathsf{R} \, a$.
So, $b \, \mathsf{R} \, a$ and $a \, \mathsf{R} \, x$. Since R is transitive, $b \, \mathsf{R} \, x$.
So, by the definition of $[b]$, $x \in [b]$.
Therefore, $[a] \subseteq [b]$.

Conversely, assume that $x \in [b]$.
By the definition of $[b]$, $b \, \mathsf{R} \, x$.
In the beginning of Claim 3, we established that $a \, \mathsf{R} \, b$.
So, $a \, \mathsf{R} \, b$ and $b \, \mathsf{R} \, x$. Since R is transitive, $a \, \mathsf{R} \, x$.
So, by the definition of $[a]$, $x \in [a]$.
Therefore, $[b] \subseteq [a]$.

The above argument shows that $[a] = [b]$. Since any equivalence class that contains a must be equal to $[a]$, a is in only one element of P.

Therefore, the set P of equivalence classes of the relation R is a partition of the original set S.

If you fully understand all the nuances, notation, and detail in the above proof, congratulations, for you have reached a new level of reasoning ability. If you can sit down and write a proof of this theorem, without having just looked at it and without trying to memorize it, using only the definitions to structure your proof and focus your thoughts, then you have elevated your logical reasoning skills to the major leagues.

On the other hand, if you do not yet fully understand this proof, don't be discouraged – keep working on it. You probably need to deepen your understanding of all the terms and notations. Definitions and symbolic notation are empty words if we do not have a personal understanding of what they represent. Each time we work through an example or a problem, we are building our personal understanding of the concepts. After you work through the exercises for this section, you may find the reasoning in this proof easier to follow.

Using the previous theorem, we can prove that if two elements are related to each other under an equivalence relation, their equivalence classes must be equal:

Theorem Let R be an equivalence relation on S and let a and b be elements in S. If $a \mathrel{R} b$, then $[a] = [b]$.

Proof Assume that $a \mathrel{R} b$.

By the definition of $[b]$, $a \in [b]$.

$a \in [b]$ and $a \in [a]$.

Since the equivalence classes form a partition,

a is in only one equivalence class. Therefore, $[a] = [b]$.

This theorem can significantly reduce our work in computing all the equivalence classes of a particular equivalence relation, For example, with congruence mod 2 on $S = \{1, 2, 3, 4, 5\}$:

$$[1] = \{1, 3, 5\}$$

Since $3 \mathrel{R} 1$, we do not have to compute $[3]$: $[3] = [1]$.

You are asked to prove the following properties of equivalence classes in (13) of the next exercise set.

If $\sim(a \mathrel{R} b)$, then $[a] \cap [b] = \emptyset$.

If $[a] \cap [b] \neq \emptyset$, then $[a] = [b]$.

Let R be an equivalence relation.

If $a \mathrel{R} b$, then $[a] = [b]$.

If $\sim(a \mathrel{R} b)$, then $[a] \cap [b] = \emptyset$.

Partition to Equivalence Relation

Given an equivalence relation on a set S, its equivalence classes form a partition of S (page 331). So each equivalence relation on a set has an associated partition. The converse is also true. Each partition on a set has an associated equivalence relation, which we define as follows. All elements in S that are in the same subset of the partition are related to each other. In other words, two elements in S are related means that there exists an X in the partition P that contains both elements. Since the elements of the partition P are subsets of S, we have to keep our verbs straight. If $X \in P$, then $X \subseteq S$.

Associated Relation

Let P be a partition of a set S. Let a and b be in S. The associated relation of P is defined as follows.

---

a R b
if and only if
a and b are in the same
element of the partition.

---

a R b
if and only if
there exists an X in P such that $a \in X$ and $b \in X$.

Before we prove that the above relation is an equivalence relation, let's think about its meaning for the following example.

⊕ Example

Let $S = \{1,2,3,4,5,6\}$. Let $P = \{\{1,2\}, \{3,4\}, \{5\}, \{6\}\}$. The partition P is illustated on the left.

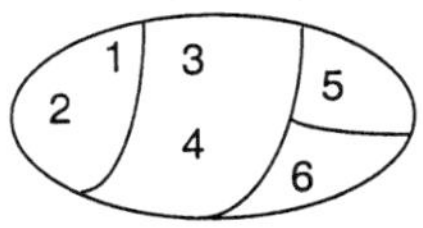

1. Find the associated relation of the partition P.
   List all of the elements that are related to each other.

   Let $X = \{1,2\}$. 1 R 2 because $1 \in X$ and $2 \in X$.
   2 R 1 because $2 \in X$ and $1 \in X$.
   1 R 1 because $1 \in X$ and $1 \in X$.
   2 R 2 because $2 \in X$ and $2 \in X$.

   Let $X = \{3,4\}$. 3 R 4 because $3 \in X$ and $4 \in X$.
   4 R 3 because $4 \in X$ and $3 \in X$.
   3 R 3 because $3 \in X$ and $3 \in X$.
   4 R 4 because $4 \in X$ and $4 \in X$.

   Let $X = \{5\}$. 5 R 5 because $5 \in X$ and $5 \in X$.

   Let $X = \{6\}$. 6 R 6 because $6 \in X$ and $6 \in X$.

2. Is R an equivalence relation?

   Since each a in the set S is also a member of some element in the partition, a R a. So R is reflexive.

Suppose that $a \mathsf{R} b$.
Then a and b are in the same member of the partition.
So, b and a are in the same member of the partition.
Hence, $b \mathsf{R} a$. Therefore, R is symmetric.

Suppose that $a \mathsf{R} b$ and $b \mathsf{R} c$.
Then there exists an X in P that contains a and b.
Likewise, there exists a Y in P that contains b and c.
Since b can be in only one member of the partition, $X = Y$.
Thus, a and c are in the same member of the partition.
So, $a \mathsf{R} c$. Hence, R is transitive.

So, R is is an equivalence relation on S.

3. Find the equivalence classes of R.

$$[a] = \{ x \mid a \mathsf{R} x \}$$

$$[1] = \{ x \mid 1 \mathsf{R} x \} = \{1,2\}$$
$$[2] = \{ x \mid 2 \mathsf{R} x \} = \{1,2\}$$

$$[3] = \{ x \mid 3 \mathsf{R} x \} = \{3,4\}$$
$$[4] = \{ x \mid 4 \mathsf{R} x \} = \{3,4\}$$

$$[5] = \{ x \mid 5 \mathsf{R} x \} = \{5\}$$

$$[6] = \{ x \mid 6 \mathsf{R} x \} = \{6\}$$

Note that the equivalence classes are the members of the original partition. Even though there are 6 elements in S, there are only 4 different equivalence classes.

$$\{ [a] \mid a \in S \} = \{ [1], [2], [3], [4], [5], [6] \}$$
$$= \{ \{1,2\}, \{3,4\}, \{5\}, \{6\} \}$$

The symmetric property of R is inherited from the wording of the relation: "*a* and *b* are in the same set" has the same meaning as "*b* and *a* are in the same set." However, the reflexive and transitive properties of R are inherited from the properties of a partition:

Each element in S is in at least one of the members of the partition, which gives R the reflexive property.

Each element in S is in only one of the members of the partition, which gives R the transitive property.

Note how these properties are used in the following proof.

Theorem Let P be a partition of a set S, and let a and b be in S. Define the relation R on S as follows:

$$a \,\mathsf{R}\, b$$

if and only if

there exists an X in P such that $a \in X$ and $b \in X$.

Then R is an equivalence relation on S.
Furthermore, the set of equivalence classes of R is equal to P:

$$P = \{\, [a] \mid a \in S \,\}$$

Proof Let a, b, and c be elements in S.

Claim 1: R is reflexive.
Since P is a partition, a is in at least one member of P.
So, there exists an X in P such that $a \in X$.
Since $a \in X$ and $a \in X$, $a \,\mathsf{R}\, a$.
Therefore, R is reflexive.

Claim 2: R is symmetric.
Assume that $a \,\mathsf{R}\, b$.
Then there exists an X in P such that $a \in X$ and $b \in X$.
So, $b \in X$ and $a \in X$. Hence, $b \,\mathsf{R}\, a$.
Therefore, R is symmetric.

Claim 3: R is transitive.
Assume that $a \,\mathsf{R}\, b$ and $b \,\mathsf{R}\, c$.
Since $a \,\mathsf{R}\, b$, for some X in P, $a \in X$ and $b \in X$.
Since $b \,\mathsf{R}\, c$, for some Y in P, $b \in Y$ and $c \in Y$.
Note that $b \in X \cap Y$.
Since P is a partition, b can be in only one element of P.
Hence, $X = Y$. So, $a \in X$ and $c \in X$. Thus, $a \,\mathsf{R}\, c$.
Therefore, R is transitive.

Hence, R is an equivalence relation on S.

Let $a \in S$. Then there exists an X in P such that $a \in X$.

$$[a] = \{\, x \mid a \,\mathsf{R}\, x \,\} = X$$

So, the set of all equivalence classes of the relation R is the original partition:

$$P = \{\, [a] \mid a \in S \,\}$$

⊕ *Example*

Let $S = \{1, 2, 3, 4, 5, 6\}$. Let $P = \{A, B, C\}$, where $A = \{1, 2\}$, $B = \{3, 4, 5\}$, and $C = \{6\}$.

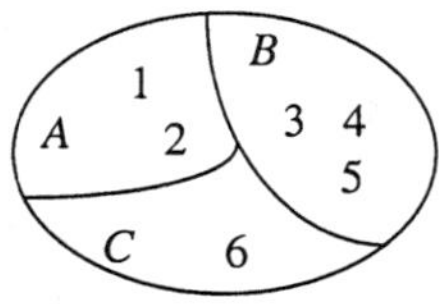

P is collection of nonempty, pairwise disjoint sets whose union is S, so P is a partition of S. The equivalence relation induced by this partition is defined as follows. Let a and b be in S.

$$a \,\mathsf{R}\, b$$

if and only if

there exists an X in P such that $a \in X$ and $b \in X$.

$1 \,\mathsf{R}\, 1$ since 1 and 1 are both in A.

$1 \,\mathsf{R}\, 2$ since 1 and 2 are both in A.

$$[1] = \{ x \mid x \,\mathsf{R}\, 1 \} = \{1, 2\}$$
$$[1] = [2] = A$$

$$[3] = \{ x \mid x \,\mathsf{R}\, 3 \} = \{3, 4, 5\}$$
$$[3] = [4] = [5] = B$$

$$[6] = \{ x \mid x \,\mathsf{R}\, 6 \} = \{6\} = C$$

Note how each equivalence class of R is a set in the original partition. Using the notation for equivalence classes, we can list the elements in P as follows:

$$P = \{ [1], [2], [3], [4], [5], [6] \}$$

In the above listing, P appears to have 6 elements.
However, some of these equivalence classes are equal.
P has only 3 elements:

$$P = \{A, B, C\}$$

As illustrated in the following examples, if P is not a partition of S, the above definition of R does not produce an equivalence relation:

If P contains subsets of S that overlap,
the relation will not have the transitive property.

If the subsets from P do not cover all of S,
the relation will not have the reflexive property.

Example

Let $S = \{1,2,3,4,5,6\}$. Let P be the given collection of subsets of S. Let a and b be in S. Define the relation R as follows:

$a \,\mathsf{R}\, b$

if and only if

there exists an X in P such that $a \in X$ and $b \in X$.

Is R an equivalence relation?

1. Let $P = \{A,B,C\}$, where $A = \{1,2,3\}$, $B = \{3,4,5\}$, $C = \{6\}$.

   Since both 1 and 3 are in A, $1 \,\mathsf{R}\, 3$.
   Since both 3 and 4 are in B, $3 \,\mathsf{R}\, 4$.
   There is no element in P that contains both 1 and 4.
   So $\sim(1 \,\mathsf{R}\, 4)$. Thus, R is not transitive.

   Hence, R is not an equivalence relation.
   Note that P is not a partition of S since $A \cap B \neq \emptyset$.

2. Let $P = \{A,B,C\}$, where $A = \{1,2\}$, $B = \{3,4\}$, $C = \{6\}$.

   There is no element of P that contains 5.
   So 5 is not related to itself.
   Therefore, R is not reflexive.

   Hence, R is not an equivalence relation. Note that P is not a partition of S since 5 is not in any of its members.

Equivalence Relation vs. Partition

We have an intimate relation between equivalence relations and partitions.

An equivalence relation on S determines a partition of S.

A partition of S determines an equivalence relation on S.

Like heads and tails, equivalence relations and partitions are different sides of the same coin. In some applications, it may be more comfortable to work with a partition; in other cases, the relation format may be more convenient. Either way, we are working with essentially the same concept.

A relation on a set embeds a structure between elements in the set. The structure embedded by an equivalence relation can be described as a partition. An equivalence relation structures a set by dividing it into nonoverlapping subsets.

Congruence mod n

$a \equiv_n b$

if and only if

n divides $a-b$.

We will now examine the relation, *congruence mod n,* in more detail (page 327). For the rest of this section, n will denote a positive integer. To prove that congruence mod n is an equivalence relation, we must work with the definition of divides:

$$a \equiv_n b \text{ if and only if } n \text{ divides } a-b.$$

$$n \text{ divides } a-b \text{ if and only if } \tfrac{a-b}{n} = k \text{ for some integer } k.$$

Theorem Congruence mod n is an equivalence relation on the set Z of integers.

Proof Let a, b and c be integers.

Claim 1: Congruence mod n is reflexive.

$\frac{a-a}{n} = 0$. So n divides $a-a$. Therefore $a \equiv_n a$.
So congruence mod n is reflexive.

Claim 2: Congruence mod n is symmetric.

Assume that $a \equiv_n b$. Then n divides $a-b$.

So $\frac{a-b}{n} = k$ for some integer k.

$$\frac{b-a}{n} = \frac{-(a-b)}{n} = -k.$$

Since $-k$ is an integer, n divides $b-a$. So $b \equiv_n a$.

Therefore, congruence mod n is symmetric.

Claim 3: Congruence mod n is transitive.

Assume that $a \equiv_n b$ and $b \equiv_n c$.

Then n divides $a-b$ and n divides $b-c$.

$\frac{a-b}{n} = k$ for some integer k. $\frac{b-c}{n} = j$ for some integer j.

$$\frac{(a-b)+(b-c)}{n} = k+j. \quad \text{So, } \frac{a-c}{n} = k+j.$$

Since k and j are integers, $k+j$ is an integer.

The above equation shows that n divides $a-c$. So $a \equiv_n c$.

Therefore, congruence mod n is transitive.

Therefore, congruence mod n is an equivalence relation on Z.

$a \equiv_n b$
if and only if
$a = b + nk$
for some integer k.

To get a different view of congruence mod n, let's rephrase it in terms of multiplication:

$a \equiv_n b$ if and only if $\frac{a-b}{n} = k$ for some integer k.

$a \equiv_n b$ if and only if $a = b + nk$ for some integer k.

Using the latter form, we can quickly compute everyone that is related to b:

$$[b] = \{ b + nk \mid k \text{ is an integer} \}$$

Congruence mod 2

For congruence mod 2, $[b] = \{ b + 2k \mid k \text{ is an integer} \}$.

$$[0] = \{ 0 + 2k \mid k \text{ is an integer} \}$$
$$[0] = \{ \ldots, -4, -2, 0, 2, 4, \ldots \}$$

$$[1] = \{ 1 + 2k \mid k \text{ is an integer} \}$$
$$[1] = \{ \ldots, -3, -1, 1, 3, 5, \ldots \}$$

Congruence mod 2 partitions the set Z of integers into two equivalence classes: the even integers and the odd integers.

Congruence mod 3

For congruence mod 3, $[b] = \{ b + 3k \mid k \text{ is an integer} \}$.

$$[0] = \{ 0 + 3k \mid k \text{ is an integer} \}$$
$$[0] = \{ \ldots, -6, -3, 0, 3, 6, 9, \ldots \}$$

$$[1] = \{ 1 + 3k \mid k \text{ is an integer} \}$$
$$[1] = \{ \ldots -5, -2, 1, 4, 7, 10, \ldots \}$$

$$[2] = \{ 2 + 3k \mid k \text{ is an integer} \}$$
$$[2] = \{ \ldots -4, -1, 2, 5, 8, 11, \ldots \}$$

In $[0]$, we list 0, then count by 3's in both directions.
In $[1]$, we list 1, then count by 3's in both directions.
In $[2]$, we list 2, then count by 3's in both directions.

Since we count by 3's in each equivalence class, the successive listing of equivalence classes keeps cycling around the above 3 equivalence classes:

$$[3] = [0] \qquad [6] = [0] \qquad [9] = [0]$$
$$[4] = [1] \qquad [7] = [1] \qquad [10] = [1]$$
$$[5] = [2] \qquad [8] = [2] \qquad [11] = [2]$$

So, congruence mod 3 partitions the set of integers into three equivalence classes: $[0]$, $[1]$, $[2]$

Congruence mod 4 partitions the set of integers into four equivalence classes: [0], [1], [2], [3]

Congruence mod n has n equivalence classes: $[0], [1], [2], \ldots, [n-1]$

$$[0] = \{ \ldots, -12, -8, -4, 0, 4, 8, 12, \ldots \}$$
$$[1] = \{ \ldots, -11, -7, -3, 1, 5, 9, 13, \ldots \}$$
$$[2] = \{ \ldots, -10, -6, -2, 2, 6, 10, 14, \ldots \}$$
$$[3] = \{ \ldots, -9, -5, -1, 3, 7, 11, 15, \ldots \}$$

In a similar manner, congruence mod n partitions the set Z of integers into n equivalence classes, which we notate as Z_n:

$$Z_n = \{ [0], [1], [2], [3], \ldots, [n-1]$$

We always list the equivalence classes in Z_n using one of the above representatives. Instead of writing $[n]$, we write $[0]$.

Division Algorithm

When we use the division algorithm to divide an integer x by n, we obtain a unique integer q (*quotient*) and a unique integer r (*remainder*) such that the following is true:

$$x = qn + r \text{ and } 0 \leq r < n$$

Since $x - r$ is a multiple of n, $x \equiv_n r$. So, $x \in [r]$.

Let x be an integer. Let r be the remainder when we divide x by n. Then $x \equiv_n r$.

Thus, each integer is in an equivalence class determined by one of the possible remainders when we divide by n. When we divide by n, there are n possible remainders:

Remainders: $0, 1, 2, 3, \ldots, n-1$

Each of these remainders represents a different equivalence class for congruence mod n. Furthermore, they represent all the equivalence classes.

When we divide an integer by 4, there are 4 different possible remainders:

Remainders: 0, 1, 2, 3

Each of these remainders represents a different equivalence class for congruence mod 4. Congruence mod 4 has 4 equivalence classes, which we notate as Z_4:

$$Z_4 = \{ [0], [1], [2], [3] \}$$

Congruence mod 5 has 5 equivalence classes, which we notate as Z_5:

$$Z_5 = \{ [0], [1], [2], [3], [4] \}$$

To determine which equivalence class in Z_n contains a given integer, we divide by n and use the remainder, as illustrated in the following examples:

Example

1. In Z_4, which equivalence class contains 34?

$$\begin{aligned}\frac{34}{4} &= 8 \text{ with a remainder of } 2\\ 34 &= 2+4(8)\end{aligned}$$

So, $34 \equiv_4 2$.

Thus, $34 \in [2]$.

2. In Z_4, which equivalence class contains 4387?

$$\begin{aligned}\frac{4387}{4} &= 1096 \text{ with a remainder of } 3.\\ 4387 &= 3 + 4(1096)\end{aligned}$$

So, $4387 \equiv_4 3$.

Thus, $4387 \in [3]$.

3. In Z_5, which equivalence class contains 47?

$$\frac{47}{5} = 9 \text{ with a remainder of } 2.$$
$$47 = 2+5(9)$$

So, $47 \equiv_5 2$.

Thus, $47 \in [2]$.

4. In Z_5, which equivalence class contains -47?

$$47 = 2+5(9)$$
$$-47 = -2+5(-9)$$

To get the remainder r in the required form, $0 \le r < 5$, we rewrite -2 in terms of 5: $-2 = 3-5$

$$-47 = (3-5)+5(-9)$$
$$-47 = 3 + 5(-1)+5(-9)$$
$$-47 = 3 + 5(-10)$$

So, $-47 \equiv_5 3$.

Thus, $-47 \in [3]$.

Exercise Set 4.2

1. Let S be the set of students at your college and let a and b be in S. Determine if the given relation is an equivalence relation on S. If so, what are its equivalence classes?
   a. a is related to b iff a and b were born in the same country.
   b. a is related to b iff a has the same birthday as b.
   c. a is related to b iff a and b are taking the same course.
   d. a is related to b iff a and b are in the same club.
   e. a is related to b iff a and b have the same major.

2. Let a and b be natural numbers. Determine if the relation R is an equivalence relation on $\mathbb{N}$. If so, find its equivalence classes.
   a. $a \mathrm{R} b$ iff a is a factor of b.
   b. $a \mathrm{R} b$ iff $a \leq b$.
   c. $a \mathrm{R} b$ iff a and b are both multiples of 3, or neither is a multiple of 3.

3. Let a and b be in the set S of all geometric figures in a plane. Determine if the relation R is an equivalence relation on S.
   a. $a \mathrm{R} b$ iff the area of a is equal to the area of b.
   b. $a \mathrm{R} b$ iff $a \cap b \neq \emptyset$.

4. Let $S = \{0, 1, 2, 3, 4, 5, 6, 7, 8, 9, 10, 11, 12, 13, 14, 15, 16\}$. Let a and b be in S. $a \mathrm{R} b$ if and only if 4 divides $a-b$.
   a. Is R an equivalence relation on S?
   b. List the elements in $[a]$ for each a in S.
   c. Let $P = \{[a] \mid a \in S.\}$ How many elements are in P? Is P a partition of S?

5. Let $S = \{1, 2, 3, 4, 5, 6, 7\}$. Let a and b be in S. Define the relation R as follows:

   $a \mathrm{R} b$ iff there exists an X in P such that $a \in X$ and $b \in X$.

   For the given set P, determine if R is reflexive, symmetric, or transitive.
   a. $P = \{A, B\}$, where $A = \{1,3,4,6\}$ and $B = \{2,5,7\}$.
   b. $P = \{A, B\}$, where $A = \{1,3,4,6\}$ and $B = \{2,3,5,7\}$.
   c. $P = \{A, B\}$, where $A = \{1,3,4\}$, and $B = \{2,5,7\}$.
   d. $P = \{A, B, C\}$, where $A = \{7\}$, $B = \{2,5\}$ and $C = \{1,4,6\}$.
   e. $P = \{A, B, C\}$, where $A = \{3,5,7\}$, $B = \{2,5\}$ and $C = \{1,4,6\}$.
   f. $P = \{A, B, C\}$, where $A = \{6\}$, $B = \{2,5\}$ and $C = \{1,3,4,7\}$.

6. List the equivalence classes of each equivalence relation in the previous exercise.

7. Let P be a partition of a set S. Define an equivalence relation on S whose equivalence classes are the elements in P.

   Let a and b be in S. $a \mathsf{R} b$ if and only if ________

8. True or false?

   a. $3 \equiv_5 8$ b. $3 \equiv_5 -8$ c. $3 \equiv_5 -7$ d. $3 \equiv_4 123$

9. a. Let a and b be integers. Define congruence mod 6: $a \equiv_6 b$

   b. Is $11 \equiv_6 5$? Is $1024 \equiv_6 2$?

   c. List the elements in the equivalence class of 5.

   d. Z_6 is the set of equivalence classes of congruence mod 6. How many elements are in Z_6? List them.

   e. Which element in Z_6 contains 2345? −38?

10. a. Let a and b be integers. Define congruence mod 7: $a \equiv_7 b$

    b. List the elements in the equivalence class of 5.

    c. List the elements in Z_7.

    d. Which element in Z_7 contains 2345? −38?

11. Prove or disprove: If $x \equiv_5 2$, then $x+1 \equiv_5 3$.

12. Let R be an equivalence relation on a set S. Let a and b be elements in S. Prove each statement.

    a. If $a \mathsf{R} b$, then $[a] = [b]$.

    b. If $\sim(a \mathsf{R} b)$, then $[a] \cap [b] = \emptyset$.

    c. If $[a] \cap [b] \neq \emptyset$, then $[a] = [b]$.

    d. If $[a] \neq [b]$, then $[a] \cap [b] = \emptyset$.

13. Let R be an equivalence relation on a set S. Without looking at the proof given in the text, prove that the set of all equivalence classes of R is a partition of S.

Activity 4.3

Let f be a function whose domain is X and let A be a subset of X. The image of A under f is defined as follows: $f(A) = \{ f(x) \mid x \in A \}$

1. Let $f(x) = 2x+1$. Compute $f(A)$. On the graph of f, use bent arrows from the x-axis to the y-axis to illustrate $f(A)$.

   a. $A = \{ -1, 1, 2 \}$ b. $A = [1,3]$ c. $A = [-2,-1] \cup [1,3]$

2. Let $f(x) = x^2 + 1$. Repeat the previous exercise.

≡ 4.3 Functions ≡

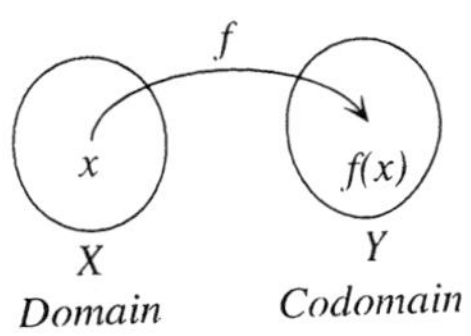

Functions are one of the most important tools in modern mathematics. They serve as a transportation vehicle that takes us from one set to another, or perhaps to another location in the same set. In calculus, we usually visualize a function in terms of its graph. In this section, though, we will focus on mapping pictures in order to build a more general understanding of a function. In a mapping representation, a function consists of two sets and a mapping that assigns each element in the first set to a unique element in the second set. The first set is called the *domain* of the function and the second set is called the *codomain*. If X is the domain and Y is the codomain, we say that f maps X *into* Y, which is notated as $f: X \rightarrow Y$ or $X \xrightarrow{f} Y$. This notation conveys three pieces of information:

---

f is a *function* from X into Y
if and only if
f maps each element in X
to a unique element in Y.

---

1) f is a function.
2) Domain $(f) = X$.
3) If $x \in X$, then $f(x) \in Y$.

A Function? No

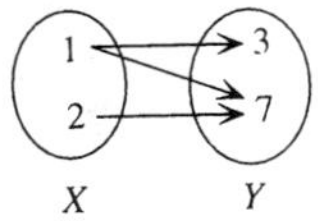

A Function? Yes

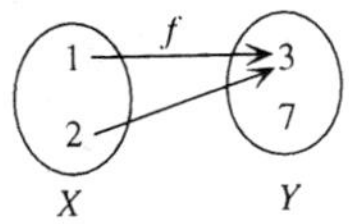

The adjacent mappings illustrate the difference between a function and a relation that is not a function. A function can map an element x to only one location, which is designated as $f(x)$. We use the visual terms "image" and "pre-image" to denote the relationship between x and $f(x)$. The *image* of x under the function f is $f(x)$. In the reverse direction, x is called a *pre-image* of $f(x)$.

For example, in the adjacent function, $f(1) = 3$ and $f(2) = 3$. The image of 1 under the function f is 3. The image of 2 is also 3. On the other hand, both 1 and 2 are pre-images of 3. Under a function, an element can have only one image, but it can have more than one pre-image. We sometimes view x as being transformed to $f(x)$, which is why functions are also called *transformations*.

When Leonhard Euler created the $f(x)$ notation in 1734, he revolutionized the way we think about functions. This simple notional device gives a concise way to represent the values of a function without using arrows or ordered pairs:

$$f(x) = y \qquad x \xrightarrow{f} y \qquad (x,y) \in f$$

The advantage of the "$f(x) = y$" notation is the equals sign which we can easily manipulate.

f versus f(x)

The notation f has a different meaning than the notation $f(x)$. The letter f represents the complete function, whereas $f(x)$ is only a part of the total function. This difference is apparent when we represent f as a set of ordered pairs. For example, consider the function f that maps $\mathbb{R}$ into $\mathbb{R}$ under the mapping, $f(x) = 2x$. $f(x)$ represents a real number, whereas f represents the following set of ordered pairs:

$$f = \{ (x, f(x)) \mid x \text{ is a real number } \}$$
$$f = \{ (x, 2x) \mid x \text{ is a real number } \}$$

When we have a set whose elements are functions, we must use f instead of $f(x)$. For example, let S be the following set:

$$S = \{ f \mid f \text{ is a function from } \mathbb{R} \text{ into } \mathbb{R} \}$$

Consider the function, $g(x) = 3$, where x is a real number. The function g maps $\mathbb{R}$ into $\mathbb{R}$. So, $g \in S$. However, $g(x) \notin S$.

Domain and Range

The definition of the domain and range of a function is the same as the definition for a relation (page 313). When a function is represented as a set of ordered pairs, the domain is the set of all the first terms and the range is the set of all the second terms. In terms of function notation, the domain of a function f is the set of all x for which $f(x)$ is defined:

---

x is in the *domain* of f
if and only if
$f(x)$ is defined.

$$\text{Domain } (f) = \{ x \mid f(x) \text{ is defined } \}$$

---

y is in the *range* of f
if and only if
there exists an x in the
domain such that $y = f(x)$.

The range of f is the set of of images of elements from the domain:

$$\text{Range } (f) = \{ f(x) \mid x \in \text{Domain } (f) \}$$

If y is an element of the range, there must exist an x in the domain such that $y = f(x)$. If X represents the domain of f, we use the notation $f(X)$ to represent the range of f:

$$f(X) = \{ f(x) \mid x \in X \}$$

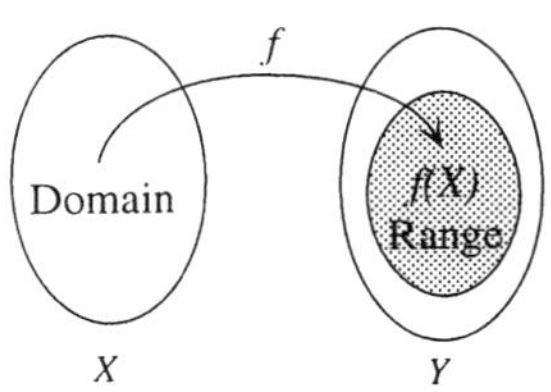

The image of a set X under a function is the set of all the individual images of elements in X.

If f maps X into Y, $f: X \rightarrow Y$, then X is the domain, but Y is not necessarily the range, as indicated in the adjacent sketch. If $x \in X$, $f(x)$ is an element of Y, so the range must be a subset of Y. However, the range is not always equal to Y.

⊕ Example

Let f be a function where the mapping rule is $f(x) = 2x$.

1. Let A be the domain of f where $A = \{1, 2, 3\}$.
   The range of f is the following set:

   $$f(A) = \{2, 4, 6\}$$

   Let $B = \{2, 4, 6\}$. For the codomain of f, we can use any set that contains B. For example, we can consider f as a mapping from A into B, or from A into the set $\mathbb{N}$ of natural numbers, or from A into the set $\mathbb{R}$ of real numbers.

   $$f: A \rightarrow B \quad f: A \rightarrow \mathbb{N} \quad f: A \rightarrow \mathbb{R}$$

   However, f does not map A into A.

2. Let X be the domain of f where $X = [1, 3]$.
   The range of f is the following set:

   $$f(X) = [2, 6].$$

   For the codomain of f, we can use any set that contains $[2, 6]$:

   $$f: X \rightarrow [2, 6] \quad f: X \rightarrow [1, 100] \quad f: X \rightarrow \mathbb{R}$$

Equal Functions

Two functions are equal if they represent the same mapping on the same domain. Functions defined by the same mapping rule are not equal if they have different domains. For example, let f be defined by the rule $f(x) = 2x$ where the domain is $\{1, 2\}$, and let $g(x) = 2x$ where the domain is $\{1, 2, 3\}$:

$$f = \{ (1, 2), (2, 4) \}$$
$$g = \{ (1, 2), (2, 4), (3, 6) \}$$

$f \neq g$ because they are different sets of ordered pairs. For two functions to be equal, they must have the same domain. Secondly, their function values must agree for each x in the domain.

Let X be the domain for the functions f and g.

$f = g$

if and only if

$f(x) = g(x)$ for all x in X.

$f = g$

if and only if

f and g have the same domain and

$f(x) = g(x)$ for all x in the domain.

In the above definition, f represents the function, whereas $f(x)$ represents the value of the function at x.

Example

Let $f(x) = x$ and $g(x) = |x|$, where $x \in X$. Is $f = g$?

1. Let $X = \mathbb{R}$, the set of real numbers.

   $f(-1) = -1$, but $g(-1) = |-1| = 1$. So $f \neq g$.

2. Let $X = \mathbb{R}^+$, the set of positive real numbers.

   If $x \in \mathbb{R}^+$, $|x| = x$. So, $g(x) = f(x)$ for all positive real numbers. Therefore, $f = g$. Even though the mapping rules look different, on the given domain, they produce the same function values.

Restricting a Function

If we restrict the domain of a function, we produce a new function. Let f be a function that maps X into Y and suppose that $A \subseteq X$. The *restriction* of f to A, notated as $f|_A$, is defined as follows:

$$f|_A(x) = f(x) \text{ for each } x \text{ in } A.$$

Since $f|_A$ and f have different domains, $f|_A \neq f$. They represent different sets of ordered pairs:

$$f|_A = \{ (x, f(x)) \mid x \in A \}$$
$$f = \{ (x, f(x)) \mid x \in X \}$$

In the other direction, we sometimes want to extend a function to a larger domain, preserving the values on the original domain. Let f be a function that maps X into Y and suppose that $X \subseteq \overline{X}$. The function g is called an *extension* of f to $\overline{X}$ if g maps $\overline{X}$ into Y and the following is true:

$$g(x) = f(x) \text{ for each } x \text{ in } X.$$

If g is an extension of f, then f is the restriction of g to X.

Identity Function

The identity function on a particular set is the function that maps each element to itself: $f(x) = x$. We often notate the identity function as e: $e(x) = x$ for all x in X. The identity function on X is different from the identity function on Y because they have different domains. When working with two different identity functions, we use subscripts to distinguish between them:

e_X denotes the identity function on X.

e_Y denotes the identity function on Y.

Defining Functions

In calculus, we usually define a function in terms of a formula, such as $f(x) = 5x^2$. However, not all functions have a simple formulaic expression. We can also define a function by giving a verbal description of the mapping, or by using cases, or by listing individual values for finite domains. Regardless of how we define a function, we must make sure that we assign each element in the domain to only one element. If our definition achieves this goal, we say that our function is *well defined*. To say that a function is well defined means that its definition produces a function. With a formula definition like $f(x) = 5x^2$, it is easy to see that the function is well defined: for each x, we have only one y. However, with other types of definitions, we must check to make sure that the function is well defined.

⊕ *Example*

For the given domain X, determine if the definition produces a well-defined function.

1. Let $X = \{1, 2, 5\}$.
   Define $f(1) = 23$, $f(2) = 23$, and $f(5) = 4$.

   For each x in X, there is only one value for $f(x)$.
   So f is a function. Thus, f is well defined.

2. Let $X = \{1, 2, 5\}$.
   Define $f(1) = 2$, $f(2) = 4$, $f(5) = 6$, and $f(1) = 8$.

   There are two values for $f(1)$, so f is not a function.
   Thus, f is not well defined.

3. Let $X = \{1, 2, 5\}$. Define $f(1) = 2$, $f(2) = 4$.

   $f(5)$ is not defined, so f is not well defined on X.

4. Let X be the set of real numbers. Define f as follows:

   $$\text{If } x < 1,\ f(x) = x + 1.$$
   $$\text{If } x \geq 1,\ f(x) = 3x.$$

   To evaluate $f(.99)$, we use the first rule: $f(.99) = 1.99$.
   To evaluate $f(1)$, we use the second rule: $f(1) = 3$.
   For each x in X, there is a unique value for $f(x)$.
   So, f is a well-defined function.

5. Let X be the set of integers. Let $f(ab) = a$.

   $f(2\cdot 3) = 2$ and $f(3\cdot 2) = 3$. Since 6 is mapped to both 2 and 3, f is not a function. So, f is not well defined.

If f is a function and $a = b$, then $f(a)$ must equal $f(b)$, for otherwise the same element is mapped to two different values. For f to be a function, the following must be true for each a and b in the domain:

$$\text{If } a = b, \text{ then } f(a) = f(b).$$

f is a *function*
if and only if
for every a and b in the domain,
if $a = b$, then $f(a) = f(b)$.

The above implication ensures that the definition of $f(a)$ is a unique element. We use this implication to determine whether or not f is welldefined so that we can then use function notation with it. If we can use function notation, the above implication is obviously true by the substitution principle.

When the definition of f is based on a representation of x, we must carefully check to make sure that the function is well defined. In the example, $f(ab) = a$, the definition of f is based on how we factor a number. This definition does not produce a function. Sometimes, though, the definition of $f(x)$ can be based on a representation of x that is not unique and still produce a function. For example, we often define a function on a set of equivalence classes in terms of a representative of the equivalence class – we define $f([a])$ in terms of a. This type of definition must be carefully checked to make sure that it is well defined, as illustrated in the following example.

⊕ *Example*

1. Let Z be the set of integers.
If $a \in Z$, define f as follows: $f(a) = 2a$
Each element a in Z is assigned to a unique element.
So, f is well defined.

2. Let Z_4 be the set of equivalence classes for Z under congruence mod 4 (page 341).
If $[a] \in Z_4$, define f as follows: $f([a]) = [2a]$

   Is f well defined? This question is more involved than in the previous example. Under congruence mod 4, $[3] = [7]$.
   Is $f([3]) = f([7])$? Is $[2 \cdot 3] = [2 \cdot 7]$?
   Since $14-6$ is divisible by 4, the answer is yes.
   Let's prove that f is well defined:

   Assume that $[a] = [b]$.
   Then, $a-b = 4k$ for some integer k.
   So, $2a-2b = 4(2k)$
   Hence, $[2a] = [2b]$.
   So, $f([a]) = f([b])$.
   Therefore, f is a well-defined function.

The following functions, which are often used in mathematics and computer science, have definitions based on cases or verbal descriptions.

⊕ *Example*

For the given domain X, is $f(x)$ well defined?

1. Let $X = \mathbb{R}$. Define f as follows:

   $f(x)$ is the greatest integer less than or equal to x.

   For each real number x, there is exactly one integer that fits the above description. So f is a well-defined function. The range of f is the set Z of integers: $f(\mathbb{R}) = Z$

   This function provides the "floor" when we round a real number x down to the next integer, so it is called the *floor function.* Its notation, $\lfloor x \rfloor$, also suggests the floor:

   $$\lfloor 2.999999 \rfloor = 2 \qquad \lfloor 3 \rfloor = 3 \qquad \lfloor \pi \rfloor = 3$$

2. Let $X = \mathbb{R}$. Define f as follows:

   $f(x)$ is the smallest integer greater than or equal to x.

   For each real number x, there is exactly one integer that fits the above description. So f is a well-defined function. The range of f is the set Z of integers: $f(\mathbb{R}) = Z$

   This function provides the "ceiling" when we round a real number x up to the next integer, so it is called the *ceiling function*. It is notated as $\lceil x \rceil$:

   $$\lceil 2.999999 \rceil = 3 \qquad \lceil 3 \rceil = 3 \qquad \lceil \pi \rceil = 4$$

3. Let X be a set and A a subset of X. Define f as follows:

   $$f(x) = 1, \text{ if } x \in A.$$
   $$f(x) = 0, \text{ if } x \notin A.$$

   For each x in X, x is assigned a unique value, so f is a well-defined function. This function, called the *characteristic function of A*, gives a way to characterize membership in a set A using the numbers 1 and 0. It is notated as χ_A.

   Let X be the set of real numbers and $A = [3, 7]$.
   Then $\chi_A(4) = 1$ and $\chi_A(8) = 0$.

4. Let X be a set of statements, where each statement is either true or false, but not both. Define f as follows:

$$f(x) = 1, \text{ if } x \text{ is true.}$$
$$f(x) = 0, \text{ if } x \text{ is false.}$$

For each statement x in X, x is assigned a unique value, so f is a well-defined function. This function has a rather descriptive name; it is called the *truth value function.*

If the domain of a function is a cross product, technically we need two sets of parentheses to denote the element to which (x,y) is mapped: $f((x,y))$. However, too many parentheses are visually confusing, so we omit the second pair and write $f(x,y)$.

⊕ *Example*

For the given domain X, determine if f is a well-defined function. If so, find its range.

1. Let $X = \mathbb{R}^2$. Define f as follows: $f(x, y) = x$.

   Each ordered pair in $\mathbb{R}^2$ is mapped to a unique element, so f is a well-defined function. This function projects each ordered pair onto its first coordinate.

   The range of f is $\mathbb{R}$: $f(\mathbb{R}) = \mathbb{R}$

2. Let $X = \mathbb{R}$. Define g as follows: $g(x) = 3$.

   g is a well-defined function.
   The range of g has only one element: $g(\mathbb{R}) = \{3\}$

3. Let $X = \mathbb{R}$. Define f as follows: $f(x) = (x, 3)$.

   Each real number is mapped to a unique ordered pair, so f is a well-defined function. The function f maps $\mathbb{R}$ into $\mathbb{R}^2$. However, $\mathbb{R}^2$ is not the range of f because no one maps to $(1, 4)$. We can describe the range as follows:

   $$f(\mathbb{R}) = \{ (x, 3) \mid x \text{ is a real number} \}$$

   Even though f is similar to the function g in the previous example, the two functions are not equal: $f(x) \neq g(x)$.

4. Let $X = \mathbb{R}^2$. Define f as follows: $f(x, y) = (y, x)$

   Each ordered pair in $\mathbb{R}^2$ is mapped to a unique ordered pair, so f is a well-defined function. The range of f is $\mathbb{R}^2$.

Binary Operations

A *binary operation* on a set S is a function that maps $S \times S$ into S. Each two elements in S must be assigned a unique value and that value must also be in S. This definition is a translation in function language of the definition given in Chapter 1, page 96.

One of the first binary operations that we learn is addition on the set $\mathbb{N}$ of natural numbers. Given two natural numbers x and y, the addition operation assigns a unique value, which can be represented as the following function:

f is a *binary operation* on a set S if and only if $f: S \times S \to S$.

$$f(x, y) = x + y$$

For example, $f(2, 3) = 5$. The function f maps $\mathbb{N} \times \mathbb{N}$ into $\mathbb{N}$, so f is a binary operation on $\mathbb{N}$.

We can represent subtraction in a similar manner as a function on $\mathbb{N} \times \mathbb{N}$:

$$g(x, y) = x - y$$

For example, $g(2, 3) = -1$. This function, though, does not map $\mathbb{N} \times \mathbb{N}$ into $\mathbb{N}$, so subtraction is not a binary operation on the set $\mathbb{N}$. However, g does map $Z \times Z$ into Z, so g is a binary operation on the set Z of integers.

Using Function Notation

To use function notation, we must read the notation carefully and substitute verbatim, as illustrated in the following examples.

Example

Let $f(x) = x^2 + 3$.

1. Compute $f(a)$, $f(a+b)$, and $f(a) + f(b)$.

   $f(a) = a^2 + 3$ ..... *Substitute a for x.*
   $f(a+b) = (a+b)^2 + 3$ ... *Substitute $a + b$ for x.*
   $f(a) + f(b) = (a^2 + 3) + (b^2 + 3)$
   .... *Substitute for $f(a)$ and $f(b)$.*

   Note that $f(a+b) \neq f(a) + f(b)$.

2. Is $f(3x) = 3\,f(x)$?

   $f(3x) = (3x)^2 + 3$ .... *Substitute $3x$ for x.*
   $3\,f(x) = 3(x^2 + 3)$ .... *Substitute $x^2 + 3$ for $f(x)$.*

   So, $f(3x) \neq 3\,f(x)$.

Onto Functions

If f maps X into Y, the range of f must be a subset of Y:

$$f(X) \subseteq Y$$

If the range of f is equal to Y, we strengthen the preposition and say that f maps X *onto* Y:

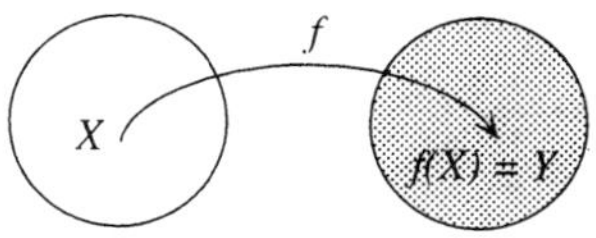

f maps X onto Y.

$$f \text{ maps } X \textit{ onto } Y \text{ if and only if } f(X) = Y.$$

An onto function is also called a *surjection*. Let $f(x) = x^2$, where x is a real number. If the codomain is the set of nonnegative real numbers, then f is a surjection. However, if the codomain is the set of real numbers, then f is not a surjection.

Let f map X into Y. To prove that f maps X onto Y, we must demonstrate the following:

> f maps X *onto* Y
> if and only if
> for every y in Y, there exists an x in X such that $f(x) = y$.

For every y in Y, there exists an x in X such that $f(x) = y$.

Earlier we examined the technique for proving that a function is onto (page 151–153). Starting with the quantifier on the left, we assume that y is an element in Y. Our job then is to find an x that maps to y, as illustrated in the following proof.

⊕ *Example*

Let $f(x) = x^2 + 1$, where x is a real number.
Prove that f maps $\mathbb{R}$ onto $[1, \infty)$.

First, note that f maps $\mathbb{R}$ into $[1, \infty)$.
If x is a real number, $x^2 \geq 0$, so $f(x) \geq 1$.

Let y be an element in $[1, \infty)$. Then $y \geq 1$.
So $y - 1 \geq 0$. Thus $\sqrt{y-1}$ is a real number.
Set $x = \sqrt{y-1}$.

$$\begin{aligned} \text{Then } f(x) &= f(\sqrt{y-1}) \\ &= (\sqrt{y-1})^2 + 1 \\ &= (y-1) + 1 \\ &= y \end{aligned}$$

So, for every y in $[1, \infty)$, there exists a real number x such that $f(x) = y$. Therefore, f maps $\mathbb{R}$ onto $[1, \infty)$.

To prove that f does not map X onto Y, we must demonstrate the negation of the definition:

There exists a y in Y such that for each x in X, $f(x) \neq y$.

⌖ *Example*

Let $f(x) = x^2 + 1$, where x is a real number.
Prove that f does not map $\mathbb{R}$ onto $\mathbb{R}$.

> Set $y_0 = 0$. Let x be any real number.
> Since $f(x) = x^2 + 1$, $f(x) \geq 1$. Thus, $f(x) \neq 0$.
> So, f does not map $\mathbb{R}$ onto $\mathbb{R}$.

One-to-One Functions

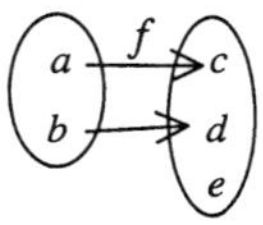

f is one-to-one

A function is *one-to-one* if different elements in the domain map to different elements in the range:

> Let a and b be elements in the domain.
> If $a \neq b$, then $f(a) \neq f(b)$.

A one-to-one function is also called an *injection*. A function that is both one-to-one and onto is called a *bijection*. A bijection is another name for a one-to-one correspondence. If a function is a bijection, then the domain and co-domain must have the same size.

The above implication can be phrased in terms of its contrapositive:

$$\text{If } f(a) = f(b), \text{ then } a = b.$$

This implication is simpler since it has no negations, so we normally use this form in proofs. However, the original form is easier to remember, so it is featured in the adjacent definition.

---

f is *one-to-one*
if and only if
for every a and b in the domain,
if $a \neq b$, then $f(a) \neq f(b)$.

⌖ *Example*

Let $f(x) = 2x + 1$ where x is a real number.
Prove that f is a one-to-one function.

> Let a and b be real numbers.
> Assume that $f(a) = f(b)$.
> $$2a + 1 = 2b + 1$$
> $$2a = 2b$$
> So, $a = b$.
> Therefore, if $f(a) = f(b)$, then $a = b$.
> So, f is a one-to-one function.

To prove that a function is not one-to-one, we must produce two different x's that map to the same value:

$$\sim[\forall x_1\ \forall x_2,\ \text{if } x_1 \neq x_2, \text{ then } f(x_1) \neq f(x_2)].$$

$$\exists x_1\ \exists x_2,\ x_1 \neq x_2 \text{ and } f(x_1) = f(x_2).$$

✧ *Example*

Let $f(x) = x^2$, where x is a real number.
Prove that f is not a one-to-one function.

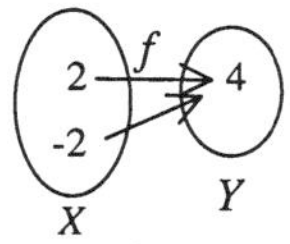

f is not one-to-one.

$$2 \neq -2$$
$$f(2) = f(-2)$$

So f is not a one-to-one function.

Note the similarity between the definition of one-to-one and the definition of a function given on page 349. Let a and b be elements in the domain of f.

Function: If $a = b$, then $f(a) = f(b)$.
One-to-One: If $f(a) = f(b)$, then $a = b$.

The definition of one-to-one is the converse of the definition of a function. Consequently, we get a reverse arrow effect in the mapping pictures:

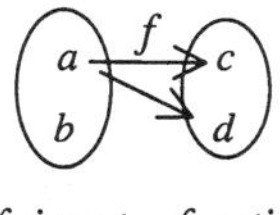

f is not a function.

To be a function, each x can be mapped to only one y.
To be one-to-one, each y can have only one x mapped to it.

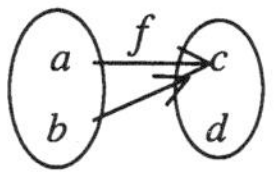

f is not one-to-one.

As illustrated in the adjacent mappings, if two different arrows have the same origin, the mapping is not a function. On the other hand, if two different arrows end up at the same place, the mapping is not one-to-one.

The second mapping is a function, but it is not one-to-one.
If we reverse the arrows, we do not get a function.

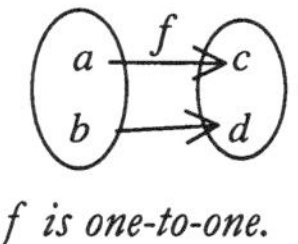

f is one-to-one.

The third mapping is a one-to-one function.
If we reverse the arrows, we get a function.

The one-to-one property determines whether or not a function has an inverse function.

Inverse Functions

Every function has an inverse relation. In the mapping picture of a function f, the inverse relation is obtained by reversing the direction of the arrows. However, the inverse relation, f^{-1}, may not be a function, which means that we cannot use function notation with it. For example, consider the following function:

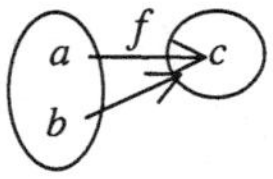

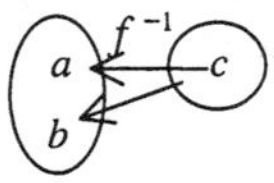

$$f = \{ (a,c), (b,c) \}$$
$$f^{-1} = \{ (c,a), (c,b) \}$$

Under f^{-1}, c is mapped to two different elements, so f^{-1} is not a function. Hence, we cannot write $f^{-1}(c)$.

If a function f is not one-to-one, its inverse relation f^{-1} will not be a function. If f is one-to-one and we reverse the arrows, each element will be mapped to only one element, so f^{-1} will be a function. Thus, we have the following equivalence:

> f^{-1} is a function if and only if f is one-to-one.

f^{-1} is a function if and only if f is one-to-one.

In the above statement, we substitute f^{-1} for f and then we substitute f for $(f^{-1})^{-1}$ (page 323):

f^{-1} is a function if and only if f is one-to-one.

$(f^{-1})^{-1}$ is a function if and only if f^{-1} is one-to-one.

f is a function if and only if f^{-1} is one-to-one.

> f is a function if and only if f^{-1} is one-to-one.

From the above equivalence, we can deduce that the inverse of a function is always one-to-one. The one-to-one property of f makes f^{-1} a function, while the function property of f makes f^{-1} one-to-one. The notions of "function" and "one-to-one" are just different directions on the same highway.

⊕ *Example*

Let $f = \{ (1,4), (2,8), (3,9) \}$.
Is f^{-1} a function? Is f^{-1} one-to-one?

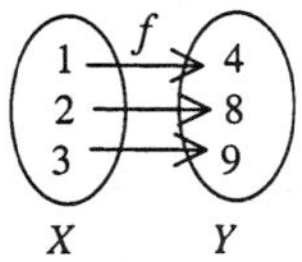

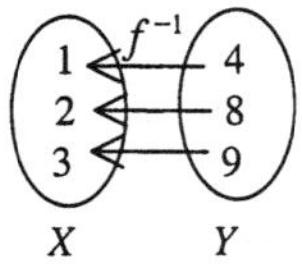

Note that f is one-to-one. Since each y has only one x mapped to it, when we reverse the mapping, we get a function. The one-to-one property of f makes f^{-1} a function.

Since f is a function, an x in X cannot map to different elements in Y. So, when we reverse the arrows, different elements in Y cannot map to the same x. The function property of f makes f^{-1} one-to-one.

Domain (f^{-1}) = Range (f)
Range (f^{-1}) = Domain (f)

As with inverse relations (page 322), the domain and range are switched for the inverse function. The range of f becomes the domain of f^{-1}, and the domain of f becomes the range of f^{-1}.

Let $f: X \to Y$. The domain of the function f is X, which is the range of f^{-1}. The range of f is not necessarily Y, so we cannot deduce that Y is the domain of f^{-1}. However, if f maps X *onto* Y, the range of f is Y, which makes Y the domain of f^{-1}.

$$\text{If } f: X \underset{onto}{\longrightarrow} Y, \text{ then } f^{-1}: Y \underset{onto}{\longrightarrow} X.$$

Theorem If f is a one-to-one function that maps X onto Y, then f^{-1} is a one-to-one function that maps Y onto X.

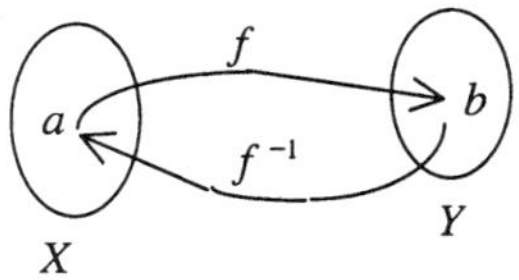

If f sends a to b, f^{-1} sends b back to a, as illustrated on the left.

$$a \xrightarrow{f} b \text{ if and only if } b \xrightarrow{f^{-1}} a.$$

$$f(a) = b \text{ if and only if } f^{-1}(b) = a.$$

$f(a) = b$
if and only if
$f^{-1}(b) = a$.

The above equivalence is the main tool for working with inverse functions. We use it to convert back and forth from a function to its inverse function:

If $f(3) = 5$, then $f^{-1}(5) = 3$.
If $f^{-1}(2) = 11$, then $f(11) = 2$.

Substitute $f(a)$ for b in the above equivalence:

$$f(a) = f(a) \text{ if and only if } f^{-1}(f(a)) = a.$$

Since the right side of the equivalence is true: $f^{-1}(f(a)) = a$.
Now substitute $f^{-1}(b)$ for a in the original equivalence:

$$f(f^{-1}(b)) = b \text{ if and only if } f^{-1}(b) = f^{-1}(b).$$

Since the right side of the equivalence is true: $f(f^{-1}(b)) = b$
These two results are summarized in the following theorem.

Theorem Let f be a one-to-one function that maps X onto Y.

For every a in X, $f^{-1}(f(a)) = a$.
For every b in Y, $f(f^{-1}(b)) = b$.

We can translate the previous theorem in terms of e_X, the identity function on X (page 347) and function composition. Let a be any element in X:

$$f^{-1}(f(a)) = a = e_X(a)$$

$$f^{-1} \circ f(a) = e_X(a)$$

$$\text{So, } f^{-1} \circ f = e_X$$

In a similar manner, we can show that $f \circ f^{-1} = e_Y$.

Composition of Functions

We can interpret the mapping picture of a function as a transportation system. When an element leaves the first set, it can be transported to only one location in the second set. After an element lands in the second set, we may choose to transport it somewhere else, which gives us the composition of two functions.

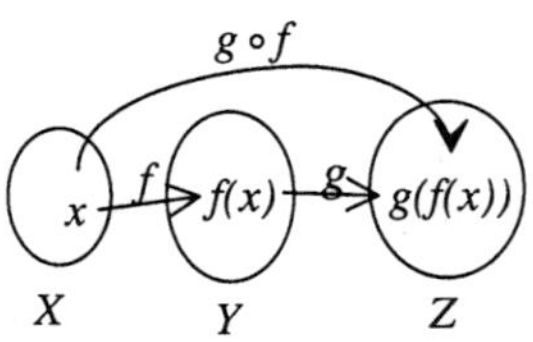

In the adjacent sketch, f maps X into Y and g maps Y into Z.

$$f \text{ maps } x \text{ to } f(x).$$

Since $f(x)$ is in the domain of g, g can transport it further:

$$g \text{ maps } f(x) \text{ to } g(f(x)).$$

The composition function, denoted as $g \circ f$, has the same end result as executing the above two mappings, one after the other.

Let f: $X \rightarrow Y$ and g: $Y \rightarrow Z$.
Let x be in X.
Then $g \circ f(x) = g(f(x))$.

$$g \circ f \text{ maps } x \text{ to } g(f(x)).$$

$$g \circ f(x) = g(f(x))$$

The notation $g \circ f$ is read as g *composition* f, but we should visualize it as f followed by g, as illustrated in the adjacent sketch. To evaluate $g \circ f(x)$, we first evaluate $f(x)$ and then evaluate $g(f(x))$.

For example, in the adjacent sketch, $f(1) = 4$ and $g(4) = 5$.

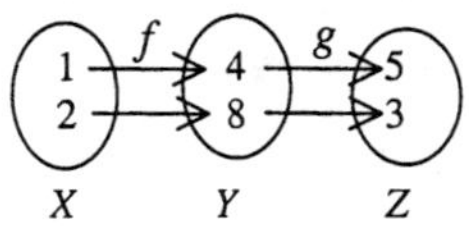

$$g \circ f(1) = g(f(1)) = g(4) = 5$$

$$g \circ f(2) = g(f(2)) = g(8) = 3$$

If $f(1)$ had been 7, we could not have formed the composition since $g(7)$ is not defined. In order to form the function $g \circ f$, the range of f must be a subset of the domain of g.

◈ *Example*

Let $f(x) = 2x$ and $g(x) = x - 1$, where x is a real number. Compute $g \circ f(x)$ and $f \circ g(x)$. Is $g \circ f = f \circ g$?

$$\begin{aligned} g \circ f(x) &= g(f(x)) \quad \ldots\ldots \quad \textit{Definition of } g \circ f \\ &= g(2x) \quad \ldots\ldots \quad \textit{Definition of } f(x) \\ &= 2x - 1 \quad \ldots\ldots \quad \textit{Definition of } g \end{aligned}$$

$$\begin{aligned} f \circ g(x) &= f(g(x)) \quad \ldots\ldots \quad \textit{Definition of } f \circ g \\ &= f(x-1) \quad \ldots\ldots \quad \textit{Definition of } g(x) \\ &= 2(x-1) \quad \ldots\ldots \quad \textit{Definition of } f \end{aligned}$$

Since $2x - 1 \neq 2(x-1)$, $f \circ g \neq g \circ f$.

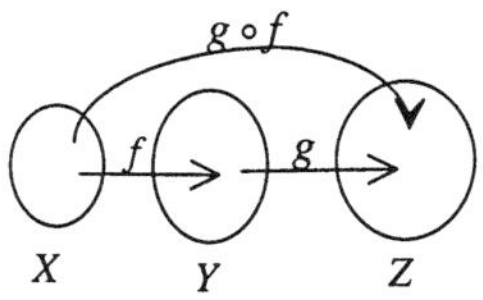

The domain of $g \circ f$ is the domain of f. The range of $g \circ f$ must be a subset of the range of g. However, they may not be equal, as illustrated in the following example.

◈ *Example*

Let $f(x) = x^2$, where x is a real number.
Let $g(x) = x - 1$, where x is a real number.
Find the range of g and the range of $g \circ f$.

$$\text{Range } (g) = (-\infty, \infty).$$
$$g \circ f(x) = g(f(x)) = g(x^2) = x^2 - 1.$$
$$\text{Range } (g \circ f) = [-1, \infty).$$

The range of $g \circ f$ is not the same as the range of g because f does not map $\mathbb{R}$ onto $\mathbb{R}$.

Composition of Onto Functions

When we form the composition of two onto functions, we produce an onto function. Let f map X into Y and g map Y into Z, as illustrated in the adjacent sketch.

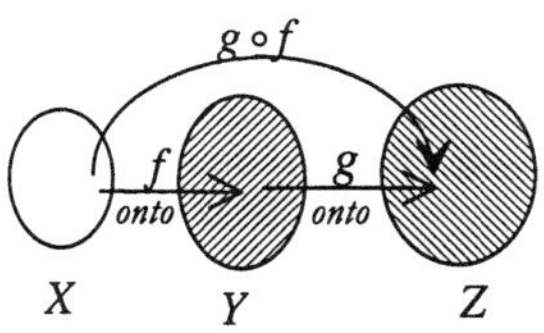

If f maps X onto Y, then the range of f is Y.

If g maps Y onto Z, then the range of g is Z.

Consequently, the range of $g \circ f$ will be Z, which means that the function $g \circ f$ maps X onto Z. To give a verbal proof of this visual reasoning, we must demonstrate the following:

For every z in Z, there exists an x in X such that $g \circ f(x) = z$.

Given a z in the third set, we must find an x in the first set that maps to z under the function $g \circ f$. To find such an x, we work backwards along the horizontal arrows in the sketch.

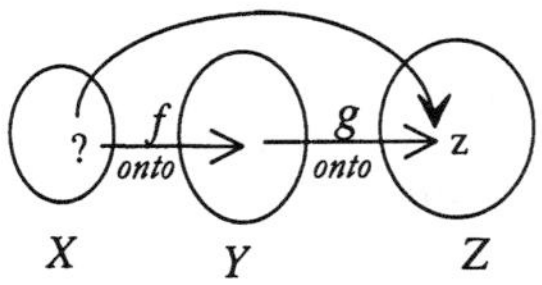

First, we use g to find a y in the second set that maps to z.

Then, we use f to find an x in the first set that maps to y.

The composition function will then send this x to the desired location. To test the development of your reasoning skills, try to write a proof of this theorem. If you have difficulty, review the construction of this proof on pages 155–157. Hopefully, it will seem simpler now.

Theorem Let f be a function that maps X onto Y and g be a function that maps Y onto Z. Then $g \circ f$ maps X onto Z.

⊕ *Example* Let $f(x) = x^3$ and $g(x) = \sin x$, where x is a real number. Find $g \circ f$ and its range.

$$g \circ f(x) = g(f(x)) = g(x^3) = \sin(x^3)$$

f maps $\mathbb{R}$ onto $\mathbb{R}$ and g maps $\mathbb{R}$ onto $[-1,1]$.

$$\mathbb{R} \xrightarrow{f} \mathbb{R} \xrightarrow{g} [-1,1]$$

Hence, $g \circ f$ maps $\mathbb{R}$ onto $[-1,1]$.

So its range is $[-1,1]$.

Composition of Injections

A one-to-one function maps different elements in the domain to different elements in the range:

If f is one-to-one and $a \neq b$, then $f(a) \neq f(b)$.

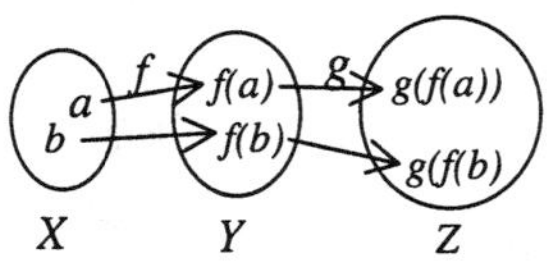

If g is also one-to-one, then g must map $f(a)$ and $f(b)$ to different elements in Z.

When both f and g are one-to-one, their composition must also be one-to-one, as demonstrated in the following proof.

Theorem Let f map X into Y and g map Y into Z. If f and g are one-to-one functions, then $g \circ f$ is a one-to-one function.

Proof Let a and b be in X. Assume that $a \neq b$.

Since f is one-to-one, $f(a) \neq f(b)$.

Since g is one-to-one, $g(f(a)) \neq g(f(b))$

So, $g \circ f(a) \neq g \circ f(b)$.

Therefore, $g \circ f$ is one-to-one.

Composition of Bijections

The previous theorem states that the composition of two injections (one-to-one functions) is an injection. We proved that the composition of two surjections (onto functions) is a surjection (page 157). Since a bijection is a function that is both one-to-one and onto, we can deduce that the composition of two bijections is a bijection.

Theorem Let f be a bijection from X onto Y and g a bijection from Y onto Z. Then $g \circ f$ is a bijection from X onto Z.

One of the fundamental relations between sets is the property of "having the same size" (page 273). Two sets have the same size if and only if there exists a bijection between them. From the above theorem, we can deduce that the relation of "having the same size" is transitive. You are asked to prove this in (19) of the next exercise set.

Inverse of a Composition

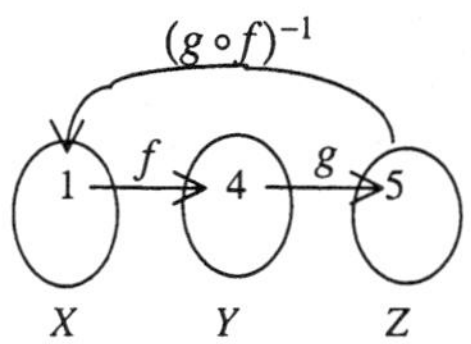

Let f be a bijection from X onto Y and g a bijection from Y onto Z. Then $g \circ f$ is a bijection from X onto Z and all three functions will have inverse functions. The inverse of $g \circ f$ can be computed from the individual inverses as illustrated in the adjacent example. Note that $(g \circ f)^{-1}$ maps Z onto X. g^{-1} also acts on elements in Z. So, we must start with g^{-1} to find the value of $(g \circ f)^{-1}(5)$:

$$\begin{aligned}(g \circ f)^{-1}(5) &= f^{-1} \circ g^{-1}(5) \\ &= f^{-1}(g^{-1}(5) \\ &= f^{-1}(4) \\ &= 1\end{aligned}$$

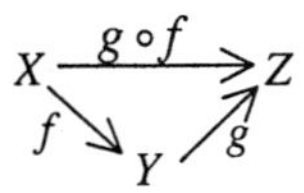

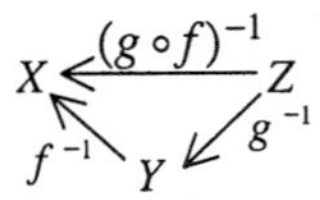

In the adjacent two diagrams, we can see why the order of f and g get reversed when we take the inverse of $g \circ f$. We are accustomed to reading from left to right. However, with composition notation, we must remember that the actual execution starts on the right. To compute a value for $g \circ f$, we start with f, not g. In the first diagram, the net result of the $g \circ f$ arrow is the same as the side path, starting with the f arrow and then followed by the g arrow.

To get the inverse of each of these 3 functions, we simply reverse each arrow, as illustrated in the second diagram. Note that the net result of the $(g \circ f)^{-1}$ arrow is the same as the side path, starting with the g^{-1} arrow and then followed by the f^{-1} arrow. So, $(g \circ f)^{-1} = f^{-1} \circ g^{-1}$.

The verbal justification of this visual picture is fairly simple, but because of all the switching going on, it can be confusing if we do not structure our reasoning and focus our thinking. We first focus on what we want to prove.

The statement of the following theorem involves the definitions of composition, inverse function, and equal functions. Working from the outside to the inside, we first focus on the definition of equal functions. Since the domain of both functions is Z, we must show that for every z in Z:

$$(g \circ f)^{-1}(z) = f^{-1} \circ g^{-1}(z).$$

This sentence sets up the outside structure of our proof. To prove this equality, we start with the left side. We set the left side equal to x so in order to apply the definition of $(g \circ f)^{-1}$. This little trick gives us the necessary notation so that we can substitute in the definition of inverse function:

$$f^{-1}(a) = b \text{ if and only if } f(b) = a.$$
$$\text{So } (g \circ f)^{-1}(z) = x \text{ if and only if } (g \circ f)(x) = z.$$

Theorem Let f be a bijection from X onto Y and g a bijection from Y onto Z. Then $(g \circ f)^{-1} = f^{-1} \circ g^{-1}$.

Proof Since f and g are bijections, $g \circ f$ is a bijection that maps X onto Z. So, f, g, and $g \circ f$ each have inverse functions. $(g \circ f)^{-1}$ maps Z onto X. Let z be an element in Z.

Set $(g \circ f)^{-1}(z) = x$.

$z = (g \circ f)(x)$. ..... *Definition of inverse of $g \circ f$*

$z = g(f(x))$ .... *Definition of composition*

$g^{-1}(z) = f(x)$ .... *Definition of inverse of g*

$f^{-1}(g^{-1}(z)) = x$ .... *Definition of inverse of f*

$(f^{-1} \circ g^{-1})(z) = x$ .... *Definition of composition*

So, for every z in Z, $(g \circ f)^{-1}(z) = f^{-1} \circ g^{-1}(z)$.

Therefore, $(g \circ f)^{-1} = f^{-1} \circ g^{-1}$. .... *Def. of equal functions*

Composition of 3 Functions

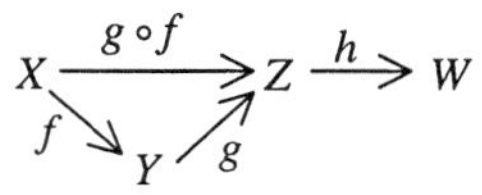

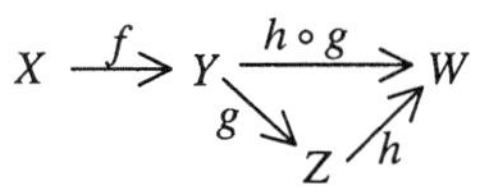

Suppose we have 3 functions whose composition can be formed. Does it matter how we group the functions when we compute the composition? Is $h \circ (g \circ f) = (h \circ g) \circ f$?

For example, let $f: X \to Y$, $g: Y \to Z$ and $h: Z \to W$. We can form the composition of these three functions in two different ways, as illustrated in the two adjacent sketches:

> If we first form $g \circ f$ and then compose this function with h, we obtain the function $h \circ (g \circ f)$.
>
> If we first form $h \circ g$ and then compose this function with f, we obtain the function $(h \circ g) \circ f$.

The computations for $h \circ (g \circ f)(x)$ are different from the computations for $(h \circ g) \circ f(x)$. However, the final outcome is the same. The proof of this theorem involves a close reading of the position of the parentheses, which indicate the two functions to which the definition of composition is applied.

Theorem Let $f: X \to Y$, $g: Y \to Z$ and $h: Z \to W$. Then $h \circ (g \circ f) = (h \circ g) \circ f$.

Proof Let x be an element in X.

$(h \circ (g \circ f))(x) = h((g \circ f)(x))$ ... *Def. of composition of h and g∘f*

$= h(g(f(x)))$ ...... *Def. of composition of g and f*

$((h \circ g) \circ f)(x) = (h \circ g)(f(x))$ .. *Def. of composition of h∘g and f*

$= h(g(f(x)))$ ..... *Def. of composition of h and g*

So, $((h \circ g) \circ f)(x) = (h \circ (g \circ f))(x)$ ..... *Transitivity of Equality*

So, $h \circ (g \circ f) = (h \circ g) \circ f$. .... *Def. of Equal Functions*

The above theorem proves that the composition of functions is associative. So, we can omit the parentheses when we write the composition of three functions since their position does not affect the outcome: $h \circ g \circ f$

Images of Sets

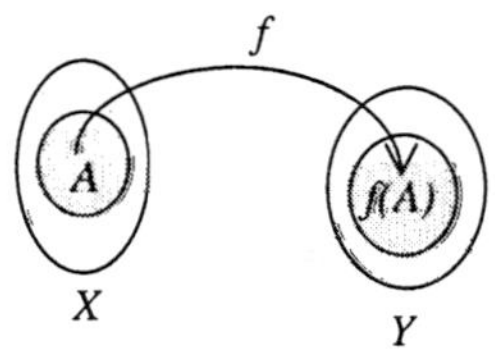

Let f be a function that maps X into Y and let A be a subset of X. We can view the function f as embedding the set X inside the set Y. Under this embedding, $f(A)$ represents the transformation of the set A. The set A could be squished to a single point, distorted in weird ways, or remain essentially the same. The set $f(A)$ is the set of all images of individual elements in A:

$$f(A) = \{ f(x) \mid x \in A \}$$

$y \in f(A)$ if and only if $y = f(x)$ for some x in A.

Note that $f(A) \subseteq Y$, whereas $f(x) \in Y$. The definition of $f(A)$ is the same as the definition of $f(X)$:

$$f(X) = \{ f(x) \mid x \in X \}$$

$y \in f(A)$
if and only if
$y = f(x)$ for some x in A.

If X is the domain of f, then $f(X)$ is the range. If $A \subseteq X$, then $f(A)$ is a subset of the range. To compute the image of a set, we compute the image of each element in the set and enclose them in a set.

Example

1. Let $f(x) = 2x$, where x is a real number.

   Let $A = \{1,2,3\}$. $f(A) = \{f(1), f(2), f(3)\} = \{2,4,6\}$

   Let $B = \{2\}$. $f(B) = \{4\}$

   Let $C = [1,3]$. $f(C) = [2,6]$, as illustrated on the left.

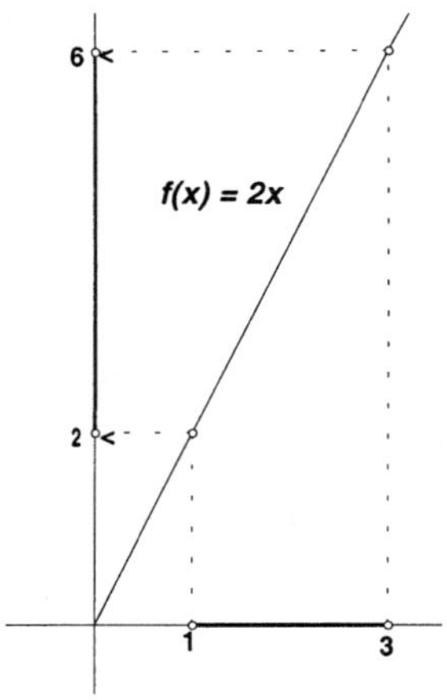

2. Let $f(x) = x^2$, where x is a real number.

   Let $A = \{2,3\}$. $f(A) = \{4,9\}$

   Let $B = \{-3,-2,3\}$. $f(B) = \{4,9\}$

   Let $C = [-2,1]$. $f(C) = [0,4]$

If $x \in A$, then $f(x) \in f(A)$. However, the converse is not true. In the above example where $f(x) = x^2$ and $A = \{2,3\}$, $f(-2) \in f(A)$, but $-2 \notin A$. The converse is true, though, when f is one-to-one. You are asked to prove the following theorem in (22) of the next exercise set.

Theorem Let f be a one-to-one function that maps X into Y and let $A \subseteq X$. If $f(x) \in f(A)$, then $x \in A$.

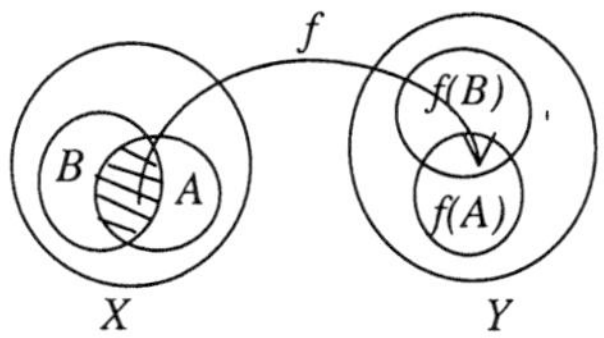

Let f map X into Y. If $A \subseteq X$ and $B \subseteq X$, what is the relation between the following two sets?

$$\text{Is } f(A \cap B) = f(A) \cap f(B)?$$

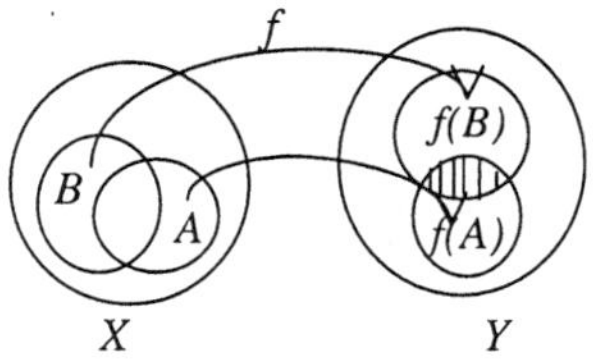

The above two sets look very similar, but the parentheses make a difference in the order in which we take the intersection and the image. To compute $f(A \cap B)$, we first intersect A and B over in X, as illustrated in the first sketch. Then we take the image of the intersection. On the other hand, to compute $f(A) \cap f(B)$, we first take the image of A and the image of B. We then intersect these two images over in Y, as illustrated in the second sketch..

It is fairly easy to prove that $f(A \cap B) \subseteq f(A) \cap f(B)$, that is, if we structure our thinking and calmly apply the definitions (see page 94). To set up the outside structure of the proof, we use the subset definition – assume y is in the left set and then demonstrate that y has to be in the right set.

Theorem Let f be a function that maps X into Y.
If $A \subseteq X$ and $B \subseteq X$, then $f(A \cap B) \subseteq f(A) \cap f(B)$.

Proof Let A and B be subsets of X.
Assume that $y \in f(A \cap B)$.
So, there exists an x_o in $A \cap B$ such that $f(x_o) = y$.
By the definition of intersection, $x_o \in A$ and $x_o \in B$.

Since $x_o \in A$, $f(x_o) \in f(A)$. .... *Def. of image of a set*
Since $x_o \in B$, $f(x_o) \in f(B)$. .... *Def. of image of a set*

But $f(x_o) = y$. So $y \in f(A)$ and $y \in f(B)$.

By the definition of intersection, $y \in f(A) \cap f(B)$.

So, $f(A \cap B) \subseteq f(A) \cap f(B)$.

Let's try to reverse the above steps and prove the following:

Claim: $f(A) \cap f(B) \subseteq f(A \cap B)$
Assume that $y \in f(A) \cap f(B)$. Then $y \in f(A)$ and $y \in f(B)$.
Since $y \in f(A)$, there exists an x_o in A such that $f(x_o) = y$.
Since $y \in f(B)$, there exists an x_1 in B such that $f(x_1) = y$.
So $f(x_o) = f(x_1)$.
. . . ? ? ? . . .
So $y \in f(A \cap B)$.

There does not appear to be anyway to build a logical bridge to the desired conclusion. In fact, this statement is not true. Before we give a counterexample, let's look at how we can finish the proof if the function f is one-to-one.

Theorem Let f be a one-to-one function that maps X into Y. If $A \subseteq X$ and $B \subseteq X$, then $f(A) \cap f(B) \subseteq f(A \cap B)$.

Proof Assume that $y \in f(A) \cap f(B)$. Then $y \in f(A)$ and $y \in f(B)$.

Since $y \in f(A)$, there exists an x_o in A such that $f(x_o) = y$.
* Since $y \in f(B)$, there exists an x_1 in B such that $f(x_1) = y$.
. . . . *Def. of image of a set*

So $f(x_o) = f(x_1)$.
Since f is one-to-one, $x_o = x_1$.

Thus, $x_o \in A$ and $x_o \in B$. Hence, $x_o \in A \cap B$.

So, $f(x_o) \in f(A \cap B)$. . . . . *Def. of image of a set*
Since $f(x_o) = y$, $y \in f(A \cap B)$.

Thus, $f(A) \cap f(B) \subseteq f(A \cap B)$.

$\exists x, p(x)$ and $\exists x, q(x)$ *is not equivalent to* $\exists x, p(x)$ and $q(x)$.

* It is easy to make a logical error in the above line marked by the asterisk. Suppose that we translate this line as follows: There exists an x_o in B such that $f(x_o) = y$. One might be tempted to deduce that $x_o \in A \cap B$. However, as we saw in Chapter 1 (page 69), the existential quantifier does not distribute across an and-statement.

If different points map to the same element under f, we can always find sets A and B such that $f(A \cap B) \neq f(A) \cap f(B)$, as illustrated in the following example.

⊕ *Example* Let $f(x) = x^2$, where x is a real number.
Let $A = \{1,2\}$ and $B = \{-1,-2\}$. Is $f(A \cap B) = f(A) \cap f(B)$?

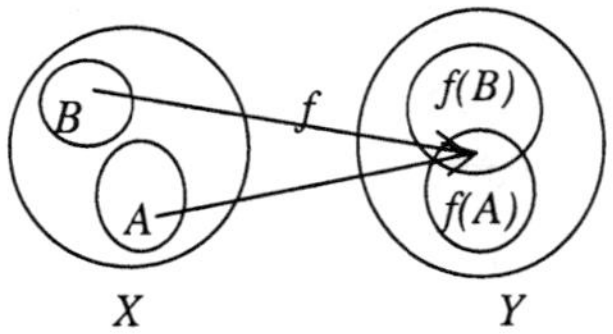

$A \cap B = \emptyset$, so $f(A \cap B) = \emptyset$
$f(A) = \{f(1), f(2)\} = \{1,4\}$
$f(B) = \{f(-1), f(-2)\} = \{1,4\}$
So, $f(A) \cap f(B) = \{1,4\}$.
Thus, $f(A \cap B) \neq f(A) \cap f(B)$.

Intersections are not always preserved under the image of a function unless the function is one-to-one. However, unions are always preserved, even if the function is not one-to-one, as demonstrated in the following proof.

Theorem Let f be a function that maps X into Y.
If $A \subseteq X$ and $B \subseteq X$, then $f(A \cup B) = f(A) \cup f(B)$.

Proof Let A and B be subsets of X.

Claim: $f(A \cup B) \subseteq f(A) \cup f(B)$.

Assume that $y \in f(A \cup B)$.

So, there exists an x_o in $A \cup B$ such that $y = f(x_o)$.

By the definition of union, $x_o \in A$ or $x_o \in B$.

So, $f(x_o) \in f(A)$ or $f(x_o) \in f(B)$.

By the definition of union, $f(x_o) \in f(A) \cup f(B)$.

Since $y = f(x_o)$, $y \in f(A) \cup f(B)$.

Thus, $f(A \cup B) \subseteq f(A) \cup f(B)$.

Claim: $f(A) \cup f(B) \subseteq f(A \cup B)$.

Assume that $y \in f(A) \cup f(B)$.

So $y \in f(A)$ or $y \in f(B)$

Case 1. Suppose that $y \in f(A)$.

There exists an x_0 such that $x_0 \in A$ and $y = f(x_0)$.

Since $x_0 \in A$, $x_0 \in A \cup B$.

So, $f(x_0) \in f(A \cup B)$.

Since $y = f(x_0)$, $y \in f(A \cup B)$.

Case 2. Suppose that $y \in f(B)$.

So, there exists an x_0 such that $x_0 \in B$ and $y = f(x_0)$.

Since $x_0 \in B$, $x_0 \in A \cup B$.

So, $f(x_0) \in f(A \cup B)$.

Since $y = f(x_0)$, $y \in f(A \cup B)$.

In both cases, $y \in f(A \cup B)$.

So, $f(A) \cup f(B) \subseteq f(A \cup B)$.

Therefore, $f(A \cup B) = f(A) \cup f(B)$.

Exercise Set 4.3

1. Draw an arrow mapping to illustrate a relation that is not a function. Use your mapping to explain why we cannot use function notation with an arbitrary relation.

2. Explain the difference between the notation f and $f(x)$.

3. Let $f(x) = x+2$, where x is a real number, and let $g(x) = x+2$, where x is a natural number. Is $f = g$? If not, why not?

4. Let a and b be natural numbers. Determine if the function is well-defined. Justify your answer.
   a. Define $f(a+b) = b$. b. Define $f(a+b) = 2(a+b)$.

5. Translate what it means for the given function to be one-to-one:
   a. $g: Y \to X$. b. $g \circ f: X \to W$.

6. Translate what it means for the given function to be onto:
   a. $g: Y \to X$. b. $g \circ f: X \to W$.

7. $\mathbb{R}$ is the set of real numbers and Z is the set of integers. Make up a function f that satisfies the following.
   a. f maps $\mathbb{R}$ onto $\{2\}$.
   b. f maps $[0,1]$ onto $[3,4]$.
   c. f maps $\mathbb{R}$ onto $\{0,1\}$.
   d. f maps $\mathbb{R}$ onto Z.
   e. f maps $\mathbb{R}\times\mathbb{R}$ into $\mathbb{R}$.
   f. f maps $\mathbb{R}$ into $\mathbb{R}\times\mathbb{R}$.

8. Find the domain and range of the function f.
   a. $f(x) = x^2$, where x is in $[-3,2]$.
   b. $f(x) = 5$, where x is a real number.
   c. $f(x, y) = x+y$, where x and y are real numbers.
   d. $f(x, y) = x+y$, where x is in $[0,2]$ and y is in $[0,3]$.
   e. $f(x, y) = (y+1, x)$, where x and y are real numbers.

9. Explain how each operation can be interpreted as a function. What would the domain and range be?
   a. Addition of natural numbers.
   b. Subtraction of natural numbers.
   c. Division of real numbers.

10. Let $f(x) = 3x+4$.
    a. Is $f(5x) = 5f(x)$? b. Is $f(x+3) = f(x)+f(3)$?

11. Prove or disprove that f maps $\mathbb{R}$ onto $\mathbb{R}$.
    a. $f(x) = 2x+7$ b. $f(x) = x^2+7$

12. Let $f: \mathbb{R} \to \mathbb{R}$. Prove or disprove that f is one-to-one.
    a. $f(x) = 2x+7$ b. $f(x) = x^2+7$

13. Let f be a one-to-one function that maps $\mathbb{R}$ onto $\mathbb{R}$ and f(3) = 5. Compute the following:
    a. $f^{-1}(5)$ b. $f^{-1}(f(7))$ c. $f(f^{-1}(2))$
14. Is the given statement true? If not, give a counterexample.
    a. If f is a one-to-one function from X into Y, then f^{-1} is a one-to-one function from Y into X.
    b. If f is a one-to-one function from X onto Y, then f^{-1} is a one-to-one function from Y onto X.
    c. For all functions f and g that map $\mathbb{R}$ into $\mathbb{R}$, $f \circ g = g \circ f$.
15. If possible, give an example of the following:
    a. A function that does not have an inverse function.
    b. A relation that does not have an inverse relation.
16. Let $f(x) = 2x$, and $g(x) = x^2$. Compute the following:
    a. $(f \circ g)(x)$ b. $(g \circ f)(x)$
17. Let $f: X \rightarrow Y$ and $g: Y \rightarrow Z$. Prove the following:
    a. If f and g are injections, then $g \circ f$ is an injection.
    b. If f and g are surjections, then $g \circ f$ is a surjection.
    c. If f and g are bijections, then $g \circ f$ is a bijection.
    d. If f is not an injection, there exist sets A and B such that $f(A \cap B) \neq f(A) \cap f(B)$.
    e. If f and g are bijections, then $(g \circ f)^{-1} = f^{-1} \circ g^{-1}$.
18. Rewrite the given equation in terms of the original function:
    a. $f^{-1}(y) = x$ b. $(g \circ f)^{-1}(z) = x$
19. Let S be a collection of sets. Let X and Y be in S.
    Define the relation $\approx$ as follows:
    $X \approx Y$ if and only if
    there exists a one-to-one function f that maps X onto Y.
    Prove that $\approx$ is an equivalence relation on S.
20. Let $f: X \rightarrow Y$ and $g: Y \rightarrow Z$. Is the statement true? If not, give a counterexample using a simple mapping picture where X, Y and Z have at most 3 points.
    a. If $g \circ f$ is a surjection, then f and g are surjections.
    b. If $g \circ f$ is an injection, then f and g are injections.
    c. Let $h: X \rightarrow Y$. If $g \circ f = g \circ h$, then $f = h$.
    (Can we cancel on the left?)
21. Let f be a one-to-one function that maps X into Y and let $A \subseteq X$. Prove the following: If $f(x) \in f(A)$, then $x \in A$.

22. Translate the given sentence by substituting in the appropriate definitions in the proper order.
    a. $y \in f(A) \cup f(B)$
    b. $y \in f(A \cup B)$
    c. $y \in f(A) \cap f(B)$
    d. $y \in f(A \cap B)$
    e. $y \in (\bigcup_{i \in I} f(A_i)$
    f. $y \in f(\bigcup_{i \in I} A_i)$
    g. $y \in \bigcap_{i \in I} f(A_i)$
    h. $y \in f(\bigcap_{i \in I} A_i)$

23. Let f be a function that maps X into Y. Let A and B be subsets of X. Prove or disprove the following.
    a. If $A \subseteq B$, then $f(A) \subseteq f(B)$.
    b. $f(A) \cup f(B) = f(A \cup B)$
    c. $f(A \cap B) \subseteq f(A) \cap f(B)$
    d. $f(A) \cap f(B) \subseteq f(A \cap B)$

24. Let $f: X \longrightarrow Y$. Let $A_i \subseteq X$ for each i in I. Prove the following.
    a. $f(\bigcup_{i \in I} A_i) = \bigcup_{i \in i} f(A_i)$
    b. $f(\bigcap_{i \in I} A_i) \subseteq \bigcap_{i \in I} f(A_i)$

25. Let $f: X \longrightarrow Y$ and let A and B be subsets of X. What is wrong with the following *fallacious* argument?
    1. Let $y \in f(A) \cap f(B)$. $y \in f(A)$ and $y \in f(B)$.
    2. So, there exists an x_o in A such that $f(x_o) = y$ and there exists an x_o in B such that $f(x_o) = y$.
    3. $x_o \in A$ and $x_o \in B$. So, $x_o \in A \cap B$.
    4. So, $f(x_o) \in f(A \cap B)$. Since $f(x_o) = y$, $y \in f(A \cap B)$.
    5. So, $f(A) \cap f(B) \subseteq f(A \cap B)$.

26. Let P be a partition of X. If $x \in X$, define f as follows: $f(x) = A$ if and only if $A \in P$ and $x \in A$. Is f well-defined?

27. Let f map X onto Y. Let $A_y = \{ x \text{ in } X \mid f(x) = y \}$.
    Let $P = \{ A_y \mid y \in Y \}$. Is P a partition of X? Justify your answer.

28. Let F be the following set of functions. $F = \{ f \mid f: \mathbb{N} \longrightarrow \{0,1\} \}$
    A member of F is a function on $\mathbb{N}$ whose only values are 0 and 1. For example, let $f(1) = 0$ and $f(n) = 1$ if $n \neq 1$. Then $f \in F$.
    a. Give 3 examples of different elements in F.
    b. Is F countable or uncountable?
       Hint: Suppose that F is countable. Arrange the functions in a sequence of rows with their values displayed. Can you construct a member of F that is not in any of the rows?
    c. Find a bijection from the power set of $\mathbb{N}$ onto F.

29. The set of all possible computer programs in a given language is countable (page 301). In a given computer language, a function is said to be computable if there is a computer program that will give any requested values of the function. Explain why there will be noncomputable functions in every computer language.

Activity 4.4

Determine if the relation is reflexive, transitive, or antisymmetric. Draw a directed graph and try to arrange the arrows so that each arrow points upward. If it is not possible to do so, explain why.

1. R = { (1,1), (1,3), (3,4), (4,3), (3,4), (4,4), (2,5) }
2. R = { (6,5), (5,7), (6,7), (5,5), (6,6), (7,7), (3,8), (3,7), (3,3), (8,8) }

≡ 4.4 Order Relations ≡

Without order, there is chaos. As we try to make sense of the world around us, we seek ways to instill a notion of order, for our reasoning faculties need some sort of order to keep things in a proper perspective. In the final section of our journey through this book, we will investigate the abstract structure of order relations.

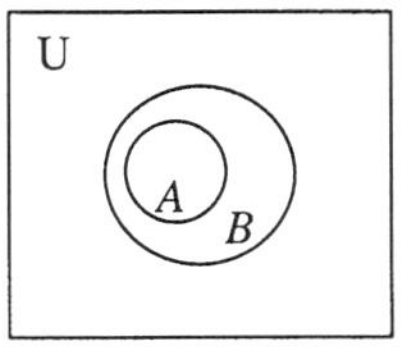

The Subset Relation

The seminal concept from which the notion of order is created is the concept of a subset. When we add a pebble to a sack with 3 pebbles, the original set is a subset of the new set. For this reason, we say that $3 \leq 4$. Using the subset principle as a guide, the counting numbers were created in an orderly manner, one after the other:

$$1\ 2\ 3\ 4\ 5\ 6\ 7\ 8\ 9\ \ldots$$

The power of the counting numbers as a quantifying tool comes from their order.

From the act of counting, we advance to the stage of measuring lengths. As we construct real numbers to measure all possible lengths, the order on the counting numbers is forever present, guiding us in our ordering of the real numbers: $\frac{3}{7} < \frac{4}{7}$ because $3 < 4$. As we tackle the more difficult task of logically explaining the set of irrational numbers, the order on the counting numbers guides us safely through the delicate task of setting up limits to represent an uncountable infinitude of irrational numbers. In the creation of the real numbers, we lose the step-by-step process of the counting numbers, but we still retain an order. The ordered set of real numbers provide us with one of the most important reasoning tools in mathematics and science. As with the counting numbers, the power of this universe of real numbers comes from their order. We will now go back to the original source of this power, the ordering properties of the subset relation.

Partial Orders

The subset relation has the following three fundamental properties. Let A, B and C be sets.

Reflexive Property: $A \subseteq A$.

Transitive Property: If $A \subseteq B$ and $B \subseteq C$, then $A \subseteq C$.

Antisymmetric Property: If $A \subseteq B$ and $B \subseteq A$, then $A = B$.

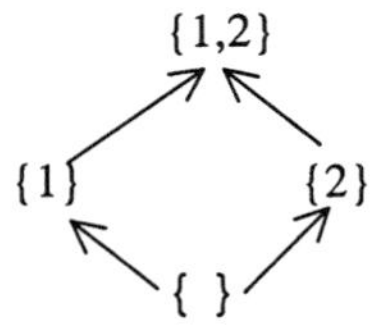

The Subset Relation

These three properties partially order any collection of sets. For example, let S be the power set of $\{1, 2\}$, which is all the subsets of $\{1, 2\}$. The adjacent diagram indicates the partial ordering of S by the subset relation. The set S is not totally ordered because we cannot compare $\{1\}$ and $\{2\}$ with the subset relation; neither is a subset of the other.

A partial order is a generalization of the subset relation. To be a partial order, a relation must possess the three basic properties of the subset relation: the reflexive, transitive, and antisymmetric properties. Let R be a relation on a set S. R is a *partial order* if and only if the following three statements are true for every a, b, and c in S:

Let R be a relation on the set S.
R is a *partial order*
if and only if
R has the following 3 properties:

- Reflexive
- Transitive
- Antisymmetric

Reflexive Property: $a \mathrel{R} a$

Transitive Property: If $a \mathrel{R} b$ and $b \mathrel{R} c$, then $a \mathrel{R} c$.

Antisymmetric Property: If $a \mathrel{R} b$ and $b \mathrel{R} a$, then $a = b$.

The antisymmetric property gives the distinguishing feature of a partial order; the only time we can reverse the order of two elements is when the elements are equal. Unlike an equivalence relation, it makes a difference in a partial order as to who comes first.

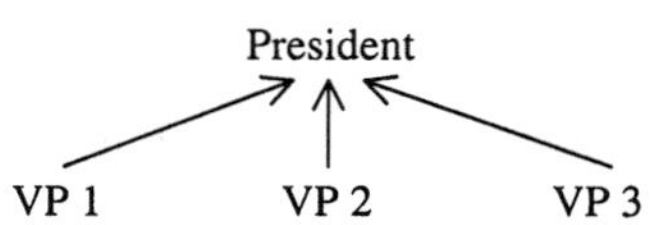

Instead of arranging elements in egalitarian equivalence classes where everyone is essentially the same, a partial order embeds a pecking order. The pecking order is not necessarily linear, but it does have a hierarchical structure like the pecking order in a corporation. Each vice-president reports to the president, but there is usually no direct authority connection between two vice-presidents. In the adjacent diagrams, notice how we can model the corporate diagram with the subset relation. We will see later on that even though we can conjure up a wide range of examples of partial orders, each and every one of them can be modeled with the subset relation on a particular collection of sets.

We use notation similar to the subset notation to represent a partial order:

$$a \preceq b \qquad a \trianglelefteq b \qquad a \ll b$$

The notation $(S, \preceq)$ represents a partially ordered set where $\preceq$ is a partial order on the set S. We often abbreviate "partially ordered set" as *poset*. A poset is a set that has a structure embedded via a partial order. The language developed for order relations describes the form of this structure and special features that it may have.

If $a \preceq b$ or $b \preceq a$, we say that a and b are *comparable*. For example, consider the poset $(S, \subseteq)$ where S is the power set of $\{1,2,3\}$. Since $\{1,2\} \subseteq \{1,2,3\}$, $\{1,2\}$ and $\{1,2,3\}$ are comparable, but $\{1,2\}$ and $\{1,3\}$ are not comparable.

Strict Order

The subset relation has a sidekick companion, the proper subset relation:

$$A \subset B \text{ if and only if } A \subseteq B \text{ and } A \neq B.$$

The $\leq$ relation has a similar companion, the $<$ relation. Similarly, each partial order has an associated strict order. If $\preceq$ is a partial order, we define the *strict order* $\prec$ as follows:

$$a \prec b \text{ if and only if } a \preceq b \text{ and } a \neq b.$$

$a \prec b$
if and only if
$a \preceq b$ and $a \neq b$.

If $a \prec b$, we say that b is a *successor* of a. To say that b is an *immediate successor* of a means that $a \prec b$ and there does not exist an x such that $a \prec x \prec b$. For example, in the set $\mathbb{N}$ of natural numbers, 4 is a successor of 2, but 3 is the immediate successor of 2.

The $\prec$ relation is transitive and antisymmetric, but it is not reflexive. In fact, it has the opposite property:

$$\text{For every } a, \sim(a \prec a).$$

A transitive and antisymmetric relation that has the above property is called a strict order. Each partial order has an associated strict order. Conversely, given a strict order $\prec$, we can construct a partial order from it as follows:

$$a \preceq b \text{ if and only if } a \prec b \text{ or } a = b.$$

$a \preceq b$
if and only if
$a \prec b$ or $a = b$.

The concepts of partial order and strict order are essentially the same, except that we insist on the reflexive property for partial orders and completely prohibit it for strict orders.

Total Order

On the set $\mathbb{R}$ of real numbers, $\leq$ is a partial order. However, the $\leq$ relation does more than partially order the real numbers; it totally orders them.

For all real numbers a and b, either $a \leq b$ or $b \leq a$.

A partial order that has this property is called a total order.
Let $\preceq$ be a partial order on the set S.

> Let $(S, \preceq)$ be a poset.
> $\preceq$ is a *total order*
> if and only if
> every two elements
> in S are comparable.

$\preceq$ is a *total order* if and only if
for any a and b in S, $a \preceq b$ or $b \preceq a$.

The $\leq$ relation is a total order on any set of real numbers. For example, the $\leq$ relation totally orders $\{3, 11, 8, 2\}$ as follows:

$$2 \rightarrow 3 \rightarrow 8 \rightarrow 11$$

A totally ordered set is also called a *chain* because the order chains the elements together in a linear manner.

In a totally ordered set, every two elements are comparable. If $\preceq$ is a total order and $\sim(a \preceq b)$, we can deduce that $b \prec a$. We cannot make this type of deduction for the subset relation on the set $\{\{1\},\{2\},\{1,2\}\}$. However the subset relation is a total order on the following set:

$$T = \{\{1\}, \{1,2\}, \{1,2,4\}, \{1,2,4,8\}\}$$

For every A and B in T, $A \subseteq B$ or $B \subseteq A$.

Example

Let $S = Z - \{0\}$. Is the divides relation a partial order on S? What about on $\mathbb{N}$? Is it a total order on either set?

Let a and b be nonzero integers. $a|b$ represents "a divides b."

$a|b$ if and only if there exists an integer k such that $b = ak$.

Reflexive: $a|a$

$a = a \cdot 1$. So, $a|a$.
Hence, the divides relation is reflexive on S and on $\mathbb{N}$.

Transitive: If $a|b$ and $b|c$, then $a|c$.

Assume that $a|b$ and $b|c$.
There exist integers k and j such that $b = ak$ and $c = bj$.
So $c = bj = (ak)j = a(kj)$. Since kj is an integer, $a|c$.
Therefore, the divides relation is transitive.

Antisymmetric: If $a|b$ and $b|a$, then $a = b$.

$3|-3$ and $-3|3$, but $3 \neq -3$. So the divides relation is not antisymmetric on S. However, it is antisymmetric on $\mathbb{N}$:

Let a and b be positive integers.
Assume that $a|b$ and $b|a$.
There exist integers k and j such that
$$b = ak \text{ and } a = bj.$$
$$a = bj = (ak)j = a(kj)$$
Since $a \neq 0$, we can divide by a. So $1 = kj$.
The only solutions to this equation are:
$$(k = 1 \text{ and } j = 1) \text{ or } (k = -1 \text{ and } j = -1).$$
But $b = ak$. Furthermore, b and a are both positive.
So k is positive. Thus, $k = 1$ and $j = 1$. Hence $b = a$.

Therefore, the divides relation is antisymmetric on the set $\mathbb{N}$ of natural numbers. Hence, it is a partial order on $\mathbb{N}$. It is not a total order on $\mathbb{N}$ because 2 and 3 are not comparable. However, the divides relation is a total order on the set $\{1,3,6,30\}$.

Hasse Graph

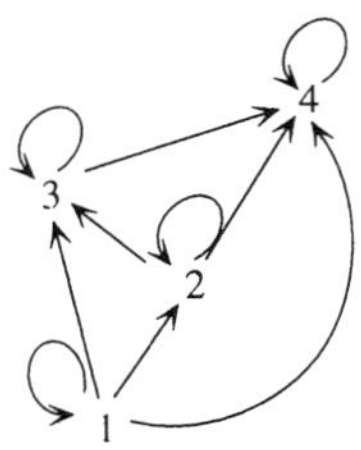

Directed Graph

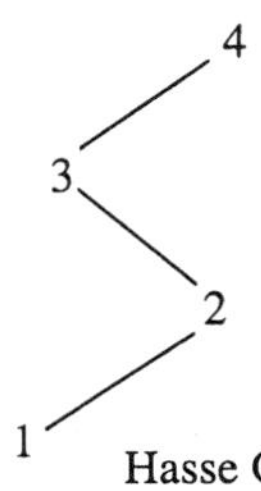

Hasse Graph

The structure embedded by an order relation on a set can be illustrated with a directed graph. For example, let $S=\{1,2,3,4\}$. Consider the structure on the poset $(S, \leq)$, which is illustrated in the adjacent directed graph. We have an arrow from 1 to 2 since $1 \leq 2$. We also have an arrow from 2 to 3 and another arrow from 1 to 3. In a directed graph, we have an arrow for each pair of related elements. However, too many arrows can obscure the view. Since the relation is transitive, we could deduce that 1 is related to 3 and not draw the arrow.

A clearer picture of the underlying structure of this relation is revealed when we remove the arrows that can be deduced by transitivity and the loop arrows that can be deduced by the reflexive property. Because of the antisymmetric property, we can always arrange the elements in a poset so that the relation arrows point upwards. If we agree to position the elements so that the relation points in an upward direction, we can remove the arrowheads. The remaining minimalist structure is called a Hasse graph. This reduced graph, which is illustrated on the left, reveals a simple structure that gives all the necessary information. Transitivity tells us that 1 is related to any element that we can get to in an upwards direction.

The *Hasse graph* of a partially ordered set is a subgraph of the directed graph in which we first position the elements so that all arrows point upwards and then we omit the following:

- the loops that can be deduced by reflexivity
- the arrows that can be deduced by transitivity
- the arrowheads

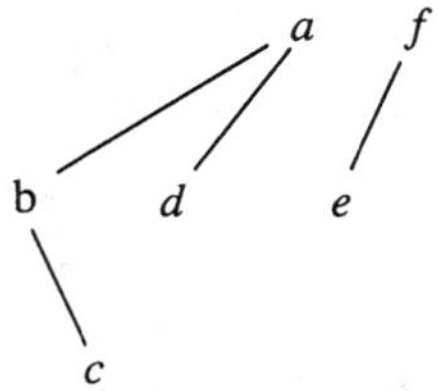

We can construct examples of partial orders by first drawing a Hasse graph. For example, the adjacent Hasse graph represents the following partial order on $S = \{ a, b, c, d, e, f \}$:

$a \preceq a$	$b \preceq b$	$c \preceq c$	$d \preceq d$	$e \preceq e$	$f \preceq f$
$c \preceq b$	$c \preceq a$	$b \preceq a$	$d \preceq a$	$e \preceq f$	

⊕ *Example*

Draw the Hasse graph of the partially ordered set.

1. $(S, \subseteq)$
   $S = \{ \emptyset, \{1\}, \{2\}, \{3\}, \{1,2\}, \{2,3\}, \{1,3\}, \{1,2,3\} \}$

   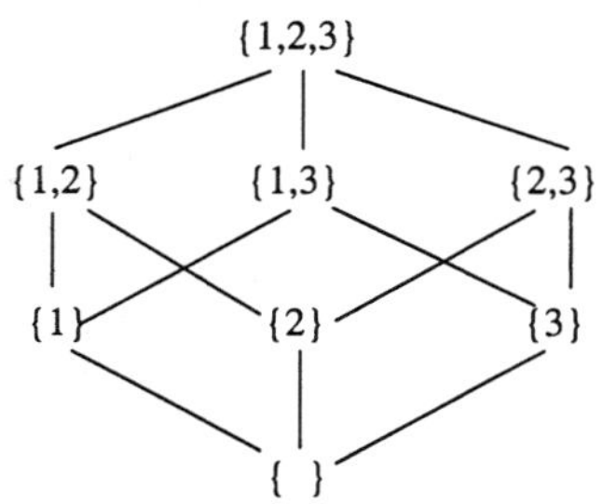

   To make the arrows point upwards, we put the empty set in the basement, the singleton sets on the first floor, the sets with 2 elements on the second floor, and the biggest set in the penthouse. We then draw the segments, keeping in mind that two elements are related if we can find an upward path that starts at the first element and ends at the second element. We do not draw a segment from $\{ 3 \}$ to $\{ 1, 2, 3 \}$ because we can get there by another path.

2. $(T, |\ )$ where $T = \{ 1, 2, 3, 5, 6, 10, 15, 30 \}$ and $|$ is divides..

   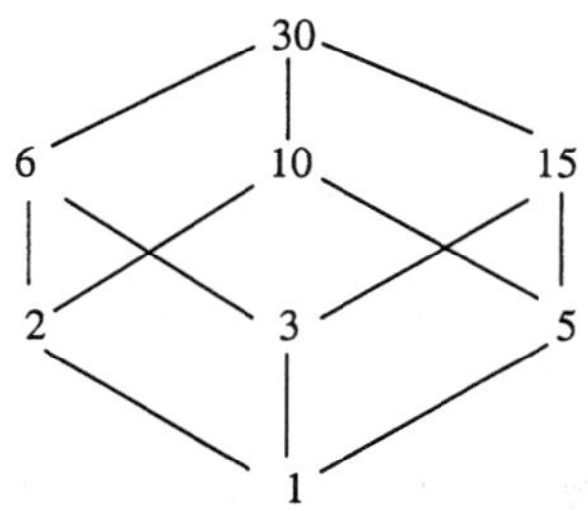

   $5|10$, so we draw an arrow from 5 to 10.
   $10|30$, so we draw an arrow from 10 to 30.

   Since 1 divides all the other numbers, we put it in the basement. At each level, we list elements that are not comparable.

Note the similarity of the above two graphs. Even though the definitions of the two relations are very different, they have identical structures. As far as structure goes, these two posets are essentially the same. They are *isomorphic*.

Isomorphisms

Now we come to one of the most important concepts in modern mathematics, that of an isomorphism. Through an isomorphic lens, different objects can look the same. The meaning of an isomorphism is built from the Greek words, *isos*, which means "same," and *morphe*, which means "form." Two isomorphic objects have the same form or structure.

Relations embed structures on sets. For example, order relations embed structures like those illustrated in the previous two examples. In abstract algebra, binary operations, like addition and multiplication, embed different types of structures on a set. In geometry, various distance functions embed different structures on a set. In analysis and topology, spatial relations which provide a notion of closeness embed different structures on a set. In each of these very diverse branches of mathematics, we try to simplify the playing field by identifying and classifying isomorphic structures.

One of the simplest visual illustrations of isomorphic structures are the two examples on the previous page. The divides relation embeds the same structure on the set of factors of 30 as does the subset relation on the set of all subsets of $\{1, 2, 3\}$. Even though the elements in the Hasse graph are labeled differently, the structure is the same. In terms of the structure, the labels applied to the elements are not significant. If we rename the elements or the relation, they will still have the same structure.

$(S, \preceq)$ is *isomorphic* to $(T, \trianglelefteq)$
if and only if
there exists a bijection f
from S onto T such that
for every a and b in S,
if $a \preceq b$, then $f(a) \trianglelefteq f(b)$.

The concept of "having the same structure" was formalized in 1910 by Alfred North Whitehead (1861–1947) and Bertrand Russell (1872–1970) in their classic text, *Principia mathematica*. To have the same structure, two structured sets must first of all have the same size (page 273), which means that there exists a bijection f from S onto T. However, there will be many different one-to-one functions that map S onto T. What we need is a bijection that also preserves the order structure. In other words, if a is related to b in the set S, then $f(a)$ must be related to $f(b)$ over in the set T, as illustrated in the adjacent sketch.

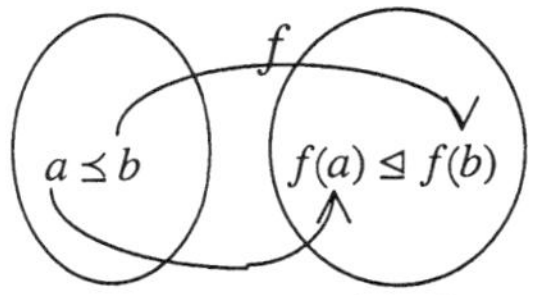

$$\text{If } a \preceq b, \text{ then } f(a) \trianglelefteq f(b).$$

This implication signifies that f preserves the order relation as it transports elements from S to T. If two elements are related in S, their images will be related in T. If f is also one-to-one and onto, then f is called an *isomorphism* and we say that $(S, \preceq)$ is *isomorphic* to $(T, \trianglelefteq)$. The relation of being isomorphic is an equivalence relation on posets (page 396), so we often represent it with an equivalence sign: $S \simeq T$.

When we construct an isomorphism between two structures, we look for a way to relabel the first structure so that we get the second structure, as illustrated in the following example.

✧ *Example*

Let $(S, \preceq)$ and $(T, \trianglelefteq)$ be the posets whose order is given in the adjacent Hasse graphs. Construct an isomorphism from S onto T.

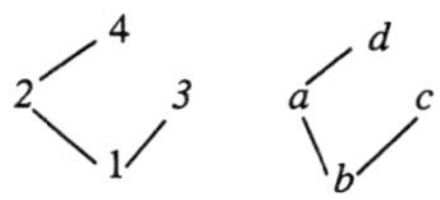

$(S, \preceq)$ $(T, \trianglelefteq)$

Define f as follows: $f(1) = b,\ f(2) = a,\ f(3) = c,\ f(4) = d$

f is a bijection from S to T. To show that f preserves the order, we need to verify the following:

For every a and b in S, if $a \preceq b$, then $f(a) \trianglelefteq f(b)$.

$1 \preceq 2$	Is $f(1) \trianglelefteq f(2)$?	Yes, $b \trianglelefteq a$.
$1 \preceq 3$	Is $f(1) \trianglelefteq f(3)$?	Yes, $b \trianglelefteq c$.
$2 \preceq 4$	Is $f(2) \trianglelefteq f(4)$?	Yes, $a \trianglelefteq d$.

By transitivity, we can deduce that the order is preserved for the remaining possibilities. Therefore, f is an isomorphism from $(S, \preceq)$ to $(T, \trianglelefteq)$. The poset $(S, \preceq)$ is isomorphic to $(T, \trianglelefteq)$, which means that they have the same order structure.

✧ *Example*

Is $(\mathbb{R}, \leq)$ isomorphic to $(\mathbb{R}, \geq)$?

Let a and b be real numbers. Define f as follows: $f(\text{a}) = -a$

Claim: f is one-to-one.

Assume that $f(a) = f(b)$.
Then $-a = -b$.
So, $-(-a) = -(-b)$.
Thus, $a = b$.
So, f is one-to-one.

Claim: f is onto.

Let y be a real number. Then $-y$ is a real number.
$f(-y) = -(-y) = y$. So, f is onto.

Claim: f preserves the order.

Assume that $a \leq b$. Then $-a \geq -b$.
So, $f(a) \geq f(b)$. So, f preserves the order.

Thus, f is an isomorphism. So, $(\mathbb{R}, \leq)$ is isomorphic to $(\mathbb{R}, \geq)$.

⊕ *Example*

1. Let $S = \{\ \{2\}, \{2,3\}, \{2,5\}\}$. As illustrated in the adjacent pair of sketches, $(S, \subseteq)$ is not isomorphic to $(S, \supseteq)$.

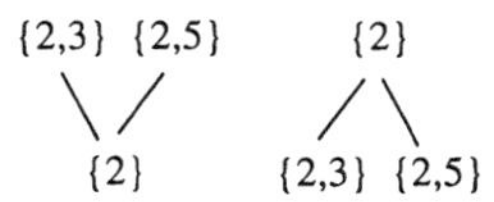

$(S, \subseteq)$ $(S, \supseteq)$

2. Let $T = \{\ \{2\}, \{2,3\}, \{2,5\}, \{2,3,5\}\}$. As illustrated in the lower pair of sketches, $(T, \subseteq)$ is isomorphic to$(T, \supseteq)$.

3. Let U be the power set of $\mathbb{R}$.
   Is $(U, \subseteq)$ isomorphic to $(U, \supseteq)$?

   Let A and B be in U. Let A' denote the complement of A in $\mathbb{R}$. Define f as follows: $f(A) = A'$

   Claim: f is one-to-one.
   Assume that $f(A) = f(B)$.
   Then $A' = B'$. So, $(A')' = (B')'$. Thus, $A = B$.
   So, f is one-to-one.

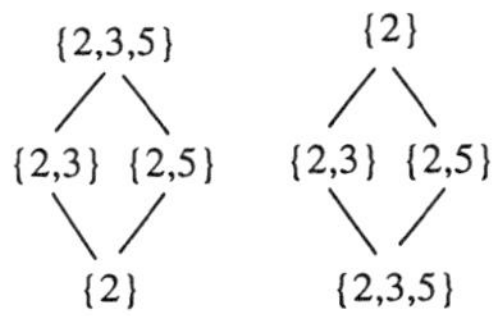

$(T, \subseteq)$ $(T, \supseteq)$

   Claim: f is onto.
   Let B be an element in U. Then $B' \in U$.
   $f(B') = (B')' = B$. So, f is onto.

   Claim: f preserves the order.
   Assume that $A \subseteq B$. Then $B' \subseteq A'$.
   So, $A' \supseteq B'$. Thus, $f(A) \supseteq f(B)$.
   So, f preserves the order.

   So, f is an isomorphism. $(U, \subseteq)$ is isomorphic to $(U, \supseteq)$.

Characterization of Partial Orders

A good stock of examples, as large as possible, is indispensable for a thorough understanding of any concept, and when I want to learn something new, I make it my first job to build one.

Paul R. Halmos

When we learn a new concept, we need to work with a wide range of examples in order to deepen our understanding of the meaning of the concept. As we explore various examples, we may wonder if we have overlooked any interesting examples that might suggest possible theorems or, in the other direction, shed light on the limitations of the system. The desired situation is to have an archetypal class of examples so that any other example is isomorphic to one of the archetypes. Then we can focus our attention on the archetypal class, for all possible structures are represented within that class.

The study of posets is fairly simple, for we do have an archetypal class of examples, namely those that are ordered by the subset relation. Given any example of a partial order on a set S, such as "divides" or $\leq$, we can produce an isomorphic copy of it using the subset relation. The technique for producing the isomorphic image is illustrated in the next example.

◈ *Example*

Let $S = \{a,b,c,d,e,f\}$. The order on S is given by the adjacent Hasse graph. Construct a collection T of sets so that $(T,\subseteq)$ is isomorphic to $(S,\preceq)$.

To construct a set that models a, note that 4 elements are related to a from its left side: $a \preceq a,\ b \preceq a,\ c \preceq a,\ d \preceq a.$
Let L_a denote this set:

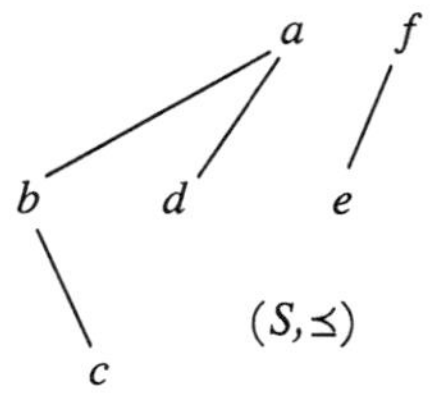

$(S,\preceq)$

$$L_a = \{\, y \mid y \preceq a \,\} = \{\, a,\ b,\ c,\ d \,\}$$

We construct similar sets to model each element in S:

$$L_b = \{\, y \mid y \preceq b \,\} = \{\, b,\ c \,\}$$
$$L_c = \{\, y \mid y \preceq c \,\} = \{\, c \,\}$$
$$L_d = \{\, y \mid y \preceq d \,\} = \{\, d \,\}$$
$$L_e = \{\, y \mid y \preceq e \,\} = \{\, e \,\}$$
$$L_f = \{\, y \mid y \preceq f \,\} = \{\, f,\ e \,\}$$

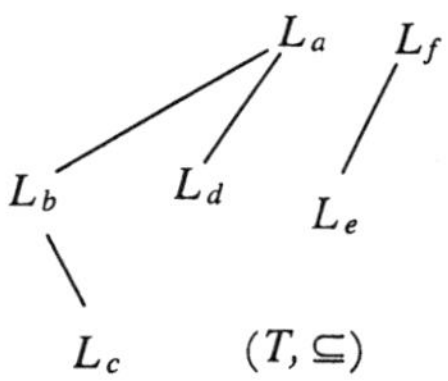

$(T,\subseteq)$

To produce an isomorphic copy of $(S,\preceq)$, we place each of the above sets in another set, which we label as T:

$$T = \{\, L_x \mid x \in S \,\} = \{\, L_a,\ L_b,\ L_c,\ L_d,\ L_e,\ L_f \,\}$$

The subset relation between sets in T clones the $\preceq$ relation on S:

$b \preceq a$	$c \preceq a$	$e \preceq f$
$L_b \subseteq L_a$	$L_c \subseteq L_a$	$L_e \subseteq L_f$

The Hasse graph of $(T,\subseteq)$, which is given on the left, has the same structure as the graph of $(S,\preceq)$.

So, $(T,\subseteq)$ is isomorphic to $(S,\preceq)$.

We will prove this statement in the following theorem.

Using the technique from the above example, we can prove that every poset is isomorphic to a poset whose order is the subset relation. Given an element x in the poset $(S,\preceq)$, we construct a set that models x by forming the subset of S that contains all elements related to x from its left side: $L_x = \{\, y \mid y \preceq x \,\}$. If you get lost in the following proof, pause and interpret the meaning for the above example.

Theorem Let $(S, \preceq)$ be a partially ordered set. There exists a collection T of sets such that the poset $(T, \subseteq)$ is isomorphic to $(S, \preceq)$.

Proof For each element x in S, define $L_x = \{ y \mid y \preceq x \}$.

Set $T = \{ L_x \mid x \in S \}$. $(T, \subseteq)$ is a partially ordered set.

Define $f: S \to T$ as follows: If $x \in S$, $f(x) = L_x$.

For each x in S, $f(x)$ is unique, so f is a function.

Claim: f is one-to-one.

Let x and z be elements in S.

Assume that $f(x) = f(z)$. So, $L_x = L_z$.

Since $x \preceq x$, $x \in L_x$. But $L_x = L_z$. So, $x \in L_z$.

By the definition of L_z, $x \preceq z$.

Similarly, $z \in L_z$, so $z \in L_x$. Hence, $z \preceq x$.

Since $\preceq$ is antisymmetric, $x = z$.

Thus, f is one-to-one.

Claim: f maps S onto T.

Let L_x be an element in T.

$f(x) = L_x$. So f maps S onto T.

Claim: f preserves the order.

Let x and z be elements in S.

Assume that $x \preceq z$.

Assume that $w \in L_x$. By the definition of L_x, $w \preceq x$.

$w \preceq x$ and $x \preceq z$. By transitivity, $w \preceq z$. So, $w \in L_z$.

So, $L_x \subseteq L_z$.

So, $f(x) \subseteq f(z)$.

Hence, f preserves the order.

Since f is a one-to-one function from S onto T that preserves the order, $(S, \preceq)$ is isomorphic to $(T, \subseteq)$.

Isomorphic posets are identical twins with different names. If a poset has a particular property, then all isomorphic posets must have the same property, as translated through the isomorphism. The above theorem tells us that the subset relation provides isomorphic copies of all partial orders. Consequently, when contemplating a conjecture about partial orders, we can restrict our focus to the subset relation. If we can prove a result for the subset relation on an arbitrary collection of sets, then the result will be true for any partial order. On the other hand, if a conjecture about partial orders is not true, there must be a counterexample for some set ordered by the subset relation.

Least Element

Some posets contain a least element. A least element of a poset $(S, \preceq)$ is the smallest element in S. Let c be in S.

c is the *least element* of S if and only if for every x in S, $c \preceq x$.

Let c be in a poset S.
c is the *least* element of S
if and only if
for every x in S, $c \preceq x$.

For example, let $S = [2,5]$ and $T = (2,5]$. Under the $\leq$ relation, 2 is the least element of S, but T does not have a least element. In a totally ordered set, we can linearly arrange the elements in a manner similar to a real number line where $a \preceq b$ means that a is to the left of b. In this type of visual, the least element is the one furthest to the left. Some sets, such as the interval (0,1], do not have an element that is furthest to the left. Unlike the real numbers, every subset of the set ℕ of natural numbers must have a least element.

In a Hasse graph of a partially ordered set, we can visually find the least element – if it exists – at the bottom of the graph.

⌖ *Example*

Does the poset $(S, \subseteq)$ have a least element?

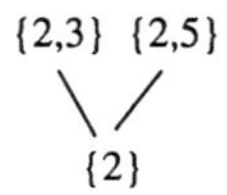

1. $S = \{\{2\}, \{2,3\}, \{2,5\}\}$

   For every set A in S, $\{2\} \subseteq A$.
   So, $\{2\}$ is the least element of $(S, \subseteq)$. $\{2\}$ is at the bottom of the adjacent Hasse graph, connected to everyone else.

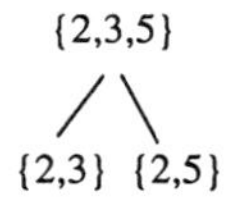

2. $S = \{\{2,3\}, \{2,5\}, \{2,3,5\}\}$

   S does not have a least element.
   Instead, S has two minimal elements.

Minimal Elements

An element of an ordered set S is a *minimal element* if there are no other elements before it. Let d be in S.

d is a *minimal element* of S if and only if
for every x in S, if $x \neq d$, then $\sim(x \preceq d)$.

Let d be in a poset S.
d is a *minimal* element of S
if and only if
for every x in S,
if $x \preceq d$, then $x = d$.

For a more positive tone, let's rephrase the above implication in terms of the contrapositive.

For every x in S, if $x \preceq d$, then $x = d$.

The definition of a least element sounds similar, but they are not equivalent, as illustrated in the following example.

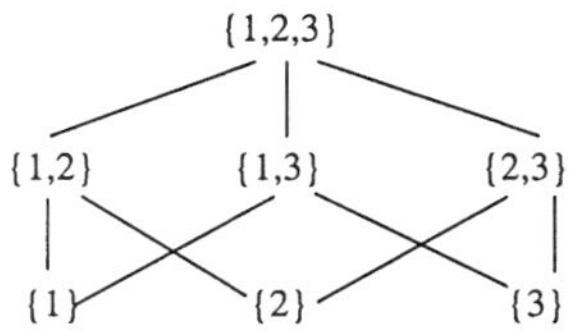

$(S, \subseteq)$

Example Let $S = \{\{1\},\{2\},\{3\},\{1,2\},\{2,3\},\{1,3\},\{1,2,3\}\}$.

Under the subset relation, does S have a least element?
Does S have any minimal elements?

In the adjacent Hasse graph of $(S,\subseteq)$, note that there is no other element that is a subset of $\{1\}$. So $\{1\}$ is a minimal element.

$\{1\}$ is not a least element because $\{1\}$ is not a subset of $\{2\}$.

Likewise, $\{2\}$ and $\{3\}$ are minimal elements of S, but neither is a least element.

S has 3 minimal elements, but it does not have a least element.

If a poset has a least element, the least element will be the only minimal element in the set, as demonstrated in the following theorem.

Theorem Let S be a partially ordered set. If b is the least element of S, then b is a minimal element of S. Furthermore, b is the only minimal element of S.

Proof Let b be the least element of S.

Claim: b is a minimal element of S.

Let x be an element in S. Assume that $x \preceq b$.

Since b is the least element of S, $b \preceq x$.

By the antisymmetric property, $x = b$.

So, for every x in S, if $x \preceq b$, then $x = b$.

Thus, b is a minimal element in S.

Claim: b is the only minimal element in S.

Suppose that c is also a minimal element in S.

Since b is the least element in S, $b \preceq c$.

Since c is a minimal element, $b = c$.

So, b is the only minimal element in S.

The converse of the previous theorem is true for totally ordered sets.

Theorem Let S be a totally ordered set. If b is a minimal element of S, then b is the least element of S.

Proof Let b be a minimal element of the totally ordered set $(S,\preceq)$.

Let x be an element in S.

Case 1: Suppose that $x = b$.
Since $\preceq$ is reflexive, $b \preceq x$.

Case 2: Suppose that $x \neq b$.
Since S is totally ordered, $b \preceq x$ or $x \preceq b$.
Since b is a minimal element and $x \neq b$, $\sim(x \preceq b)$.
So, $b \preceq x$.

Thus, b is the least element in S.

If a poset has a least element, it can have only one, as demonstrated in the following proof. To prove that a least element is unique, we assume that there are two least elements and then demonstrate that they must be equal.

Theorem A partially ordered set can have at most one least element.

Proof Let $(S,\preceq)$ be a partially ordered set.

Suppose that a and b are both least elements of S.

Since a is a least element, for every x in S, $a \preceq x$.
In particular, let $x = b$. So $a \preceq b$.

Since b is a least element, for every x in S, $b \preceq x$.
In particular, let $x = a$. So $b \preceq a$.

Since $\preceq$ is antisymmetric, $a = b$.

Some posets do not have a minimal element. For example, the interval $(1,2]$ with the $\leq$ relation does not have a minimal element. If a poset is finite, though, it must have a minimal element. This statement is visually obvious in a Hasse graph of a finite set. There must be an element at the bottom that has no elements below it.

To logically verify this visual observation, we could set up a step-by-step procedure analogous to starting at an arbitrary point on the Hasse graph and working our way down to the bottom. We first pick an arbitrary element a_1 in the Hasse graph and then we look at all the elements that are before it:

$$S_1 = \{\ x \text{ in } S \mid x \prec a_1\ \}$$

Since $a_1 \notin S_1$, S_1 must have less elements than S. If a_1 is not a minimal element, then $S_1 \neq \emptyset$. We then pick an element in S_1 and repeat the same procedure, at each stage creating a subset S_k that has less elements than the previous set. Since S is finite, we must eventually run out of elements. The last element left standing will be a minimal element. This process can be used in a computer program to locate minimal elements. We could use a similar argument to prove the following theorem, but a proof by induction is not nearly as tedious. This theorem can also be proved using a proof by contradiction, which you are asked to do in (13) of the next exercise set.

Theorem A finite, nonempty poset must have a minimal element.

Induction Proof Let $(S, \preceq)$ be a poset.
Let $p(n)$: If S has n elements, then S has a minimal element.

$p(1)$ is true.

Let n be a natural number. Assume that $p(n)$ is true.
Assume that S has $n+1$ elements. S can be represented as follows: $S = \{a_1, a_2, a_3, \ldots, a_{n+1}\}$. Let $T = \{a_1, a_2, a_3, \ldots, a_n\}$.
Since $p(n)$ is true, T has a minimal element. Call it c.

Case 1: Assume that $c \preceq a_{n+1}$.
Then c is a minimal element of S.

Case 2: Assume that $a_{n+1} \preceq c$.
Let x be in T. Assume that $x \preceq a_{n+1}$.
By transitivity, $x \preceq c$.
Since c is a minimal element of T, $x = c$.
Since $a_{n+1} \preceq c$, $a_{n+1} \preceq x$.
By antisymmetry, $x = a_{n+1}$.
Therefore, a_{n+1} is a minimal element of S.

Case 3: Assume that c and a_{n+1} are not comparable.
Then c is a minimal element of S.

Thus, S has a minimal element. So, $p(n) \Rightarrow p(n+1)$.

By mathematical induction, $p(n)$ is true for all natural numbers n.

Topological Sorting

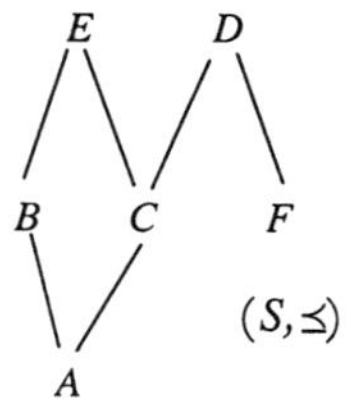

Using the previous theorem, we can extend a partial order on a finite set to a total order. For example, suppose that we have a list of 6 tasks to do and some tasks must be performed before others, as indicated in the adjacent graph. Task A must be performed before B and C, B and C must each be performed before E, and F must be performed before D. Let S represent the set of 6 tasks and $\preceq$ the partial order on S. We could schedule the tasks as follows:

$$A \ B \ C \ E \ F \ D$$

This listing of the elements in S gives a total order of S, which we will notate as $\trianglelefteq$. This new order has the following property:

For every x and y in S, if $x \preceq y$, then $x \trianglelefteq y$.

Let $\preceq$ be a partial order on S.
Let $\trianglelefteq$ be a total order on S.

$\trianglelefteq$ is a *topological sorting* for $\preceq$
if and only if
for every x and y in S,
if $x \preceq y$, then $x \trianglelefteq y$.

A total order of a poset that preserves the above implication is called a *topological sorting*. A topological sorting embeds a partially ordered set in a totally ordered set so that the original order is preserved. However, the converse of the above implication is not true: $E \trianglelefteq F$, but $E \npreceq F$.

To construct a topological sort of a finite poset, we select a minimal element at each stage and then remove it from the list.

- Select the first element to be a minimal element in S. Then remove it from S and call the new set S_1.
- Select the 2nd element to be a minimal element in S_1. Then remove it from S_1 and call the new set S_2.

Continuing this process produces a topological sorting of the original poset, as illustrated in the following example

---

✦ *Example*

Let $S = \{\{1,2,3\}, \{1\}, \{1,2\}, \{1,3\}, \{2\}, \{3\}, \{2,3\}\}$.
Find a topological sorting for $\subseteq$ on S.

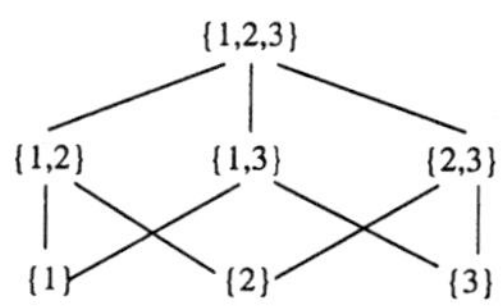

1. In the adjacent Hasse graph for S, we have 3 minimal elements. We may select either $\{1\}$, $\{2\}$, or $\{3\}$ for the first element in our sort. Let's choose $\{3\}$.
2. Remove $\{3\}$ from the graph. We now have 2 minimal elements: $\{1\}$, $\{2\}$. Let's choose $\{2\}$.

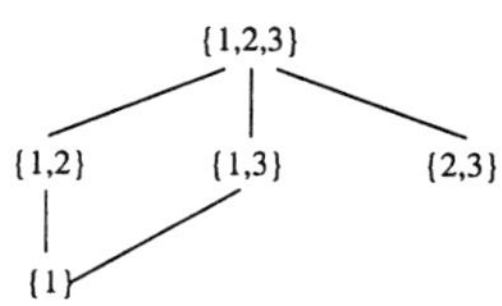

3. Remove $\{2\}$ from the graph. As illustrated in the adjacent graph, we now have 2 minimal elements: $\{1\}$, and $\{2,3\}$. Let's choose $\{2,3\}$.
4. Remove $\{2, 3\}$ from the graph. We only have 1 minimal element, so we must choose $\{1\}$.

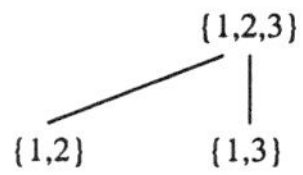

5. Remove $\{1\}$ from the graph. As illustrated in the adjacent graph, we now have 2 minimal elements: $\{1,2\}$ and $\{1,3\}$. Let's choose $\{1,2\}$.
6. When we remove $\{1, 2\}$ from the graph, we must first choose $\{1,3\}$ and then $\{1,2,3\}$.

We now have a topological sorting for $\subseteq$ on S:

$$\{3\}\ \{2\}\ \{2,3\}\ \{1\}\ \{1,2\}\ \{1,3\}\ \{1,2,3\}$$

The above listing gives a total order that has the required property: If $A \subseteq B$, then A is before B in the above list.

The following arrangement is not a topological sorting because $\{2,3\}$ is before $\{2\}$:

$$\{3\}\ \{2,3\}\ \{2\}\ \{1\}\ \{1,2\}\ \{1,3\}\ \{1,2,3\}$$

We can also do a topological sorting by listing the sets according to their size:

$$\{1\}\ \{2\}\ \{3\}\ \{1,2\}\ \{1,3\}\ \{2,3\}\ \{1,2,3\}$$

Greatest & Maximal

Let c be in a poset S.

c is the *greatest* element in S
if and only if
for every x in S, $x \preceq c$.

c is a *maximal element* in S
if and only if
for every x in S,
if $c \preceq x$, then $x = c$.

We have analogous terms for least and minimal at the other end of the spectrum. The opposite of least is greatest. Let c be in a partially ordered set S.

c is the *greatest* element in S
if and only if
for every x in S, $x \preceq c$.

The opposite of a minimal element is a maximal element. An element is a maximal element if there are no other elements that come after it.

c is a *maximal element* in S
if and only if
for every x in S, if $c \preceq x$, then $x = c$

For example, let $S = [1,4]$ and $T = [1,4)$. Under the $\leq$ relation, 4 is the greatest element in S, but T does not have a greatest element or a maximal element.

⊕ *Example*

Under the subset relation, does the given set have a greatest element, a least element, any maximal elements, or any minimal elements?

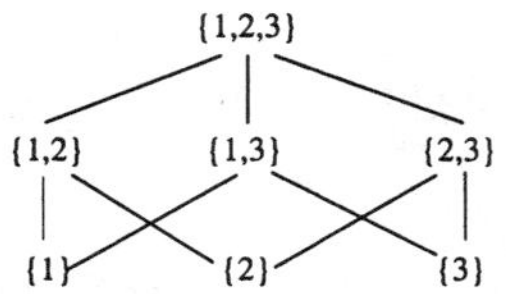

1. $S = \{ \{1\}, \{2\}, \{3\}, \{1,2\}, \{2,3\}, \{1,3\}, \{1,2,3\} \}$.

   $\{1,2,3\}$ is the greatest element of S.
   S does not have a least element.

   $\{1,2,3\}$ is a maximal element.
   S has 3 minimal elements: $\{1\}$, $\{2\}$, $\{3\}$

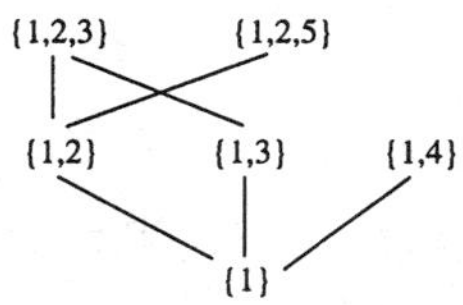

2. $T = \{ \{1\},\{1,2\},\{1,3\},\{1,4\},\{1,2,3\},\{1,2,5\} \}$.

   T does not have a greatest element.
   $\{1\}$ is the least element.

   S has 3 maximal elements: $\{1,2,3\}$, $\{1,2,5\}$, $\{1,4\}$
   $\{1\}$ is a minimal element.

As you probably suspect, for each theorem on minimal and least elements, we have a dual theorem for maximal and greatest elements. You are asked to prove the following theorems in the next exercise set.

Theorem

Let S be a partially ordered set.

1. S can have at most one greatest element.
2. If b is the greatest element of S, then b is the only maximal element of S.
3. If S is totally ordered, a maximal element of S must be the greatest element of S.
4. A finite, nonempty poset must have a maximal element.

Lower & Upper Bounds

The concepts of least element and lower bound are very similar. However, the least element of a set S must be in S, whereas a lower bound of S does not have to be in S. Since we are going outside of S, we need to specify a universal set U. Let $(U, \preceq)$ be a partially ordered set. Let $S \subseteq U$ and $b \in U$.

b is a *lower bound* for S if and only if for every x in S, $b \preceq x$.

b is an *upper bound* for S if and only if for every x in S, $x \preceq b$.

Let U be the set of real numbers ordered by $\leq$. Set $S = (5,7]$. On the real number line, 3 is a lower bound of S. 5 is also a lower bound of S. S has an infinite set of lower bounds. 5 is the greatest of all the lower bounds, so it is called the *greatest lower bound.* The opposite terms – greatest and lower – make this concept a little confusing. However, if we focus on the set of all lower bounds, we can easily visualize the greatest element in that set. The greatest lower bound of (5,7] is not a member of the set. The greatest lower bound of [5,7] is also 5 and it is a member of the set.

Let U be a poset.
$S \subseteq U$ and $b \in U$.

b is a *lower bound* for S
if and only if
for every x in S, $b \preceq x$.

b is an *upper bound* for S
if and only if
for every x in S, $x \preceq b$.

At the other end of the spectrum, we can look at the set of all upper bounds of a set S. The least element in this set is called the *least upper bound* of S. Let $S = (5,7]$. 21 is an upper bound for S. 100 is an upper bound for S. The least upper bound is 7. In some posets, the least upper bound or greatest lower bound may not exist. For example, let U be the set of rational numbers. Let $S = \{ x \mid x \text{ is rational and } x^2 > 2 \}$. Since $\sqrt{2}$ is not in our universal set, we cannot use it for the greatest lower bound for S. On a calculator, we can get decimal approximations to $\sqrt{2}$, such as 1.414213. 1.414213 is a rational number, so it is in U. This number is a lower bound for S, but it is not the greatest lower bound. S does not have a greatest lower bound in the set of rational numbers.

A subset of a finite poset may not have a greatest lower bound, as illustrated in the next example.

Example

$U = \{ \{1\}, \{1,2\}, \{1,3\}, \{1,2,3,4\}, \{1,2,3,5\}, \{1,2,3,4,5\} \}$

In the universal poset $(U, \subseteq)$, find the greatest lower bound and least upper bound for the given set.

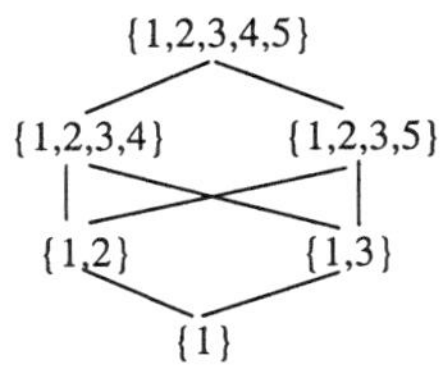

1. $A = \{ \{1,2\}, \{1,3\} \}$

   $\{1\}$ is the only lower bound for A,
   so it is the greatest lower bound.

   A has 3 upper bounds: $\{1,2,3,4\}, \{1,2,3,5\}, \{1,2,3,4,5\}$
   However, A does not have a least upper bound.

2. $B = \{ \{1,2,3,4\}, \{1,2,3,5\} \}$

   B has 3 lower bounds: $\{1,2\}$, $\{1,3\}$ and $\{1\}$.
   However, B does not have a greatest lower bound.

   $\{1,2,3,4,5\}$ is the only upper bound for B,
   so it is the least upper bound.

Well-Ordered Set

A well-ordered set S is a partially ordered set that has the following additional property:

Every nonempty subset of S has a least element.

Let S be a poset.
S is *well-ordered*
if and only if
every nonempty subset of S
has a least element.

The positive rational numbers do not have a least element, so the set of rational numbers is not well-ordered. Neither is the set Z of integers nor the set $\mathbb{R}$ of real numbers. Each of these examples is a totally ordered set. Hence, a totally ordered set does not have to be well-ordered.

On the other hand, a totally ordered set S that is finite must be well-ordered. Let A be a nonempty subset of S. Since A is a finite poset, A must have a minimal element (page 385). In a totally ordered set, a minimal element must be the least element (page 384). Therefore, A has a least element. So, every finite totally ordered set is well-ordered. We will now prove that every well-ordered set must be totally ordered.

Theorem Let $(S, \preceq)$ be a partially ordered set.
If S is well-ordered, then S is totally ordered.

Proof Let S be a well-ordered poset.
Let a and b be elements in S.
  Let $C = \{ a,b \}$.
  Since S is well-ordered, C must have a least element.
    If the least element is a, then $a \preceq b$.
    If the least element is b, then $b \preceq$ a.
  Hence, $a \preceq b$ or $b \preceq a$.
Therefore, S is totally ordered.

Let S be a poset.
b is an *immediate successor* to a
if and only if
$a \prec b$ and there does not
exist an x such that $a \prec x \prec b$.

The archetypal example of an infinite well-ordered set is the set $\mathbb{N}$ of natural numbers. Every nonempty subset of $\mathbb{N}$ has a least element. This property of $\mathbb{N}$ is inherited from the step-by-step manner in which the natural numbers are constructed:

$$0, 1, 2, 3, \ldots .$$

Each natural number has an immediate successor. For b to be an *immediate successor* to a, b must be greater than a and there cannot exist an element x between a and b.

In $\mathbb{N}$, the immediate successor of 1 is 2.

In $\mathbb{R}$, 1 does not have an immediate successor.

Since 1 does not have an immediate successor in $\mathbb{R}$, the following set does not have a least element:

$$A = \{\ x \text{ in } \mathbb{R} \mid 1 < x\ \}$$

Suppose that c is the least element in A.
Since $c \in A$, $1 < c$. Since 1 does not have an immediate successor, there exists a real number b such that $1 < b < c$.
Since $1 < b$, $b \in A$. Since $b < c$, c is not the least elements in A. Contradiction!

Using a similar argument, we can prove that if an element, other than the greatest element, does not have an immediate successor, the set cannot be well-ordered. This result is stated in contrapositive form in the following proof: if a set is well-ordered, each element, other than the greatest element, will have an immediate successor. To find the immediate successor of a, we look at the set of all elements that are greater than a and take its least element.

Theorem Let $(S, \preceq)$ be a well-ordered set. If $a \in S$ and a is not the greatest element in S, then a must have an immediate successor.

Proof Assume that a is not the greatest element in S.
Then there exists a d such that $\sim(d \preceq a)$.
Since S is totally ordered, $a \prec d$.

Set $T_a = \{x \mid a \prec x\}$. $d \in T_a$, so $T_a \neq \emptyset$.

Since S is well-ordered and $T_a \neq \emptyset$, T_a has a least element.
Let b denote the least element in T_a. Since $b \in T_a$, $a \prec b$.

Claim: b is the immediate successor of a.
Suppose that $x \in S$ and $a \prec x \prec b$.
Since $a \prec x$, $x \in T_a$.
Since b is the least element in T_a, $b \preceq x$.
By antisymmetry, $c = b$. Contradiction!
Thus, it is not true that there exists a x in S such that $a \prec x \prec b$. So b is the immediate successor of a.

Well-ordering implies immediate successors, but the converse is not true. For example, in the following totally ordered set, each element has an immediate successor, but S is not well-ordered:

$$S = \{\ \tfrac{1}{n} \mid n \in \mathbb{N}\ \} = \{\ \ldots, \tfrac{1}{5}, \tfrac{1}{4}, \tfrac{1}{3}, \tfrac{1}{2}, \tfrac{1}{1}\ \}$$

In a well-ordered set, we cannot have an infinite accumulation of elements in the downwards direction, but we can have an infinite accumulation of elements in the upwards direction. For example, the following set is well-ordered:

$$T = \{ 1 - \tfrac{1}{n} \mid n \in \mathbb{N} \} = \{ 0, \tfrac{1}{2}, \tfrac{2}{3}, \tfrac{3}{4}, \tfrac{4}{5}, \ldots \}$$

Well-Ordering Principle

Well-Ordering Principle
Every nonempty subset of $\mathbb{N}$ has a least element.

Every nonempty subset of the set $\mathbb{N}$ of natural numbers has a least element. This property of the set of natural numbers is called the Well-Ordering Principle. There are an uncountable number of different subsets of $\mathbb{N}$ (page 298). Thus, the Well-Ordering Principle applies to a vast universe of sets, which is the reason for its great power. In advanced math courses, you will develop an appreciation for the powerful reasoning tool provided by this deceptively simple sounding principle: every nonempty subset of $\mathbb{N}$ has a least element. The opposite extreme, though, is not true. A nonempty subset of $\mathbb{N}$ does not have to possess a greatest element.

In the construction of numbers, the set of natural numbers is our first contact with the mysterious realm of the infinite. Even though the natural numbers are unbounded at the top, the Well-Ordering Principle gives us a firm grip on its lower side, which is the power of the Well-Ordering Principle. We may not have a largest element in a set of natural numbers, but we will always have a smallest element. It is surprising how useful that can be.

Using the Well-Ordering Principle, we can derive the Principle of Mathematical Induction, and other powerful tools, such as the Division Algorithm and the Fundamental Theorem of Arithmetic. Actually, the Well-Ordering Principle is equivalent to the Principle of Mathematical Induction. In a book devoted to the art of logical reasoning, it is appropriate that we close with a proof that two of the most powerful reasoning tools are logically equivalent.

Theorem The Well-Ordering Principle is equivalent to the Principle of Mathematical Induction.

Proof Assume that the Well-Ordering Principle is true.

Let $p(n)$ be an open statement about n.
Assume $p(1)$ is true.
Assume that for every positive integer n, $p(n) \Rightarrow p(n+1)$.

Claim: For every positive integer n, $p(n)$ is true.

Let $T = \{ j \text{ in } \mathbb{N} \mid p(j) \text{ is false} \}$

* Assume that T is not empty.

By the Well-Ordering Principle, T has a least element.

Let c denote the least element in T.
Since $c \in T$, $p(c)$ is false.

Since $p(1)$ is true, $1 \notin T$. So $1 \neq c$.
Thus, $c-1$ is a natural number.

Since c is the least element in T, $c-1 \notin T$.
So, $p(c-1)$ is true.

By the second assumption, $p(c-1) \Rightarrow p(c)$.
Thus $p(c)$ is true. Contradiction!

Thus, the * assumption is false: T is empty.
Thus, for every integer n, $p(n)$ is true.

Hence, the Principle of Mathematical Induction is true.

Converse Assume that the Principle of Mathematical Induction is true.

Let A be a subset of $\mathbb{N}$.
Assume that A does not have a least element.
We will now demonstrate that A must be empty.

Let $p(n)$: For every positive integer j, if $j \leq n$, then $j \notin A$.

Since A does not have a least element, $1 \notin A$. So $p(1)$ is true.

Let n be a positive integer.
Assume that $p(n)$ is true.
Then $1 \notin A$, $2 \notin A$, . . . , and $n \notin A$.
If $n+1 \in A$, then $n+1$ is the least element in A.
But we assumed that A does not have a least element.
Thus, $n+1 \notin A$. So $p(n+1)$ is true.

So for every positive integer n, $p(n) \Rightarrow p(n+1)$.

By mathematical induction, $p(n)$ is true for every integer n.
Thus, for every integer n, $n \notin A$. So $A = \emptyset$.

We have proved the following implication:

If A does not have a least element, then $A = \emptyset$.

So, if A is not empty, then A must have a least element.
Hence, the Well-Ordering Principle is true.

Conclusion Therefore, the Well-Ordering Principle is equivalent to the Principle of Mathematical Induction.

Exercise 4.4

1. Let R denote a relation on the set S. Define the following terms.
   a. R is a partial order on S.
   b. R is a total order on S.
   c. (S, R) is well-ordered.
2. On the set S, determine if $\leq$ is a partial order, a total order, or a well-ordering.
   a. $S = \{\pi, \frac{8}{11}, 1.2, 3.4, 2.0000004\}$
   b. $S = \mathbb{N}$
   c. $S = \{-1, -2, -3, -4, \ldots\}$
   d. $S = \mathbb{R}^+$
   e. $S = \{2 - \frac{1}{n} \mid n \in \mathbb{N}\}$
   f. $S = \{2 + \frac{1}{n} \mid n \in \mathbb{N}\}$
3. On the set S, determine if the subset relation is a partial order, a total order, or a well-ordering.
   a. $S = \{\{5\}, \{3,4\}, \{3,4,5\}, \{3,4,5,6\}\}$
   b. $S = \{[-1,1], [-2,1], [-3,1], [-4,1], \ldots\}$
   c. $S = \{\{0\}, \{0,1\}, \{0,1,2\}, \{0,1,2,3\}, \ldots\}$
   d. $S = P(\mathbb{N})$, the power set of $\mathbb{N}$.
4. On the given set, determine if the divides relation is a partial order, a total order, or a well-ordering.
   a. $\mathbb{N}$ b. Z c. $\{2, 4, 6, 8, 12, 16\}$ d. $\{2, 4, 8, 16, 32\}$
5. Let $S = \{a, b, c, d, e\}$ If possible, give examples of the following:
   a. A partial order on S that is not a total order.
   b. A total order on S.
   c. A total order on S that is not well-ordered.
6. Is the given statement true? If not, give a counterexample.
   a. Every finite, partially ordered set is totally ordered.
   b. Every finite, nonempty, partially ordered set has a greatest element.
   c. Every nonempty, partially ordered set has a maximal element.
   d. Every finite, nonempty, partially ordered set has a maximal element.
   e. Every totally ordered set is well-ordered.
   f. Every well-ordered set is totally ordered.

7. A partial order on the indicated set is given by the Hasse graph.

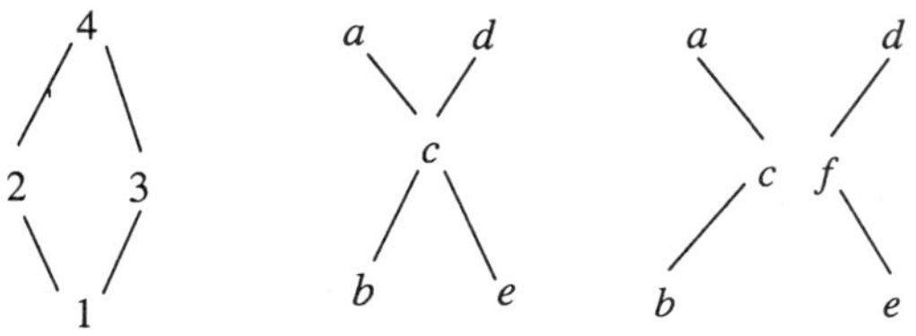

$(A, \preceq)$ $(B, \trianglelefteq)$ $(C, \ll)$

   a. List the ordered pairs in each relation.
   b. Find the least, greatest, minimal, and maximal elements.
   c. If possible, do two different topological sortings of each poset.
   d. For each poset, find a collection T of sets so that the poset is isomorphic to $(T, \subseteq)$.

8. Order the set S with the subset relation. Then determine if $(S, \subseteq)$ is isomorphic to any of the posets in the previous exercise.
   a. $S = \{ \{7\}, \{7,11\}, \{7,11,8\}, \{7,8\} \}$
   b. $S = \{ \{1,2\}, \{1,3\}, \{1,2,5\}, \{1,2,5,9\}, \{1,3,4,7,8\} \}$
   c. $S = \{ \{1,2\}, \{1,3\}, \{1,2,5\}, \{1,2,5,9\}, (1,3,4\}, \{1,3,4,7,8\} \}$

9. Draw a Hasse graph of a poset S with the given property.
   a. S has a greatest element, but no least element.
   b. S has a minimal element, but no least element.
   c. S has a maximal element, but no greatest element.
   d. S has 3 minimal elements and 4 maximal elements.

10. Order the following sets with the subset relation.

$$S = \{ \{2\}, \{2,3\}, \{2,3,5\}, \{2,3,5,9\} \}$$
$$T = \{ \{2\}, \{2,3\}, \{2,3,6\}, \{4,5,6\}, \{2,3,5,6\} \}$$
$$V = P(\{1,2,3,4\}) \text{ (} V \textit{ is the power set.)}$$

   a. Draw a Hasse graph of each poset.
   b. Find all least, greatest, minimal, and maximal elements.
   c. If possible, do two different topological sortings of each poset.
   d. Determine if the relation is a total order or a well-ordering.

11. Order the following sets with the divides relation.

$$S = \{ n \text{ in } \mathbb{N} \mid n \text{ divides } 18 \}$$
$$T = \{ n \text{ in } \mathbb{N} \mid n \text{ divides } 32 \}$$
$$W = \{ n \text{ in } \mathbb{N} \mid n \text{ divides } 48 \}$$

   a. Draw a Hasse graph of each poset.
   b. Find all least, greatest, minimal, and maximal elements.
   c. If possible, do two different topological sortings of each poset.
   d. For each poset, find a collection T of sets so that the poset is isomorphic to $(T, \subseteq)$.

12. Prove the following.
    a. A poset can have at most one least element and at most one greatest element.
    b. If a poset has a greatest element, it is the only maximal element in the set.
    c. In a totally ordered set, a minimal element must be the least element in the set and a maximal element must be the greatest element in the set.
13. Prove the following theorem using a proof by contradiction.

    Theorem: A finite nonempty poset must have a minimal element.
14. Find all lower and upper bounds for the set S in $\mathbb{R}$. Then determine if S has a greatest lower bound or a least upper bound..

    a. $S = (2,3]$ b. $S = [1,5)$ c. $S = \{ x \mid x \text{ is rational and } x^2 > 2 \}$
15. Let $S = \{ \{1,3,4\}, \{1,2,3\} \}$. Find all lower and upper bounds for S in the given set U with the subset relation. Then determine if S has a greatest lower bound or a least upper bound.
    a. $U = \{ \emptyset, \{1\}, \{2\}, \{3\}, \{1,2\}, \{1,3\}, \{1,2,3\}, \{1,3,4\} \}$
    b. $U = \{ \emptyset, \{1\}, \{2\}, \{3\}, \{1,2\}, \{1,2,3\}, \{1,3,4\}, \{1,2,3,4,7\} \}$
16. Let $\preceq$ be a partial order on S. Consider ways to define a partial order $\trianglelefteq$ on $S \times S$ that is related to $\preceq$. Let a, b, c, and d be in S.
    a. Define: $(a,b) \trianglelefteq (c,d)$ if and only if $a \preceq c$ or $b \preceq d$.
       Is $\trianglelefteq$ a partial order on $S \times S$? Justify your answer.
    b. Using $\preceq$, define a partial order $\trianglelefteq$ on $S \times S$ by modeling the lexicographic ordering used in a dictionary.
       $(a,b) \trianglelefteq (c,d)$ if and only if ____. Justify your answer.
    c. Define another partial order on $S \times S$. Justify your answer.
17. $(U, \preceq)$ is a partially ordered set, $S \subseteq U$, and $T \subseteq U$.
    Check your language skills by translating the following.
    a. S has a least element. T does not have a least element.
    b. S has a maximal element. T does not have a maximal element.
    c. S has a lower bound in U. T does not have a lower bound in U.
    d. S has a greatest lower bound in U. T does not have a greatest lower bound in U.
    e. S has an upper bound in U.
    f. S has a least upper bound in U.
18. Let U be a set of posets, and let $(S, \preceq)$ and$(T, \trianglelefteq)$ be partially ordered sets that are in U. Prove that the isomorphic relation is an equivalence relation on U.

    $$S \simeq T \text{ if and only if } (S,\preceq) \text{ is isomorphic to } (T,\trianglelefteq).$$

Review

Relation Embeds a structure between two sets by giving a connection between various elements. A relation R from X into Y can be described as a mapping where some of the elements in X are mapped to some of the elements in Y. It can also be viewed as a set of ordered pairs whose first terms come from X and whose second terms come from Y. Mathematical activity has always focused on relations. In the reasoning process, we are usually trying to figure out how various objects may be related to each other. When we work with a set, we do not individually analyze its elements; instead, we compare the set with other sets by looking for relations between them. Within the grand house of mathematics, there are many diverse areas of study, but within each area, the focus is on relations. Mathematics can be described as the study of relations.

Domain The set of all first terms of the ordered pairs in a relation R.

Range The set of all second terms of the ordered pairs in a relation R.

Inverse relations A relation obtained by reversing the order of a given relation.

$$a \, \mathsf{R}^{-1} \, b \text{ if and only if } b \, \mathsf{R} \, a.$$

Every relation has an inverse relation.

n-ary relation A set of ordered n-tuples. A subset of a cross product of n sets.

Graph A visual representation of a relation where we use points in a coordinate plane to represent the ordered pairs in a function.

Directed graph A visual representation of a relation where the mapping is represented by arrows, with each member of the domain and range listed only once. Consequently, some of the arrows may be chained together.

Matrix A rectangular array used to represent a finite relation. Matrices have a wide range of applications.

Reflexive relation A relation R on a set S that has the following property: for every a in S, $a \, \mathsf{R} \, a$. Each element is related to itself.

Transitive relation A relation R on a set S that has the following property: for every a, b, and c in S, if $a \, \mathsf{R} \, b$ and $b \, \mathsf{R} \, c$, then $a \, \mathsf{R} \, c$.

Symmetric relation A relation R on a set S that has the following property: for every a and b in S, if $a \, \mathsf{R} \, b$, then $b \, \mathsf{R} \, a$. The order of the elements does not affect the relation.

Equivalence relation A relation on a set S that is reflexive, transitive, and symmetric. The set of equivalence classes of an equivalence relation partitions the set S into nonoverlapping subsets. An equivalence relation identifies a property that makes elements essentially the same with respect to that property, such as the property of "having the same size and shape." "Is congruent to" and "is similar to" are important equivalence relations between figures. Congruence mod n is an important equivalence relation between integers. "Has the same size" is a very important equivalence relation between sets. "Is isomorphic to" is an extremely important equivalence relation between structured sets, such as partially ordered sets.

Equivalence class – $[a]$ The set of elements related to a by an equivalence relation R on a set S. Let a be in S. $[a] = \{ x \text{ in } S \mid a \, \mathsf{R} \, x \}$. Related elements have the same equivalence class: if $a \, \mathsf{R} \, b$, then $[a] = [b]$. Elements in the same equivalence class are considered as essentially the same with respect to the relation.

Partition A subdivision of a set into nonoverlapping subsets. A partition P of a set S is a collection of nonempty subsets of S where each element in S is in one and only one of the subsets. Each partition of a set has an associated equivalence relation.

Congruence mod n An equivalence relation on the set Z of integers. $a \equiv_n b$ if and only if n divides $a - b$. If r is the remainder when we divide x by n, then $x \equiv_n r$. Congruence mod n partitions the set Z of integers into n equivalence classes, which we notate as Z_n:

$$Z_n = \{ [0], [1], [2], [3], \ldots, [n-1] \}$$

Function f is a *function* from X into Y if and only if f maps each element in X to a unique element in Y. If $a = b$, then $f(a) = f(b)$.

Function notation $f(x)$ denotes the value assigned to x by the function f. The following have the same meaning: $f(x) = y$, $x \xrightarrow{f} y$, $(x,y) \in f$.

Domain The set of elements for which a function f is defined. x is in the domain of f means that $f(x)$ is defined.

Range The set of images of elements in the domain. y is in the range of a function f if and only if there exists an x in the domain such that $y = f(x)$.

Equal functions Two functions that have the same domain and the same function values for each element in the domain. $f = g$ if and only if $f(x) = g(x)$ for all x in the domain.

Into function f maps X into Y if and only if for each x in X, $f(x)$ is in Y.

Onto function	f maps X onto Y if and only if the range of f is Y. For each y in Y, there must exist an x in X such that $f(x) = y$.
One-to-one function	A function that maps different elements to different images. f is a one-to-one function if and only if for every a and b in the domain of f, if $a \neq b$, then $f(a) \neq f(b)$. A one-to-one function has an inverse function.
Injection	A one-to-one function.
Surjection	An onto function.
Bijection	A one-to-one and onto function.
Image of a set	The set of images of individual elements in a set under a function. $f(A) = \{ f(x) \mid x \in A \}$. $y \in f(A)$ if and only if there exists an x in A such that $f(x) = y$.
Inverse function	The inverse relation of a one-to-one function. Let f be a one-to-one function that maps X onto Y. Then f^{-1} is a one-to-one function that maps Y onto X. $f^{-1}(a) = b$ if and only if $f(b) = a$. For every a in X, $f^{-1}(f(a)) = a$. For every b in Y, $f(f^{-1}(b)) = b$.
Composition of functions	A function formed from two functions f and g where f maps X into Y and g maps Y into Z. If x is in X, $g \circ f(x) = g(f(x))$. The composition of two injections is an injection. The composition of two surjections is a surjection. The composition of two bijections is a bijection. If f and g have inverse functions, $(g \circ f)^{-1} = f^{-1} \circ g^{-1}$. Composition of functions is associative: $(h \circ g) \circ f = h \circ (g \circ f)$.
Restriction of a function	A function whose domain is restricted to a subset of the original domain, but using the same mapping.
Extension of a function	A function whose domain is extended beyond the domain of the original function, while preserving the original mapping.
Identity function	A function that maps each element to itself: $e(x) = x$. The identity function is dependent on its domain. e_A denotes the identity function on the set A.
Binary operation on S	A function that maps $S \times S$ into S. A binary operation maps each pair of elements in S to an element in S.
Well-defined	A definition that is logically acceptable. A "well-defined function" means that the definition produces a function. A "well-defined set" means that the definition produces a legitimate set, one whose members can be determined.

Antisymmetric relation — A relation R on a set S that has the following property: For every a and b in S, if $a \mathrel{R} b$ and $b \mathrel{R} a$, then $a = b$. The only time we can reverse the order is when the elements are equal.

Partial order — A relation that is reflexive, transitive, and antisymmetric. Instead of arranging elements in egalitarian equivalence classes where everyone is essentially the same, a partial order embeds a hierarchical structure on a set. Given any example of a partial order on a set S, such as "divides" or $\leq$, we can produce an isomorphic copy of it using the subset relation.

Poset — A partially ordered set. A set that has a partial order relation defined on its elements.

Strict order — A transitive and antisymmetric relation that has no element related to itself. Every partial order ($\preceq$) has an associated strict order: $a \prec b$ if and only if $a \preceq b$ and $a \neq b$.

Total order — A partial order in which each pair of elements are comparable. For every a and b in the set S, $a \preceq b$ or $b \preceq a$.

Hasse graph — A minimalist graph of a partially ordered set in which its directed graph is positioned so that all arrows point upwards, then we omit the arrowheads, the loops that can be deduced by reflexivity, and the arrows that can be deduced by transitivity.

Isomorphic structures — Two structures that have the same form, which means that one structure can be relabeled to produce the other structure. For posets, $(S, \preceq)$ is isomorphic to $(T, \trianglelefteq)$ if and only if there exists a bijection f from S onto T that preserves the order on S:

For every a and b in S, if $a \preceq b$, then $f(a) \trianglelefteq f(b)$.

The function f is called an *isomorphism*. An isomorphism preserves the order relation. If two elements are related in S, their images must be related in T.

Least element — The smallest element in a poset $(S, \preceq)$. b is the *least* element of S if and only if $b \in S$ and for every x in S, $b \preceq x$. A partially ordered set can have at most one least element. If b is the least element in S, b is the only minimal element of S. In a totally ordered set S, a minimal element is also the least element in S.

Greatest element — The largest element in a partially ordered set $(S, \preceq)$. b is the greatest element of S if and only if $b \in S$ and $x \preceq b$ for all x in S. A partially ordered set can have at most one greatest element. If b is the greatest element in S, b is the only maximal element of S. In a totally ordered set S, a maximal element is also the greatest element in S.

Minimal element — An element in a poset S that has no other elements before it. b is a minimal element of S if and only if $b \in S$ and for every x in S, if $x \preceq b$, then $x = b$. A nonempty poset that is finite must have at least one minimal element.

Maximal element — An element in a partially ordered set S that has no other elements after it. b is a maximal element of S if and only if $b \in S$ and, for every x in S, if $b \preceq x$, then $x = b$. A nonempty poset that is finite must have at least one maximal element.

Lower bound — b is a lower bound for S if and only if for every x in S, $b \preceq x$. Unlike a minimal element, a lower bound does not have to be in S. The *greatest lower bound* of S is the greatest of all the lower bounds of S.

Upper bound — b is an upper bound for S if and only if for every x in S, $x \preceq b$. Unlike a maximal element, an upper bound does not have to be in S. The *least upper bound* of S is the least of all the upper bounds of S.

Topological sorting — The embedding of a poset in a totally ordered set. We start with a partial order and construct a total order from it, preserving all the relations of the partial order. Let $\preceq$ be a partial order on S and $\trianglelefteq$ be a total order on S. $\trianglelefteq$ is a topological sorting for $\preceq$ if and only if for every x and y in S, if $x \preceq y$, then $x \trianglelefteq y$.

Immediate successor — An element that comes after another element with no other elements between them. In a poset, b is an immediate successor to a if and only if $a \prec b$ and there does not exist an x such that $a \prec x \prec b$.

Well-ordered set — A partially ordered set in which every nonempty subset has a least element. A well-ordered set must be totally ordered. In a well-ordered set, every element, except for the greatest element, has an immediate successor. The set of natural numbers is well-ordered, but the set of real numbers is not.

Well-ordering principle — Every nonempty subset of the set $\mathbb{N}$ of natural numbers has a least element.

The ability to reason depends on the ability to both visualize and verbalize. The human thought processes seem to thrive on visuals which give us ideas that we can then verbalize and develop in a careful and logically correct manner. We conclude this book with a brief visual/verbal summary of the three special types of relations that are used throughout all of mathematics.

- A relation can be viewed as a mapping between two sets where $a \; \mathsf{R} \; b$ indicates that a is mapped to b.
- Functions allow us to compare sizes and structures of sets, and move from one set to another via the $f(x)$ notation.
- Equivalence relations enable us to divide a set into nonoverlapping subsets.
- Order relations enable us to arrange the elements in a set in a hierarchical structure.

Mathematics is the study of relations.

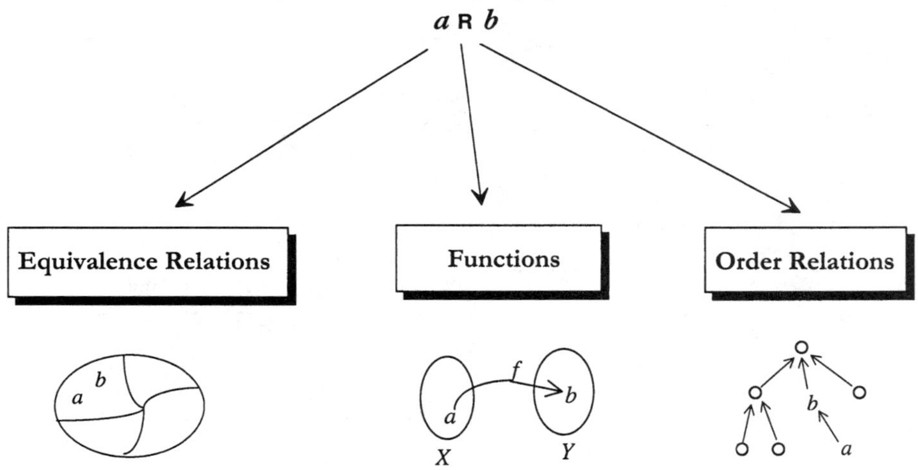

$a \; \mathsf{R} \; b$ can be interpreted as "a is essentially the same as b with respect to R."

$a \; \mathsf{R} \; b$ is interpreted as $f(a) = b$.

$a \; \mathsf{R} \; b$ can be interpreted as "a is before b."

Chapter Review

1. Define the following terms, give examples of each, and determine if a given example satisfies the definition.
   a. relation, domain, range, inverse relation
   b. reflexive, transitive, symmetric, and antisymmetric properties
   c. equivalence relation, equivalence classes, partition
   d. function, domain, range, equal functions, image of a set
   e. onto function, one-to-one function, injection, surjection, bijection, inverse function, composition of two functions
   f. partial order, total order, well-ordered set, isomorphic posets
   g. least & greatest elements, minimal & maximal elements, upper & lower bounds, greatest lower bound, least upper bound
2. Given a specific relation, illustrate it as a mapping, a set of ordered pairs, a graph, a directed graph, and a matrix.
3. Explain why the following are true:
   a. Every relation has an inverse relation, but some functions do not have an inverse function.
   b. For every relation R, domain (R^{-1}) = range (R)
   c. For every relation R, $(R^{-1})^{-1} = R$.
4. Let R be an equivalence relation on a set S. Let a and b be elements in S. Prove the following:
   a. If $a \mathrel{R} b$, then $[a] = [b]$.
   b. If $\sim(a \mathrel{R} b)$, then $[a] \cap [b] = \emptyset$.
   c. $[a] = [b]$ or $[a] \cap [b] = \emptyset$
   d. The set of all equivalence classes of R is a partition of S.
5. Given a partition of a set, define an equivalence relation on S whose equivalence classes are the members of the partition.
6. Define the relation R on the set $\mathbb{N}$ of natural numbers as follows:
   $$a \mathrel{R} b \text{ if and only if 7 divides } a-b.$$
   a. Prove that R is an equivalence relation on S.
   b. What are the equivalence classes of R?
7. Give an example of a relation that is not a function. Explain why function notation cannot be used with your example.
8. Discuss the following:
   a. The difference between the notation f and the notation $f(x)$.
   b. What does it mean to say that a function is well-defined?
   c. Under what conditions does a function have an inverse function? Explain why.

9. Explain how addition, multiplication, and division can each be interpreted as a function. Give the domain and range of each.

10. Use a mapping picture to illustrate the following:
    a. A function that is not one-to-one.
    b. A function that is not onto.
    c. The composition of two functions.
    d. Two functions whose compositions cannot be formed.
    e. If f and g have inverse functions, then $(g \circ f)^{-1} = f^{-1} \circ g^{-1}$.

11. Given a specific function f that maps X into Y, prove or disprove that f is one-to-one or that f maps X onto Y.

12. Let $f: X \rightarrow Y$ and $g: Y \rightarrow Z$. Prove the following:
    a. If f and g are surjections, then $g \circ f$ is a surjection.
    b. If f and g are injections, then $g \circ f$ is an injection.
    c. If f and g are bijections, then $g \circ f$ is a bijection.

13. Let $S = \{f \mid f: \mathbb{R} \rightarrow \mathbb{R}\}$. Is the composition of functions commutative on S? Justify your answer.

14. Given $f: X \rightarrow Y$, compute $f(A)$ where $A \subseteq X$.

15. Let $f: X \rightarrow Y$ and let A and B be subsets of X. Prove or disprove:
    a. If $A \subseteq B$, then $f(A) \subseteq f(B)$.
    b. $f(A \cup B) = f(A) \cup f(B)$
    c. $f(\bigcup_{i \in I} A_i) = \bigcup_{i \in I} f(A_i)$
    c. $f(A \cap B) \subseteq f(A) \cap f(B)$
    d. $f(A) \cap f(B) \subseteq f(A \cap B)$

16. Given a partially ordered set $(S, \preceq)$:
    a. Draw a Hasse graph of the given relation on S.
    b. Find all least, greatest, minimal, and maximal elements, and lower and upper bounds.
    c. Construct a topological sorting of S.
    d. Find a collection T of sets so that $(T, \subseteq)$ is isomorphic to S.

17. Prove the following.
    a. Every partially ordered set is isomorphic to a set of sets that are ordered by the subset relation.
    b. A partially ordered set can have at most one greatest element.
    c. A greatest element in a poset is its only maximal element.
    d. In a totally ordered set, a maximal element is the greatest element in the set.

Beneath the effort directed toward the accumulation of worldly goods lies all too frequently the illusion that this is the most substantial and desirable end to be achieved; but there is fortunately, a minority composed of those who recognize early in their lives that the most beautiful and satisfying experiences open to human kind are not derived from the outside, but are bound up within the development of the individual's own feeling, thinking and acting. The genuine artists, investigators and thinkers have always been persons of this kind. However inconspicuously the life of these individuals runs its course, none the less the fruits of their endeavors are the most valuable contributions which one generation can make to its successors.

Albert Einstein

In Memory of Emmy Noether, 1935

Appendix A

Selected Answers

Exercise Set 1.1

1. $x(a+b) = xa + xb$
3. 3 is less than x and x is less than 5. *nouns*: x, 3, 5; *verb*: is less than; *logical operator*: and
5. a. Yes b. Yes c. No d. Yes
7. a. Statement b. Open statement c. Neither d. Statement e. Statement f. Neither
9. a. False b. True c. True d. False
11. a. True b. False c. False d. True

Exercise Set 1.2

1. a. False, true b. False, true c. True – stronger, true d. True – stronger, true
3. a. $A = \{1,2\}$, $B = \{3,4\}$, $C = \{4,6\}$
5. $p(x,y)$: $x + y = 3$
   a. True b. True c. True d. False
7. a. False. There exists a real number x such that $3x \neq 4$.
   b. False. There is a real number x such that $3x = 4$.
   c. False. For every real number x, $x^2 \neq -1$.
   d. False. There exists a complex number x such that $x^2 = -1$.
   e. False. For every real number y, there exists an x such that $x + y \neq 4$.
   f. False. For every integer y, there exists a real number x such that $y \geq x$.
9. a. There exists an integer y such that for every real number x, $g(x) \neq y$.
   b. There exists a y in B such that for every x in A, $f(x) \neq y$.
   c. There exist sets A and B such that for every function f, f does not map A onto B.

Exercise Set 1.3

1. a. $\sim q \Rightarrow \sim p$ b. p and $\sim q$ c. $\sim p \Rightarrow q$, $\sim q \Rightarrow p$. d. $\sim p$ and $\sim q$
3. a. If $x \notin B$, then $x \notin A$. $x \notin A$ or $x \in B$.
   b. If $x \notin B$, then $x \in A$. $x \in A$ or $x \in B$.
5. a. True b. False c. True d. True e. False f. False.
7. a. If not r, then s. b. If r, then s. c. If $x \notin C$, then $x \in D$. d. If $x \in C$, then $x \in D$.
9. a. $3 \geq x$ or $x \geq 7$. b. $x \geq 3$ and $x \leq 7$
   c. $x \in A$ and $x \notin B$.
   d. There exists an x such that $x < 7$ and $x \geq 3$.
   e. There exists an x such that $x \in C$ and $x \notin B$.
   f. There exists an x such that $|x - 1| < 4$ and $|f(x) - f(1)| \geqq 3$.
11. a. $p \Rightarrow q$ is false only when p is true and q is false.
   b. Suppose that p or q is true. Then if p is not true, q has to be true.
13. *components–cases*: 2–4, 3–8, 4–16, 5–32, n–2^n
15. a. 10010 b. 11011 c. 01001 d. 1011

Exercise Set 1.4

1. a. $* \quad (p \Rightarrow q) \Leftrightarrow (\sim q \Rightarrow \sim p)$

p	q	$\sim q$	$\sim p$	$\sim q \Rightarrow \sim p$	$p \Rightarrow q$	*
T	T	F	F	T	T	T
T	F	T	F	F	F	T
F	T	F	T	T	T	T
F	F	T	T	T	T	T

3. a. If x is not in A, then x is not in C.
   b. If x is not in C, then x is in B.
5. a. $x \in A$ and $x \notin B$ b. $x \in C$ and $x \in A$
   c. $z \in A$ or $z \in B$ d. $x \in C$ and $x \notin D$
   e. $|x-2| < \delta$ and $|f(x)-f(2)| \geqq \varepsilon$
   f. $(x \notin D$ and $x \in B)$ or $(x \in D$ and $x \notin B)$
7. a. $z \notin X$, or $z \notin Y$ and $z \notin Z$
   b. There exists an x such that $x \notin C$ and $x \in D$.
   c. There exists an x such that $x \notin C$ and there exists an x such that $x \in D$.
   d. For all x, x is not in A or x is not in B.
   e. For all x, x is not in A or for all x, x is not in B.
9. a. Yes b. No. $p-$F, $q-$T c. Yes
   d. No. $p-$T, $q-$F e. Yes f. No. $p-$F, $q-$T
11. No. The 1st sentence is false and the 2nd is true.
13. a. No. $A = \{1\}, B = \{2\}$ b. Yes c. Yes
   d. No. Let $A = \{1\}$ and B be all real numbers except for 1.

Exercise Set 1.5

1. a. 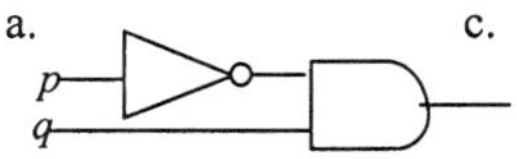

   c. See page 79
3. a. p b. $\sim q$ c. $q \wedge r$ d. $\sim(p \wedge q)$

Exercise Set 1.6

1. a. There exists an x such that x is in A and x is in B.
   b. For every x, if x is in A, then x is in B.
   c. There exists an x such that x is not in A or x is not in B.
   d. x is in A and x is not in C.
   e. For every x, if x is in B, then x is in A.
   f. If x is in A, then x is in B.
3. a. mn is even if and only if there exists an integer k such that $mn = 2k$.
   b. $m+n$ is even if and only if there exists an integer k such that $m+n = 2k$.
   c. m^2 is even if and only if there exists an integer k such that $m^2 = 2k$.
5. a. g maps X onto Z if and only if for each z in Z, there exists an x in X such that $g(x) = z$.
   b. f maps Y onto X if and only if for each x in X, there exists a y in Y such that $f(y) = x$.
7. a. x is an *odd* number if and only if there exists an integer k such that $x = 2k+1$.
   b. a is a *factor* of b if and only if there exists an integer k such that $b = ak$.
   c. Same as part (b). d. Same as part (b).
9. a. For all distinct real numbers a and b, there exists a real number c such that $a < c < b$ or $b < c < a$.
   b. For all integers a and b, if a is even and b is even, then $a+b$ is even.
   c. For all x, if $|x-3| < \delta$, then $|f(x)-f(3)| < \in$.
   d. For all real numbers a and b, if $a < b$, then $f(a) < f(b)$.
   e. There exists a c in S such that for every x in S, $x \leq c$.
   f. There exists a c in S such that for every x in S, $c \leq x$.
11. a. There exists an element u such that for every x in S, $x \leq u$.
   b. For every element u, there exists an x in S such that $x > u$.
   c. For every upper bound u of S, $m \leq u$.
13. a. For all integers x, y and z, $x+(y+z) = (x+y)+z$.
   b. There exists an integer c such that for every integer x, $x+c = x$ and $c+x = x$.
   c. For every integer x, there is an integer b such that $x+b = c$ and $b+x = c$.
15. a. $x \mathrm{R} x$ (*Reflexive Property*)
   b. If $x \mathrm{R} y$, then $y \mathrm{R} x$. (*Symmetric Property*)
   c. If $x \mathrm{R} y$ and $y \mathrm{R} z$, then $x \mathrm{R} z$. (*Trans. Property*)

Chapter Review

3. A statement is either true or false. An open statement has variables and is neither true nor false, but it does become either true or false when a substitution is made for each variable. Not all sentences with variables are open statements. If each variable is bound with a quantifier, the sentence is a statement.

7. The truth values for the contrapositive are identical to the truth values of for the implication; however, the truth values for the converse are different. To say that $p \Rightarrow q$ is true means that if p is true, then q has to be true. So, if q is false, then p has to be false. Thus, $\sim q \Rightarrow \sim p$ is true.
9. a. For all integers x, if x is not even, then x is odd.
   b. If x is not in D, then x is in E.
11. Yes.
13. a. x is not *even* $\Leftrightarrow$ for every integer n, $x \neq 2n$.
   b. x is not *rational* $\Leftrightarrow$ for all integers p and q, $x \neq \frac{p}{q}$.
   d. $x \notin A \cup B \Leftrightarrow x \notin A$ and $x \notin B$
   e. $x \notin A \cap B \Leftrightarrow x \notin A$ or $x \notin B$
   f. $A \not\subseteq B \Leftrightarrow$ there exists an x such that $x \in A$ and $x \notin B$.
   i. S does not have a largest element $\Leftrightarrow$ for every m in S, there exists an x in S such that $x \geq m$.
   j. f is not a *function* $\Leftrightarrow$ there exists a and b in the domain of f such that $a = b$ and $f(a) \neq f(b)$.
   k. f is not a *one-to-one* function $\Leftrightarrow$ there exist a and b in the domain of f such that $a \neq b$ and $f(a) = f(b)$.
   l. f does not map X into $Y \Leftrightarrow$ there exists an x in X such that $f(x)$ is not in Y.
   m. f does not map X *onto* $Y \Leftrightarrow$ there exists a y in Y such that for all x in X, $f(x) \neq y$.
   n. $y \notin f(A) \Leftrightarrow$ for all x in A, $f(x) \neq y$.
   o. The function f is not increasing on $[a, b] \Leftrightarrow$ there exists c and d in $[a,b]$ such that $c < d$ and $f(c) \geq f(d)$.
   p. The function f is not continuous at $x = a \Leftrightarrow$ there exists a positive ε such that for every positive δ, there exists an x such that $|x-a| < \delta$ and $|f(x)-f(a)| \geq \varepsilon$.
15. a. x^2 is even $\Leftrightarrow$ there exists an integer n such that $x^2 = 2n$.
   c. h maps Y onto $X \Leftrightarrow$ for each x in X, there exists a y in Y such that $h(y) = x$.
   d. $g \circ f$ maps X onto $Z \Leftrightarrow$ for each z in Z, there exists an x in X such that $g \circ f(x) = z$.
   e. $x \in g(C) \Leftrightarrow$ there exists a c in C such that $g(c) = x$.
19. a. There exists an integer k such that $n = 2k$.
   b. There exists an integer k such that $n = 2k+1$.
   c. There exists an integer k such that $m = nk$.
   d Same as part (c). f. There exists a b in S such that for every x in S, $x \leq b$.

Exercise Set 2.1

1. a. Not valid b. Valid c. Not valid
   d. Not valid e. Not valid f. Valid
   g. Valid h. Valid i. Valid

3. a. Not valid b. Valid c. Valid d. Not valid
5. a. Not valid.
   b. Valid
   c. Valid
   d. Not valid

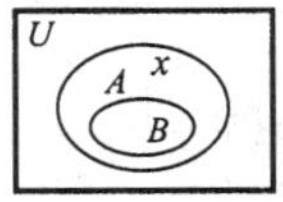

7. a. Therefore, if $x \in A$, then $x \in C$.
   b. Therefore, $x \notin C$ or $x \notin D$.
   c. Therefore, $x \notin A$ and $x \notin B$.
   d. Therefore, x is in D.
   e. Therefore, $x \notin A$.
   f. Therefore, $x \notin A$.
   g. Therefore, $x \notin D$ and $x \notin E$.
   h. Therefore, x is in A if and only if x is in B.
11. a. No b. No c. Yes d. No e. No.
   f. Yes g. Yes h. Yes i. No
13. a. There exists positive integers a and b such that $n = ab$, $a \neq 1$, $a \neq n$. Also, $b \neq 1$ and $b \neq n$.
   b. $a = 1$ or $a = n$.
15. $f(n)$ represent the number of eggs on the nth day.
   a. $f(n) = 2n$ $\quad f(90) = 180$
   b. $f(n) = 2n+1$ $\quad f(90) = 181$
   c. $f(n) = 2^n$ $\quad f(90) = 2^{90}$
   d. $f(n) = 2^n - 1$ $\quad f(90) = 2^{90} - 1$

Exercise Set 2.2

1. a. Let m and n be integers.
   Direct: Assume m and n are odd. . . .
   Then, mn is odd.
   Indirect: Assume mn is not odd.
   . . . Then, m is not odd or n is not odd.
   b. Let m and n be integers.
   Direct: Assume mn is odd. . . .
   So, m is odd and n is odd.
   Indirect: Assume m is not odd or n is not odd.
   . . . Then, mn is not odd.
   c. *Direct*: Assume that a is not a factor of $b+c$ . . .
   So, a is not a factor of b or a is not a factor of c.
   Indirect: Assume a is a factor of b and a is a factor of c. . . . So, a is a factor of $b+c$.
   d. *Direct*: Assume that x and y are rational. . . .
   So, $x+y$ is rational.
   Indirect: Assume that $x+y$ is irrational.

. . . Then x is irrational or y is irrational.

3. Let x and y be numbers.
   a. Assume that $x \notin A$. . . .
      Then, $x \in B$. So, $x \in A$ or $x \in B$.
   b. Let x be an integer. Assume that x is not even.
      . . . Then, x^2 is odd. So, x is even or x^2 is odd.
5. a. Assume that x and y are even. Then there exist integers j and k such that $x = 2j$ and $y = 2k$.
      So, $x+y = 2(j+k)$. Set $m = j+k$.
      Since j and k are integers, m is an integer.
      $x+y = 2m$. So $x+y$ is even.
   c. If an integer is not even, it must be odd. We can use the contrapositive of 5a, phrasing it as follows: If $x+y$ is odd, then x is odd or y is odd.

Exercise Set 2.3

1. a. There exists an integer k such that $x = 2k+1$.
   b. There exists an integer k such that $xy = 2k+1$.
   c. There exists an integer k such that $b = ka$.
   d. There exists an integer k such that $b+c = ka$.
3. *Structure*: Let a and b be real numbers.
   Assume that $a < b$. . . .
   So, $a < \frac{a+b}{2}$ and $\frac{a+b}{2} < b$. Thus, $a < \frac{a+b}{2} < b$.
   Middle Part: Add a to both sides of the inequality
   Since $a < b$, $a+a < a+b$.
   So, $2a < a+b$. Thus $a < \frac{a+b}{2}$.
   Since $a < b$, $a+b < b+b$.
   So, $a+b < 2b$. Thus $\frac{a+b}{2} < b$.

Exercise Set 2.4

1. a. There exist integers a and b such that $x = \frac{a}{b}$.
   b. There exist integers a and b such that $x+y = \frac{a}{b}$.
   c. There exist integers a and b such that $\frac{x}{y} = \frac{a}{b}$.
   d. For all integers a and b, $x \neq \frac{a}{b}$.
3. a. True: Let y be a real number. Set $x_o = 7-2y$.
      Then $x_o + 2y = (7-2y)+2y = 7$.
   b. False: Let x be a real number. Set $y_o = \frac{8-x}{2}$.
      Then $x+2y_o = x+2\cdot\frac{8-x}{2} = x+8-x = 8$.
      So, $x+2y_o \neq 7$. Thus, for every real number x, there exists a real number y such that $x+2y \neq 7$.
   c. False: Let x be a real number.
      Set $y_o = x-1$. Then $x > x-1$. So, $x > y_o$.
      Thus, for every real number x, there exists a real number y such that $x \geq y$.
   d. True: Set $x_o = -2$. Let y be a natural number.
      Then $x_o < y$.
5. False: Set $y = -1$. Let x be a real number.
   Then $x^2 \geq 0$. So $f(x) \neq -1$.
7. a. False: Set $x = -1$. $2(-1) > 3(-1)$.
   b. True: Let $x > 0$. Set $y = \frac{x}{2}$.
      Since x is positive, y is positive. $\frac{x}{2} < x$.
      So, there is no smallest positive real number.
   c. True: Let x be a number in the interval $(3,5)$.
      Set $y = \frac{x+3}{2}$. y is the average of 3 and x, which is halfway between x and 3 (see 8b).
      So, $3 < y < x$. Thus, the interval $(3,5)$ does not have a smallest element.
9. Let a and b be rational numbers with $a < b$.
   Set $x_o = \frac{a+b}{2}$. Since a and b are rational, $a+b$ is rational. The quotient of two rational numbers is rational, so $\frac{a+b}{2}$ is rational.
   Claim: $a < x_o$.
   Since $a < b$, $a+a < a+b$. So $2a < a+b$.
   Thus $a < \frac{a+b}{2}$.
   Claim: $x_o < b$.
   Since $a < b$, $a+b < b+b$. So $a+b < 2b$.
   Thus $\frac{a+b}{2} < b$.
   So, x_o is between a and b.
11. a. Yes. Let a and b be distinct rational numbers. There exists a rational number x_0 between a and b. There exists a rational number x_1 between a and x_0. There exists a rational number x_2 between a and x_1. . . .
    b. No. $\sqrt{2}$ is not a rational number (page 170).

Exercise Set 2.5

1. a. Suppose that $(x+2)(x-4) > 0$.
      Both factors are positive or both are negative.
      Case 1: Assume $x+2 > 0$ and $x-4 > 0$.
      So, $x > -2$ and $x > 4$.
      This is equivalent to $x > 4$.
      Case 2: Assume $x+2 < 0$ and $x-4 < 0$.
      So, $x < -2$ and $x < 4$.
      This is equivalent to $x < -2$.
      Since either Case 1 or Case 2 must occur, the solution to the original inequality is:
      $x > 4$ or $x < -2$.
   b. Assume $(x-3)(x-4) < 0$. One factor must be positive and the other one negative.
      Case 1: Assume that $x-3 < 0$ and $x-4 > 0$.
      So, $x < 3$ and $x > 4$. This is impossible, so this case will not occur.
      Case 2: Assume that $x-3 > 0$ and $x-4 < 0$.
      So, $x > 3$ and $x < 4$. Hence the solution for

this case is $3 < x < 4$.
Since Case 1 or Case 2 must occur, the solution to the original inequality is: $3 < x < 4$.

c. Assume that $x^2 - 9 > 0$. So $(x-3)(x+3) > 0$.
Both factors are positive or both are negative.
Case 1: Assume $x - 3 > 0$ and $x + 3 > 0$.
Then $x > 3$ and $x > -3$.
This is equivalent to $x > 3$.
Case 2: Assume $x - 3 < 0$ and $x + 3 < 0$.
Then $x < 3$ and $x < -3$.
This is equivalent to $x < -3$.
Since Case 1 or Case 2 must occur, the solution to the original inequality is: $x < -3$ or $x > 3$.

3. a. Let n be an integer.
Then n is even or n is odd.
Case 1: Assume n is even. Then there exists an integer k such that $n = 2k$. So,
$n^2 - n = (2k)^2 - 2k = 2(2k^2 - k)$.
Set $j = 2k^2 - k$. Then $n^2 - n = 2j$.
Since k is an integer, j is an integer.
Therefore, $n^2 - n$ is even.
Case 2: Assume that n is odd. Then there exists an integer k such that $n = 2k + 1$.
So, $n^2 - n = (2k+1)^2 - (2k+1)$
$= 4k^2 + 4k + 1 - 2k - 1 = 2(2k^2 + k)$
Set $j = 2k^2 + k$. Then $n^2 - n = 2j$.
Since k is an integer, j is an integer.
Therefore, $n^2 - n$ is even.
Since either Case 1 or Case 2 must occur, $n^2 - n$ is always an even number.

Exercise Set 2.6

1. a. Assume that $1 + \sqrt{2}$ is rational.
Since the sum of two rational numbers is rational (page 157), $(1 + \sqrt{2}) + (-1)$ is rational.
So $\sqrt{2}$ is rational. Contradiction!
Thus, $1 + \sqrt{2}$ is not rational.

c. *Hint*: Prove the contrapositive.
Assume that $x + y$ is rational.
. . . (*How do you prove an or-statement?*)
So, x is irrational or y is rational.

d. *Hint*: Is $-\sqrt{2}$ irrational?

3. a. Yes. Yes. No.

7. a. Let r be a line through the point P that is perpendicular to the line t. Suppose that s is a line through P perpendicular to t. Let A denote the intersection of t with r and B denote the intersection of t with s. Suppose that $A \neq B$. Then ΔABP has more than 180 degrees. Contradiction! Therefore, $A = B$, and hence, $r = s$.

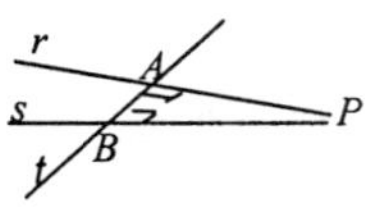

Exercise Set 2.7

1. b. Let $p(n)$: $2 + 4 + 6 + ... + 2n = n(n+1)$.
The proof is similar to the proof on page 178.

d. $\sum_{i=1}^{n} 2i = 2 + 4 + 6 + ... + 2n$

3. a. 3, 11, 19, 27, 35 b. $s_n = 3 + (n-1)8 = 8n - 5$
c. Similar to the proof on page 183.

7. a. $\sum_{i=1}^{n} 2^i$ b. $\sum_{i=1}^{n} \frac{1}{2^i}$ c. $\sum_{i=1}^{n} \frac{1}{i(i+1)}$

9. a. Let $S_n = 2 + 2^2 + 2^3 + ... + 2^n$.
$2S_n = \quad 2^2 + 2^3 + ... + 2^n + 2^{n+1}$
So, $2S_n - S_n = 2^{n+1} - 2$. Hence, $S_n = 2^{n+1} - 2$.

11. b. Similar to the proof on page 180, or the proof in (13).

13. Let $p(n)$ represent the following sentence: $2^n < n!$
Let n be an integer such that $n \geq 4$.
Assume $p(n)$ is true: $2^n < n!$
$(n+1)2^n < (n+1)\,n!$
Since $n > 1$, $2 < n+1$. So, $2\,(2^n) < (n+1)2^n$.
By transitivity, $2\,(2^n) < (n+1)\,n!$.
So, $2^{n+1} < (n+1)!$ Therefore, $p(n+1)$ is true.
So, $p(n) \Rightarrow p(n+1)$ for $n \geq 4$.

15. Let $p(n)$: n is even or n is odd.
Let n be an integer. Assume $p(n)$ is true.
Then n is even or n is odd
Case 1: Assume n is even. Then $n = 2k$ for some integer k. So, $n + 1 = 2k + 1$. Thus, $n + 1$ is odd.
So, $n + 1$ is even or $n + 1$ is odd.
Hence, $p(n+1)$ is true.
Case 2: Assume n is odd. Then $n = 2k+1$ for some integer k. So, $n+1 = 2k+2 = 2(k+1)$.
Thus, $n+1$ is even. So, $n+1$ is even or $n+1$ is odd. Thus $p(n+1)$ is true.
In either case, $p(n+1)$ is true. So, $p(n) \Rightarrow p(n+1)$.

Exercise Set 2.8

13. Undefined terms: *not, or*
Definition: *p and q* means *not* (*not p* *or* *not q*).
Definition: *p implies q* means (*not p*) *or* *q*.
Definition: *p is equivalent to q* means *p implies q* and *q implies p*.

Exercise Set 3.1

1. a. $S = \{3, 5, 7, 9, \ldots\}$.
   $x \in S \Leftrightarrow x = 2n + 1$ for some natural number n.
   $x \notin S \Leftrightarrow x \neq 2n + 1$ for every natural number n.
   b. $S = \{2, 4, 8, 16, \ldots\}$.
   $x \in S \Leftrightarrow x = 2^n$ for some natural number n.
   $x \notin S \Leftrightarrow x \neq 2^n$ for every natural number n.
   c. $S = \{3, 6, 9, 12, \ldots\}$.
   $x \in S \Leftrightarrow x = 3n$ for some natural number n.
   $x \notin S \Leftrightarrow x \neq 3n$ for every natural number n.
   d. $S = \{a, 2a, 3a, 4a, \ldots\}$.
   $x \in S \Leftrightarrow x = na$ for some natural number n.
   $x \notin S \Leftrightarrow x \neq na$ for every natural number n.
   e. $S = \{1, 2^n, 3^n, 4^n, \ldots\}$.
   $x \in S \Leftrightarrow$ if $x = y^n$ for some natural number y.
   $x \notin S \Leftrightarrow x \neq y^n$ for every natural number n.
3. a. Yes b. No c. No d. Yes e. No f. No.
5. a. 2, 5, 8 b. Yes, let $x = -8$ and $y = 5$.
7. a. For some x, ($x \in C$ and $x \notin D$) or ($x \in D$ and $x \notin C$).
   For some x, $x \in D$ and $x \notin C$.
9. a. False: Let $A = \{1,2\}$. Then $A \notin A$.
   b. True: Let A be a set. Let x be any element.
   If $x \in A$, then $x \in A$. So, $A \subseteq A$.
   c. True: Let A and B be sets.
   Assume $A \subseteq B$ and $x \notin B$.
   Since $A \subseteq B$, if $x \in A$, then $x \in B$.
   So, if $x \notin B$, then $x \notin A$. But $x \notin B$. So $x \notin A$.
   Thus, if $A \subseteq B$ and $x \notin B$, then $x \notin A$.
   d. False: Let $A = \{1\}$. Then $\emptyset \notin A$.
   e. False: Let $A = \{1,2\}$ and $B = \{1,2,3\}$.
   $A \subseteq B$, but $A \notin B$.
   f. False: Let $A = \{1,2\}$ and $B = \{\{1,2\}\{3\}\}$.
   $A \in B$, but A is not a subset of B.
   g. False: Let $A = \{1, 2\}$, $B = \{A, 3\}$, and
   $C = \{B, 4\}$. $A \in B$ and $B \in C$, but $A \notin C$.
11. No. Yes.
13. a. No. P has overlapping sets.
    b. No. P does not cover all of S.
15. a. $S_2 = \{00,01,10,11\}$, so $|S_2| = 4$.
    $S_3 = \{000,001,010,100,011,110,101,111\}$
    $|S_3| = 8$. To list all elements in S_4, we can append each bit string in S_3 with a 1 and then with a 0. So S_4 has 16 elements.
    Theorem: For every positive integer n,
    S_n has 2^n elements.
    Proof: Let $p(n)$: S_n has 2^n elements. Let n be a positive integer. Assume $p(n)$ is true:
    S_n has 2^n elements. Let a be an element of S_n.
    Let $a1$ denote the bit string obtained by appending 1 at the end of a.
    Let $a0$ denote the bit string obtained by appending 0 at the end of a.
    Since $a1$ and $a0$ are bit strings of length $n + 1$, they are both elements of S_{n+1}. Each element in S_{n+1} can be expressed in this form.
    Thus, S_{n+1} has twice as many elements as S_n.
    So the number of elements in S_{n+1} is $2 \cdot 2^n$, which is 2^{n+1}. Thus, $p(n + 1)$ is true.
    So, for all positive integers n, $p(n) \Rightarrow p(n + 1)$.
    Since $S_1 = \{0, 1\}$, $p(1)$ is true.
    Therefore, by mathematical induction, for every positive integer n, S_n has 2^n elements.
    b. $P(A)$: $\{a\}$ $\{ab\}$ $\{ac\}$ $\{b\}$ $\{c\}$ $\{bc\}$ $\{abc\}$ $\{\emptyset\}$
    $\downarrow$ $\downarrow$ $\downarrow$ $\downarrow$ $\downarrow$ $\downarrow$ $\downarrow$ $\downarrow$
    S_3: 100 110 101 010 001 011 111 000
    c. Suppose A has n elements: $A = \{a_1, a_2, a_3, \ldots a_n\}$.
    Let $X \subseteq A$. Form a bit string in S_n as follows:
    In the 1st position: write 1, if $a_1 \in X$
    write 0, if $a_1 \notin X$
    In the ith position: write 1, if $a_i \in X$
    write 0, if $a_i \notin X$
    In the nth position: write 1, if $a_n \in X$
    write 0, if $a_n \notin X$
    The mapping of each subset of A to the above bit string is a one-to-one correspondence between the subsets of A and the bit strings in S_n.
    So, $P(A)$ has the same size as S_n. Since we proved in part (a) that S_n has 2^n elements, $P(A)$ has 2^n elements.

Exercise Set 3.2

1. a. $A \cup B = \{1,2,4,6\}$
   b. $A \cap B = \{2,4\}$
   c. $A - B = \{1\}$
   d. $(A \cap B)' = \{1,3,5,6\}$
   e. $A' \cup B' = \{1,3,5,6\}$

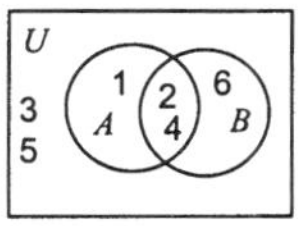

3. a. $[1,5]$ b. $[2,3]$ c. $[1,2)$ d. $[6, \infty)$
   e. $(7, \infty)$ f. $[6,7]$ g. $(-\infty, \infty)$ h. $[1,2)$
   i. $(-\infty, 1) \cup [2, \infty)$
5. a. A b. A c. A d. $A \cup B$ e. A f. A
7. Region 1 = $A \cap B \cap C$ Region 2 = $(B \cap C) - A$
   Region 3 = $(A \cap B) - C$ Region 5 = $C - (A \cup B)$
   Region 6 = $B - (A \cup C)$ Region 8 = $U - (A \cup B \cup C)$
9. a. True b. True c. False d. False
   e. True f. True g. False h. True

11. a. Yes b. No c. Yes d. Yes
e. No f. Yes

13. a. Similar to the proof on page 239.
b. Assume $A \subseteq B$. By the definition of subset, if $x \in A$, then $x \in B$. So, if $x \notin B$, then $x \notin A$. Using the definition of complement, we can translate this implication as follows:
If $x \in B'$, then $x \in A'$. So, $B' \subseteq A'$.
d. From the proof on page 242, we know that
if $A \subseteq B$, then $A \cup B = B$.
Conversely, assume that $A \cup B = B$.
Assume that $x \in A$. Then $x \in A$ or $x \in B$.
So, $x \in A \cup B$. Since $A \cup B = B$, we can deduce that $x \in B$. Thus $A \subseteq B$.
Hence, $A \subseteq B$ if and only if $A \cup B = B$.

15. c. Assume that $A \subseteq C$ and $B \subseteq C$.
Claim: $A \cup B \subseteq C$
Assume that $x \in A \cup B$.
Then $x \in A$ or $x \in B$.
Case 1: Assume that $x \in A$.
Since $A \subseteq C$, x must be in C.
Case 2: Assume that $x \in B$.
Since $B \subseteq C$, x must be in C.
In both cases, $x \in C$. So, $A \cup B \subseteq C$.

17. a. 17 b. 13 c. 8 d. 5 e. 13 f. 0

19. a. Yes, $A \triangledown B = B \triangledown A$ c. $(A \triangledown B) \triangledown C = A \triangledown (B \triangledown C)$

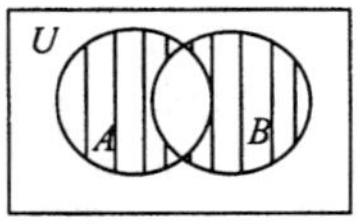

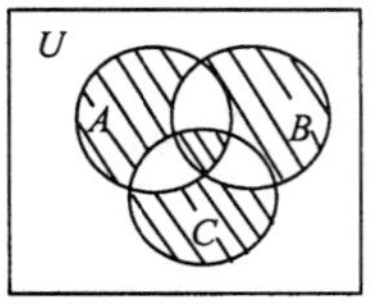

It looks like $\triangledown$ is associative. If you're willing to work through the algebra, you can prove it.
e. Yes

21. a. $(\frac{1}{16}, \infty)$ b. $(\frac{1}{2}, \infty)$ c. $(1, \infty)$ d. $(\frac{1}{16}, \infty)$
e. $(\frac{1}{\pi}, \infty)$ f. $(1, \infty)$

Exercise Set 3.3

1. a. $\bigcup_{i \in I} A_i = [-2,4]$ b. $\bigcup_{i \in N} A_i = [-2, \infty)$
c. $\bigcap_{i \in I} A_i = [-2,1]$ d. $\bigcap_{i \in N} A_i = [-2,1]$

3. $\bigcup_{i \in I} A_i = [5\frac{1}{4}, 6]$ b. $\bigcup_{i \in N} A_i = (5,6]$
c. $\bigcap_{i \in I} A_i = \{6\}$ d. $\bigcap_{i \in N} A_i = \{6\}$

5. a. $\bigcup_{A \in F} A = \{1,2,3,5,7,9\}$ b. $\bigcap_{A \in F} A = \emptyset$

7. a. True. b. False. Let $A_i = (-\frac{1}{i}, \frac{1}{i})$ c. True.
d. False. Counterexample: $A_i = (-\frac{1}{i}, \frac{1}{i})$

9. a. *Claim*: $A \cap (\bigcup_{i \in I} B_i) \subseteq \bigcup_{i \in I} (A \cap B_i)$. Page 254.
Claim: $\bigcup_{i \in I} (A \cap B_i) \subseteq A \cap (\bigcup_{i \in I} B_i)$
Assume that $x \in \bigcup_{i \in I} (A \cap B_i)$. $x \in A \cap B_{i_0}$ for some i_0 in I. (*Multiple union def.*)
$x \in A$ and $x \in B_{i_0}$. (*Intersection def.*)
Since $x \in B_{i_0}$, $x \in \bigcup_{i \in I} B_i$. (*Multiple union def.*)
So $x \in A$ and $x \in \bigcup_{i \in I} B_i$. (*Valid Argument*)
Thus, $x \in A \cap (\bigcup_{i \in I} B_i)$. (*Intersection def.*)
Therefore, $\bigcup_{i \in I} (A \cap B_i) \subseteq A \cap (\bigcup_{i \in I} B_i)$
c. *Claim*: $(\bigcap_{i \in I} A_i)' \subseteq \bigcup_{i \in I} A_i'$. Page 256.
Claim: $\bigcup_{i \in I} A_i' \subseteq (\bigcap_{i \in I} A_i)'$
Assume that $x \in \bigcup_{i \in I} A_i'$.
For some i_0 in I, $x \in A_{i_0}'$. (*Multiple union def.*)
$x \notin A_{i_0}$. (*Complement def.*)
So, ~(For each i, $x \in A_i$.).
~($x \in \bigcap_{i \in I} A_i$) (*Multiple intersection def.*)
Hence, $x \in (\bigcap_{i \in I} A_i)'$. (*Complement def.*)

Exercise Set 3.4

1. a. $\{3\} \times A$ b. $A \times \{3\}$ c. $A \times B$

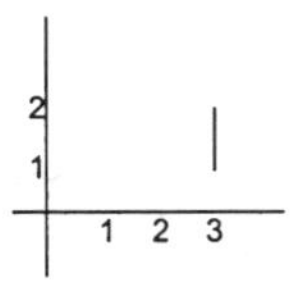

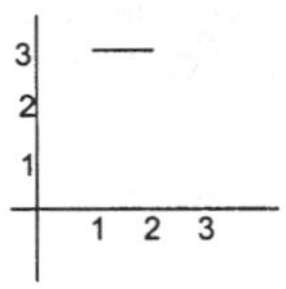

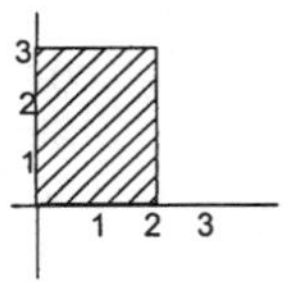

3. (0,3,0,3), (0,4,1,3).

5. Let A be the set of horses in the 1st race,
B the set of horses in the 2nd race, and
C the set of horses in the 3rd race.
$A \times B \times C$ is a model of the ways that you can place the three bets. $|A \times B \times C| = 6 \cdot 8 \cdot 5 = 240$, so there are 240 possibilities. The probability that you will win in all 3 races is $\frac{1}{240}$.

7. a. $[0,2] \times [0,2] \times [0,2]$ b. $[1,3] \times [2,4] \times [0,2]$

11. Let a and b be real numbers with $a < b$.
a. *Case* 1. Assume that $0 \leq a < b$.
Since $a \geq 0$, the distance from a to 0 is a.

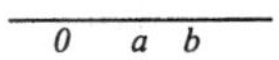

Since $b > 0$, the distance from b to 0 is b.
The distance from a to b is the distance from b to 0 minus the distance from a to 0. which is $b - a$.

Case 2. Assume that $a<0\leq b$.
Since $a<0$, the distance from a to 0 is $-a$.

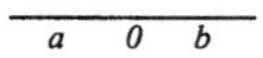

Since $b\geq 0$, the distance from b to 0 is b.
The distance from a to b is the distance from a to 0 plus the distance from b to 0, which is $b+(-a)$.

Case 3. Assume that $a<b<0$.
Since $a<0$, the distance from a to 0 is $-a$.

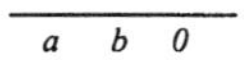

Since $b<0$, the distance from b to 0 is $-b$.
The distance from a to b is the distance from a to 0 minus the distance from b to 0, which is $(-a)-(-b)$.
In each case, the distance from a to b is $b-a$.

c. Use part (a) and the Pythagorean Theorem.

Exercise Set 3.5

1. a. Yes. Let $f(1)=a, f(2)=b, f(3)=c, f(4)=d$.
   b. No.
3. a. Yes b. No
5. a. Yes b. Yes
7. a. No. See (1a).
   b. No. See (2b).
   c. Both answers will be "yes."
9. a. Yes. b. f is not one-to-one. c. f is not onto.
11. Let S be a set and n a natural number.
    a. There exists a one-to-one function that maps $\{1,2,3,\ldots,n\}$ onto S.
    b. If f maps $\{1,2,3,\ldots,n\}$ into S and f is one-to-one, then f does not map onto S.

Exercise Set 3.6

1. a. Yes: $f(n)=3n$ b. Yes: $f(n)=3n-1$
   c. Yes: $f(n)=3n-2$ d. Yes: $f(n)=3n+3$
   e. Yes: $f(n)=2^n$ f. Yes: $f(n)=2^n-1$
3. a. True. b. False. Let $S=\{-1,-2,-3,-4,\ldots\}$.
   c. False. Let $A=\{2,3,4,\ldots\}$ and $B=\mathbb{N}$.
5. a. $R_1=\{\frac{1}{1},\frac{2}{1},\frac{3}{1},\ldots\}$
   $R_2=\{\frac{1}{2},\frac{2}{2},\frac{3}{2},\ldots\}$
   $R_3=\{\frac{1}{3},\frac{2}{3},\frac{3}{3},\ldots\}$
   b. List the elements of R_1 on the first row, R_2 on the second row, etc. Use the same diagonal counting technique as in the proof on page 292.
   c. f is not one-to-one.
   $f(1)=\frac{1}{1},\ f(2)=\frac{2}{1},\ f(3)=\frac{1}{2},\ f(8)=\frac{3}{2}$.
7. a. False. Let $X=Y=\mathbb{N}$ and $f(x)=2x$.
   f is one-to-one, but f does not map $\mathbb{N}$ onto $\mathbb{N}$.
   b. False. Let $X=Y=\mathbb{N}$. Let $f(1)=1$ and $f(n)=n-1$ if $n>1$. Since $f(1)=f(2)=1$, f is not one-to-one. However, f does map $\mathbb{N}$ onto $\mathbb{N}$.
9. a. $1\rightarrow\{1,3\}$ a. f is one-to-one.
   $2\rightarrow\{1,3,4\}$ b. f is one-to-one.
   $3\rightarrow\{2\}$ c. f is not one-to-one.
   $4\rightarrow\{1,2,4\}$
10. a. $C=\{2,3\}$. Not in the range of f.
    b. $C=\emptyset$. Not in the range of f.
    c. $C=\{2,3,4\}$. Not in the range of f.
11. a. $f(x)=\{x\}$ b. $\{x \text{ in } S \mid x\notin f(x)\}$ (page 298)
13. a. $|S_n|=2^n$. $|S|=\aleph_0$
    b. $|S_n|=3^n$. $|S|=\aleph_0$
    c. $|S_n|=x^n$. $|S|=\aleph_0$
15. a. If $A\in P(V)$, A is a set, so $A\in V$.
    Hence, $A\subseteq P(V)$.
    b. Yes. Part (a), 1st theorem on page 297.
    c. Yes. Theorem on page 298.
    d. Since $|P(V)|\leq|V|$ and $|V|<|P(V)|$, by the Schröder-Bernstein Theorem (page 296), $|V|=|P(V)|$. But $|V|<|P(V)$. Contradiction!
    e. Yes. V does not satisfy the Property Axiom for defining a set (page 279). So, we cannot apply the above cited theorems.
17. Set $n=.9999\ldots$.
    $10n=9.999\ldots$
    $\underline{-n=\ .999\ldots}$
    So, $9n=9.000\ldots$ Hence, $n=1$.

Chapter Review

3. a. There exists an x such that ($x\in A$ and $x\notin B$) or ($x\in B$ and $x\notin A$).
   b. There exists an x such that $x\in A$ and $x\notin B$.
   c. $A\not\subseteq B$ or $A=B$. d. $x\notin A$ and $x\notin B$
   e. $x\notin A$ or $x\notin B$ f. For every n in I, $x\notin A_n$.
   g. For some i in I, $x\notin A_i$.
5. b. $(A-B)-C$

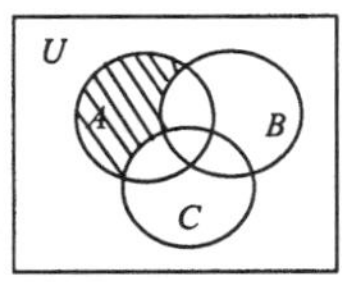

7. a. B b. B' c. C d. C' e. $\emptyset$ f. $\emptyset$
9. a. True b. False: $A=\{1\}$ c. True
   d. False: $A=\{1\}$ e. True f. True
11. a. No. Let $A=\{1,2\}$ and $B=\{2,3\}$

b. No. Let $A = \{1,2\}$ and $B = \{2,3\}$ c. Yes.

13. a. True b. False: Let $A = \{1,3\}$, $B = \{2\}$, $C = \{2,3\}$. $A-(B-C) = \{1,3\}$, $(A-B)-C = \{1\}$
    c. True d. False. Draw a Venn diagram.
17. The power set of a set S is the collection of all subsets of S. If S has n elements, S has 2^n subsets.
19. a. For every natural number n and every one-to-one function f, if f maps $\{1,2,3, \ldots, n\}$ into S, then f does not map $\{1,2,3, \ldots, n\}$ onto S.

Exercise Set 4.1

1. a. A circle – radius of 1 unit – center at origin.
   Domain: $\{ x \mid -1 \le x \le 1 \}$
   Range: $\{ y \mid -1 \le y \le 1 \}$
   b. The circle described in part (a) and its interior. It has the same domain and range.
   c. Domain: $\{ x \mid 0 \le x < \infty \}$ A parabola.
   Range: $\{ y \mid -\infty < y < \infty \}$
   d. Domain: $\{ x \mid 0 \le x \le 1 \}$
   Range: $\{ y \mid 0 \le y \le 1 \}$

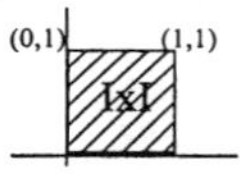

   e. Domain: $\{ x \mid 0 \le x \le 2 \}$
   Range: $\{ y \mid 0 \le y \le 4 \}$

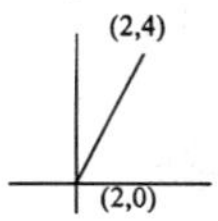

   f. Domain: $\{ x \mid 1 \le x \le 3 \}$ Line segment from
   Range: $\{ y \mid -2 \le y \le 4 \}$ $(1,-2)$ *to* $(3,4)$.
3. f must be one-to-one.
5. The graph of R^{-1} is the reflection of the graph of R about the line, $y = x$, that bisects the first quadrant. The matrix of R^{-1} is the matrix of R with the rows and columns swapped – the first row of R^{-1} is the first column of R.
7. $\supseteq$ is the inverse relation for the $\subseteq$ relation.
   $\ge$ is the inverse relation for the $\le$ relation.
9. a. Transitive, antisymmetric.
   b. Reflexive, symmetric.
   c. Reflexive, transitive, antisymmetric.
   d. Transitive, antisymmetric.
   e. Reflexive, transitive, antisymmetric.
   f. Transitive, antisymmetric.
13. R symmetric if and only if $R = R^{-1}$.
15. a. It must contain all points on the line $y = x$ where x is in the domain of the relation. From a point on the graph, move vertically to the line $y = x$ and that point must be on the graph.
    b. If (a,b) is on the graph, its reflection about the line $y = x$ must also be on the graph.
    c. If (a,b) is on the graph and $a \ne b$, then the reflection of (a,b) about the line $y = x$ cannot be on the graph.

Exercise Set 4.2

1. a. Yes. $[a]$ is the set of people born in the same country as a.
   b. Yes. $[a]$ is the set of people that have the same birthday as a.
   c. No. The relation is not transitive.
   d. No. The relation is not transitive.
   e. The relation is not transitive if someone has a double major.
3. a. Yes. b. No. The relation is not transitive.
5. a. Reflexive, symmetric, and transitive.
   b. Reflexive and symmetric.
   c. Symmetric and transitive.
   d. Symmetric and transitive.
   e. Reflexive and symmetric
   f. Reflexive, symmetric, and transitive.
7. $a \mathrm{R} b$ iff there exists an X in P such that a and b are both in X.
9. a. $a \equiv_6 b$ if and only if 6 divides $a - b$.
   b. Yes. No. c. $[5] = \{\ldots -7, -1, 5, 11, 17, \ldots\}$
   d. $Z_6 = \{ [0], [1], [2], [3], [4], [5] \}$
   e. $2345 \in [5]$, $-38 \in [4]$
11. Assume that $x \equiv_5 2$. Then 5 divides $x - 2$.
    So $x - 2 = 5k$ for some integer k.
    $x + 1 - 3 = 5k$.
    Since 5 divides $(x+1) - 3$, $x + 1 \equiv_5 3$.
    Therefore, if $x \equiv_5 2$, then $x + 1 \equiv_5 3$.

Exercise Set 4.3

3. a. No. They have different domains.
5. a. Let a and b be elements in Y.
   If $a \ne b$, then $g(a) \ne f(b)$.
   b. Let a and b be elements in X.
   If $a \ne b$, then $g \circ f(a) \ne g \circ f(b)$.
7. a. $f(x) = 2$ b. $f(x) = x + 3$.
   c. $f(0) = 0$, $f(x) = 1$, if $x \ne 0$. d. $f(x) = 3$, if x is not an integer. $f(x) = x$, if x is an integer.
   e. $f(x,y) = 2x + y$ f. $f(x) = (x, x^2)$
9.

	Function	Domain	Range
a.	$f(x,y) = x + y$	$\mathbb{N} \times \mathbb{N}$	$\mathbb{N}$
b.	$f(x,y) = x - y$	$\mathbb{N} \times \mathbb{N}$	Z
c.	$f(x,y) = \frac{x}{y}$	$\mathbb{R} \times (\mathbb{R} - \{0\})$	$\mathbb{R}$

11. a. Let y be a real number. Set $x_0 = \frac{y-7}{2}$.
Since y is a real number, x_0 is a real number.
$f(x_0) = 2x_0 + 7 = 2(\frac{y-7}{2}) + 7 = y$.
So, for every real number y, there exists a real number x such that $f(x) = y$.
Therefore, f maps $\mathbb{R}$ onto $\mathbb{R}$.
b. Set $y = 0$. 0 is a real number. Let x be a real number. Since $x^2 \geq 0, f(x) > 0$.
So, $f(x) \neq 0$. Thus, f does not map $\mathbb{R}$ onto $\mathbb{R}$.
13. a. 3 b. 7 c. 2
15. a. $f(x) = x^2$, where x is a real number.
b. Not possible. Every relation has an inverse relation.
17. a. See page 361. b. See page 153.
c. Use part (a) and part (b).
d. See example on page 366.
19. Let A, B and C be sets that are members of S.
a. Let $f(x) = x$, where x is in A. Then f is a one-to-one function that maps A onto A.
So, $A \cong A$. Thus, $\cong$ is reflexive.
b. Assume that $A \cong B$. So there exists a one-to-one function f that maps A onto B.
By a previous theorem, f^{-1} is one-to-one function that maps B onto A. So, $B \cong A$.
Thus, $\cong$ is symmetric on S.
c. Assume that $A \cong B$ and $B \cong C$.
Since $A \cong B$, there exists a one-to-one function f that maps A onto B. Since $B \cong C$, there exists a one-to-one function g that maps B onto C.
By a previous theorem, $g \circ f$ is a one-to-one function f that maps A onto C. So, $A \cong C$.
Thus $\cong$ is transitive.
Since $\cong$ is reflexive, symmetric and transitive, it is an equivalence relation.
21. Assume that $f(x) \in f(A)$. By the definition of the image of a set, there exists a c in A such that $f(c) = f(x)$. Since f is one-to-one, $c = x$. So $x \in A$.
23. a. Assume that $A \subseteq B$. Let $y \in f(A)$.
Then $y = f(x)$ for some x in A. Since $A \subseteq B$, x is also in B. So $y = f(x)$ for some x in B. Thus $y \in f(B)$. So $f(A) \subseteq f(B)$.
b. See page 367. c. See page 365.
d. See page 366.
25. Step #3. You cannot factor the existential quantifier across *and*.
27. a) Suppose that $y \in Y$. There exists an x in X such that $f(x) = y$. So $x \in A_y$. Thus, $A_y \neq \emptyset$. So, the elements in P are nonempty subsets of X.
b) Now, let x be an element in X. Then $f(x) \in Y$. So $x \in A_{f(x)}$. Thus, each element in X is in one of the sets in P.
c) Assume that $x \in A_y \cap A_z$. Then $f(x) = y$ and $f(x) = z$. So, $y = z$. Thus, $A_y = A_z$. So, each element in X is in only one of the sets in P.
29. The set of all functions from N to {0,1} is uncountable (see exercise 28). However, the set of all possible computer programs in a given language is countable (see (14), page 301). There are far more functions than possible computer programs, so some functions must be noncomputable.

Exercise Set 4.4

3. a. partial order b. total order and a well-ordering
c. total order and a well-ordering d. partial order
5. a.

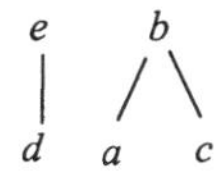

b. $a \to b \to c \to d \to e$
c. Not possible. Every finite totally ordered set is well-ordered.
7. a. $\preceq = \{ (1,1), (2,2), (3,3), (4,4), (1,2), (1,4), (1,3), (2,4), (3,4) \}$
b.

Poset	Greatest	Least	Maximal	Minimal
A	4	1	4	1
B			a, d	b, e
C			a, d	b, e

c. A: 1 2 3 4 B: $b\ e\ c\ a\ d$ C: $b\ c\ a\ e\ f\ d$
1 3 2 4 $e\ b\ c\ d\ a$ $b\ e\ c\ f\ d\ a$
d. A: $T = \{\{1,2,3,4\}, \{2,1\}, \{1\}, \{3,1\}\}$
B: $T = \{\{a,c,b,e\}, \{d,c,b,e\}, \{c,b,e\}, \{b\}, \{e\}\}$
C: $T = \{\{a,c,b\}, \{d,f,e\}, \{c,b\}, \{b\}, \{f,e\}, \{e\}\}$
9. a & b.

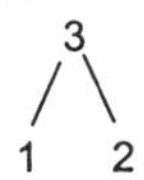

c.

d.

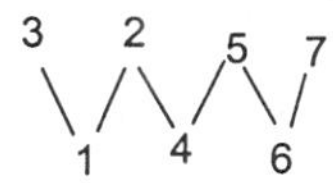

11. a. $(S, |)$

$(T, |)$

$(W, |)$

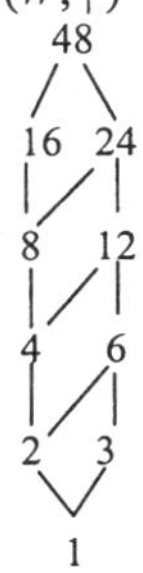

c. S: 1 2 3 6 9 18
1 3 2 9 6 18
T: 1 2 4 8 16 32 (*only possible sort*)
W: 1 2 3 4 6 8 12 16 24 48
1 3 2 4 6 8 12 16 24 48

d. $S \simeq \{ \{18,9,6,3,2,1\}, \{9,3,1\}, \{6,3,2,1\}, \{3,1\}, \{2,1\}, \{1\} \}$
$T \simeq \{ \{32,16,8,4,2,1\}, \{16,8,4,2,1\}, \{8,4,2,1\}, \{4,2,1\}, \{2,1\}, \{1\} \}$
$W \simeq \{ \{1,2,3,4,6,8,12,16,24,48\}, \{1,2,3,4,6,8,12,24\}, \{1,2,4,8,16\}, \{1,2,3,4,6,8,12\}, \{1,2,4,8\}, \{1,2,3,6\}, \{1,2,4\}, \{1,2\}, \{2\}, \{1\} \}$

13. *Proof*: Let $(S, \preceq)$ be a finite nonempty poset with n elements. Suppose that S does not have a minimal element. Pick an element a_1 in S.
Since a_1 is not a minimal element, there exists an a_2 in S such that: $a_2 \prec a_1$
Since a_2 is not a minimal element, there exists an a_3 in S such that: $a_3 \prec a_2 \ldots$
Since a_n is not a minimal element, there exists an a_{n+1} in S such that: $a_{n+1} \prec a_n$
Suppose that $i<j$. By transitivity of $\prec$: $a_j \prec a_i$
So $a_j \neq a_i$. Hence, we have a list of $n+1$ different elements in S. Contradiction!
So, S must have a minimal element.

15. a. Lower bounds: $\emptyset$, $\{1\}$, $\{3\}$, $\{1,3\}$
Greatest lower bound: $\{1,3\}$
S has no upper bounds.
b. Lower bounds: $\emptyset$, $\{1\}$, $\{3\}$
S has no greatest lower bound.
$\{1,2,3,4,7\}$ is the only upper bound of S, so it is the least upper bound.

17. a. There exists a b in S such that for every x in S, $b \preceq x$. For every b in T, there exists an x in T such that $\sim(b \preceq x)$.
b. There exists a b in S such that for every x in S, if $b \preceq x$, then $b = x$. For every b in T, there exists an x in T such that $b \preceq x$ and $b \neq x$.
c. There exists a b in U such that for every x in S, $b \preceq x$. For every b in U, there exists an x in T such that $\sim(b \preceq x)$.
d. There exists a b in U such that b is a lower bound of S and for every lower bound x of S, $x \preceq b$. For every lower bound b of S, there exists a lower bound x of S such that $\sim (x \preceq b)$.
e. There exists a b in U such that for every x in T, $x \preceq b$.
f. There exists a b in U such that b is an upper bound of T and for every upper bound x of T, $b \preceq x$.

◈ Appendix B

Glossary

$\aleph_0$ The number of elements in the set $\mathbb{N}$ of natural numbers. $|S| = \aleph_0$ if and only if there exists a one-to-one function that maps $\mathbb{N}$ onto S.

Abstract compound statement A compound statement where the component statements are represented by variables such as p and q.

Abstraction The merging of concrete examples under the rubric of a concept that expresses a property the examples have in common. An abstraction exists as an idea with no material existence. For example, the abstract number 3 describes a property that various physical sets have in common, but the number 3 has no physical existence.

Antisymmetric relation A relation R on a set S that has the following property: for every a and b in S, if $a \mathsf{R} b$ and $b \mathsf{R} a$, then $a = b$. The only time we can reverse the order is when the elements are equal.

Argument A list of sentences called hypotheses followed by a sentence called the conclusion. See Valid argument.

Associative property Let $*$ be a binary operation on a set S. $*$ is associative if and only if for every a, b, and c in S, $a * (b * c) = (a * b) * c$. The logical operators, *and* and *or*, are associative. Likewise, the set operations, union and intersection are associative. The number operations, addition and multiplication, are associative. The operation of function composition is also associative.

Axiom A statement that is assumed true in an axiomatic system, requiring no proof.

Axiomatic system A list of undefined terms, a list of axioms, and a proof procedure for deriving theorems in the system. Definitions are built from the undefined terms and previously defined terms. Theorems are derived from the axioms, previous theorems, and definitions using the proof procedure. The axioms, definitions, and theorems must each be sentences according to the grammar for the system.

Bijection A one-to-one and onto function.

Binary Refers to two. A binary operation, such as + or $\cup$, operates on two elements in a set and produces a new element in the set. A binary relation, such as $\leq$ or $\subseteq$, gives a relation between two elements. A binary decimal system has a base of two.

Binary operation on S A function that maps $S \times S$ into S. $*$ is a *binary operation* on set S if and only if for every a and b in S, $a * b$ is defined and $a * b \in S$.

Cardinal number of a set The number of elements in a set A, notated as $|A|$. Let n be a natural number. $|A| = n$ if and only if there exists a one-to-one function f that maps $\{1, 2, 3, \ldots, n\}$ onto A. Two sets that have the same size have the same *cardinality*. $|A| = |B|$ if and only if there exists a one-to-one function that maps A onto B. $|A| \leq |B|$ if and only if there exists a one-to-one function that maps A into B. $|A| \geq |B|$ if and only if there exists a function that maps A onto B. Let A and B be finite sets. If $A \subset B$, then $|A| < |B|$.

Commutative property Let $*$ be a binary operation on a set S. $*$ is commutative if and only if for every a and b in S, $a * b = b * a$. The logical operators, *and* and *or*, are commutative. Likewise, the set operations, union and intersection are commutative. The number operations, addition and multiplication, are commutative. The operation of function composition is not commutative.

Complement laws The complement of a union is the intersection of the complements. The complement of an intersection is the union of the complements. Let A_i be a set for each i in I:

$$(A_1 \cup A_2)' = A_1' \cap A_2' \quad \ldots \quad (\bigcup_{i \in I} A_i)' = \bigcap_{i \in I} (A_i')$$

$$(A_1 \cap A_2)' = A_1' \cup A_2' \quad \ldots \quad (\bigcap_{i \in I} A_i)' = \bigcup_{i \in I} (A_i')$$

Complement A unary operation on a set which produces a new set composed of all the elements in the universal set that are not in the original set: $A' = \{ x \mid x \notin A \}$. $A' = U - A$.

Complete system An axiomatic system in which every well-formed statement can be either proved or disproved. Every sentence is decidable.

Complex number A number that can be represented in the form $x + yi$ where x and y are real numbers and $i = \sqrt{-1}$. The visual picture of the complex numbers is the points in a plane, where $x + yi$ is identified with the point (x, y).

Composition of functions A function formed from two functions f and g where f maps X into Y and g maps Y into Z. If x is in X, $g \circ f(x) = g(f(x))$.
The composition of two injections is an injection.
The composition of two surjections is a surjection.
The composition of two bijections is a bijection.
If f and g have inverse functions, $(g \circ f)^{-1} = f^{-1} \circ g^{-1}$.
Composition of functions is associative: $(h \circ g) \circ f = h \circ (g \circ f)$.

Compound statement A sentence composed of statements connected with logical operators.

Conclusion See Implication and see Valid Argument.

Congruence mod n — An equivalence relation on the set Z of integers. $a \equiv_n b$ if and only if n divides $a-b$. If r is the remainder when we divide x by n, then $x \equiv_n r$. Congruence mod n partitions the set Z of integers into n equivalence classes, which we notate as Z_n:

$$Z_n = \{ [0], [1], [2], [3], \ldots, [n-1] \}$$

Conjecture — A statement someone thinks is true, but no one has proved it.

Conjunction — A compound statement of the form: *p and q*. For an and-statement to be true, both parts must be true.

Consistent system — An axiomatic system that contains no contradictions.

Contradiction — An abstract compound statement that is always false, like *p and ~p*. A negation of a law of logic is a contradiction. See also Proof by contradiction.

Contrapositive — The contrapositive of $p \Rightarrow q$ is $\sim q \Rightarrow \sim p$. The contrapositive has the same meaning as the original implication.

Converse — The converse of $p \Rightarrow q$ is $q \Rightarrow p$. The converse does not have the same meaning as the original implication.

Countable — A set that is either finite or can be placed in a one-to-one correspondence with the set $\mathbb{N}$ of natural numbers. The union of a countable collection of countable sets is countable.

Countably infinite — An infinite set that is countable. S is countably infinite if and only if $|S| = \aleph_0$. The set of natural numbers, the set of integers, and the set of rational numbers are countably infinite sets.

Cross product — $A \times B$ is the set of all ordered pairs whose first term is in A and whose second term is in B. $A \times B = \{ (a, b) \mid a \in A \text{ and } b \in B \}$ $A \times B \times C = \{ (a, b, c) \mid a \in A \text{ and } b \in B \text{ and } c \in C \}$. The number of elements in a cross product is the product of the number of elements in the individual sets.

Decidable — A sentence that can be either proved or disproved.

Deductive reasoning — Type of reasoning used when we derive a conclusion through valid arguments from other sentences that we accept as true.

Direct proof — A proof of an implication by assuming the hypothesis is true and then deriving that the conclusion must be true.

Directed graph — A visual representation of a relation where the mapping is represented by arrows, with each member of the domain and range listed only once. So, some of the arrows may be chained together.

Disjunction — A compound statement of the form: *p or q*. For an or-statement to be true, at least one part must be true, but both could be true.

Disprove a statement — Prove its negation.

Distributive property — Gives a relation between two operations. Multiplication distributes over addition, *and* distributes over *or*, *or* distributes over *and*, intersection distributes over union, and union distributes over intersection: $p \times (q + r) = (p \times q) + (p \times r)$

$$p \text{ and } (q \text{ or } r) \Longleftrightarrow (p \text{ and } q) \text{ or } (p \text{ and } r)$$

$$p \text{ or } (q \text{ and } r) \Longleftrightarrow (p \text{ or } q) \text{ and } (p \text{ or } r)$$

Distributive property (cont.) Let A be a set and B_i be a set for each i in I:
$A \cap (B_1 \cup B_2) = (A \cap B_1) \cup (A \cap B_2)$ .... $A \cap (\bigcup_{i \in I} B_i) = \bigcup_{i \in I} (A \cap B_i)$
$A \cup (B_1 \cap B_2) = (A \cup B_1) \cap (A \cup B_2)$ .... $A \cup (\bigcap_{i \in I} B_i) = \bigcap_{i \in I} (A \cup B_i)$

Divides Let a and b be integers. a divides b if and only if $b = ak$ for some integer k, which means that a is a factor of b.

Domain The set of elements that can be substituted for a variable. The set of all first terms of the ordered pairs in a relation R. The set of elements for which a function f is defined. x is in the domain of f means that $f(x)$ is defined.

Empty set A set that has no elements. The empty set is analogous to an empty box, which exists even though it has nothing in it.

Equal functions Two functions with the same domain and the same function values for each element in the domain. $f = g$ if and only if $f(x) = g(x)$ for all x in the domain.

Equals relation Two sets are equal if they contain the same elements. $A = B$ if and only if for every x, $(x \in A \Rightarrow x \in B)$ and $(x \in B \Rightarrow x \in A)$. The equals relation is reflexive, transitive and symmetric.

Equivalence A compound statement of the form: *p is equivalent to q*. For an equivalence to be true, either both parts are true or both parts are false. If two abstract compound statements composed of the same component statements are equivalent, they have the same meaning and can be used interchangeably. For frequently used equivalences, see page 103.

Equivalence relation A relation on a set S that is reflexive, transitive, and symmetric. The set of equivalence classes of an equivalence relation partitions the set S into nonoverlapping subsets. An equivalence relation identifies a property that makes elements essentially the same with respect to that property, such as the property of "having the same size and shape." "Is congruent to" and "is similar to" are important equivalence relations between figures. Congruence mod n is an important equivalence relation between integers. "Has the same size" is a very important equivalence relation between sets. "Is isomorphic to" is an extremely important equivalence relation between structured sets, such as partially ordered sets.

Equivalence class – [a] The set of elements related to a by an equivalence relation R on a set S. Let a be in S. $[a] = \{ x \text{ in } S \mid a \, \mathrm{R} \, x \}$. Two related elements have the same equivalence class: if $a \, \mathrm{R} \, b$, then $[a] = [b]$. Elements in the same equivalence class are considered as essentially the same with respect to the relation.

Even a is *even* if and only if $a = 2n$ for some integer n.

Exclusive or A logical operator that joins two statements with the exclusive or: *p XOR q*. This compound statement is true only when one statement is true and the other one false.

Existential quantifier	Asserts that at least one substitution of an element from the domain of the variable converts an open statement into a true statement. $\exists x,\ p(x)$ is true if and only if there exists at least one x in the domain of x such that $p(x)$ is true.
Extension of a function	A function whose domain is extended beyond the domain of the original function, while preserving the original mapping.
Finite set	A set S is finite if and only if S is the empty set or $\|S\| = n$ for some natural number n.
Function	f is a function from X into Y if and only if f maps each element in X to a unique element in Y. This uniqueness property allows us to use the function notation, $f(x)$. To demonstrate that the same element has not been assigned two different values, we prove the following implication: if $a = b$, then $f(a) = f(b)$.
Function notation	$f(x)$ denotes the value assigned to x by the function f. The following have the same meaning: $f(x) = y$, $x \xrightarrow{f} y$, $(x,y) \in f$.
Fundamental theorem of arithmetic	Every natural number, other than 1, can be represented in a unique manner as a product of prime numbers, with smaller factors written to the left of larger factors.
Graph	A visual representation of a relation where we use points in a coordinate plane to represent the ordered pairs in a function.
Greatest element	The largest element in a partially ordered set $(S, \preceq)$. b is the greatest element of S if and only if $b \in S$ and $x \preceq b$ for all x in S. A partially ordered set can have at most one greatest element. The greatest element in S will be the only maximal element of S. In a totally ordered set S, a maximal element is the greatest element.
Hasse graph	A minimalist graph of a partially ordered set in which its directed graph is positioned so that all arrows point upwards, then we omit the arrowheads, the loops that can be deduced by reflexivity, and the arrows that can be deduced by transitivity.
Hypothesis	See Implication and see Valid Argument.
Identity function	A function that maps each element to itself: $e(x) = x$. e_A denotes the identity function on the set A.
Image of a set	The set of images of individual elements in a set under a function. $f(A) = \{ f(x) \mid x \in A \}$. $y \in f(A)$ if and only if there exists an x in A such that $f(x) = y$.
Immediate successor	An element that comes after another element with no elements between them. In a poset, b is an immediate successor to a if and only if $a \prec b$ and there does not exist an x such that $a \prec x \prec b$.
Implication	A compound statement of the form: p *implies* q. p is called the hypothesis or premise and q is called the conclusion. The only case in which an implication is false is when the hypothesis is true and the conclusion is false. To say that $p \Rightarrow q$ is true means that if p is true, then q must be true. The implication $p \Rightarrow q$ has the same meaning as its contrapositive: $\sim q \Rightarrow \sim p$.

Independent axioms A set of axioms in which none of the axioms can be derived from the others.

Indirect proof A proof of an implication by assuming the conclusion is false and then deriving that the hypothesis must be false. An indirect proof of an implication is a direct proof of its contrapositive.

Inductive reasoning Type of reasoning used when we discover a general relation from specific examples or experiences.

Infinite set A set that is not finite.

Injection A one-to-one function.

Integers $\ldots -3, -2, -1, 0, 1, 2, 3, \ldots$

Intersection A binary operation on two sets that produces a new set from their common elements: $A \cap B = \{ x \mid x \in A \text{ and } x \in B \}$.

Into function f maps X into Y if and only if for each x in X, $f(x)$ is in Y.

Inverse function The inverse relation of a one-to-one function. Let f be a one-to-one function that maps X onto Y. Then f^{-1} is a one-to-one function that maps Y onto X. $f^{-1}(a) = b$ if and only if $f(b) = a$.
For every a in X, $f^{-1}(f(a)) = a$. For every b in Y, $f(f^{-1}(b)) = b$.

Inverse relations A relation obtained by reversing the order of a given relation. Every relation has an inverse relation. $a \mathsf{R}^{-1} b$ if and only if $b \mathsf{R} a$.

Irrational number A real number that is not rational. Unlike the rational numbers, the decimal form of an irrational number does not have a repeating cycle. The number of irrational numbers is a higher level of infinity than the number of rational numbers.

Is an element of A relation between the members of a set and the collective unit to which the members belong. In formal set theory, "is an element of" is an undefined term since there are no simpler concepts with which to define it.

Isomorphic structures Two structures that have the same form, which means that one structure can be relabeled to produce the other structure. For posets, $(S, \preceq)$ is isomorphic to $(T, \trianglelefteq)$ if and only if there exists a bijection f from S onto T that preserves the order on S: for every a and b in S, if $a \preceq b$, then $f(a) \trianglelefteq f(b)$. The function f is called an *isomorphism*. An isomorphism preserves the order relation. If two elements are related in S, their images must be related in T. Isomorphic posets are identical twins with different names. If a poset has a particular property, then all isomorphic posets must have the same property, as translated through the isomorphism.

Law of contraposition A valid argument whose hypothesis has the form, $p \Rightarrow q$ and $\sim q$, and whose conclusion is $\sim p$.

Law of detachment A valid argument whose hypothesis has the form, $p \Rightarrow q$ and p, and whose conclusion is q. Also known as modus ponens.

Law of logic An abstract compound statement that is always true, regardless of the truth values of its component statements. A law of logic is also called a tautology. See page 103 for frequently used laws of logic.

Least element The smallest element in a poset $(S, \preceq)$. b is the least element of S if and only if $b \in S$ and for every x in S, $b \preceq x$. A partially ordered set can have at most one least element. The least element in S is the only minimal element of S. In a totally ordered set S, a minimal element is also the least element in S.

Logic A formal study of the art of reasoning and the principles for making valid deductions.

Logical operators Connectives used to form a compound sentence from given component sentences: *and*, *or*, *implies*, *is equivalent to*, and *negation*.

Lower bound b is a lower bound for S if and only if for every x in S, $b \preceq x$. Unlike a minimal element, a lower bound does not have to be in S. The *greatest lower bound* of S is the greatest of all lower bounds of S.

Matrix A rectangular array used to represent a finite relation. Matrices have a wide range of applications.

Maximal element An element in a partially ordered set S that has no other elements after it. b is a maximal element of S if and only if $b \in S$ and, for every x in S, if $b \preceq x$, then $x = b$. A finite, nonempty poset must have at least one maximal element.

Minimal element An element in a poset S that has no other elements before it. b is a minimal element of S if and only if $b \in S$ and for every x in S, if $x \preceq b$, then $x = b$. A finite, nonempty poset must have at least one minimal element.

Model An example of an undefined term that has the properties specified by the axioms. A model of an axiomatic system contains a model of each undefined term in the system.

Multiple intersections A set formed from the elements that are in each member of a collection of sets. $\bigcap_{i \in I} A_i = \{ x \mid \text{for every } i \text{ in } I, x \in A_i \}$

Multiple unions A set formed by combining the elements in a collection of sets. $\bigcup_{i \in I} A_i = \{ x \mid \text{for some } i \text{ in } I, x \in A_i \}$

n-ary relation A set of ordered n-tuples. A subset of a cross product of n sets.

Natural numbers 1, 2, 3, 4, 5, 6, . . . The natural numbers were created as a standardized reference set for comparing sizes of finite sets.

Negation A logical operator that reverses the truth value of a statement. The negation of p is true if and only if p is false. The rules for negating quantifiers and logical operators are summarized on page 103. These rules are often used in the process of logical reasoning.

Odd a is *odd* if and only if $a = 2n + 1$ for some integer n.

One-to-one correspondence A one-to-one and onto function between two sets.

One-to-one function A function that maps different elements in the domain to different elements in the range. For every element a and b in the domain of f, if $a \neq b$, then $f(a) \neq f(b)$. If f is a one-to-one function that maps X into Y, then $|X| \leq |Y|$. A one-to-one function has an inverse function.

Onto function — f maps X onto Y if and only if the range of f is Y. For each y in Y, there must exist an x in X such that $f(x) = y$. If f maps X onto Y, then $|X| \geq |Y|$. If $|X| = |Y|$ and both sets are finite, f is onto is equivalent to f being one-to-one.

Open statement — A sentence with variables that is not a statement but becomes a statement whenever substitutions are made for the variables. An open statement can be converted to a statement by substituting for each variable or by binding each variable with a quantifier, such as $\forall x \, \exists y, p(x,y)$.

Ordered pair — A pairing of elements where the order affects the meaning. $(a,b) = (c,d)$ if and only if $a = c$ and $b = d$.

Partial order — A relation that is reflexive, transitive, and antisymmetric. Instead of arranging elements in egalitarian equivalence classes where everyone is essentially the same, a partial order embeds a hierarchical structure on a set. Given any example of a partial order on a set S, such as "divides" or $\leq$, we can produce an isomorphic copy of it using the subset relation.

Partition — A subdivision of a set into nonoverlapping subsets. A partition P of a set S is a collection of nonempty subsets of S where each element in S is in one and only one of the subsets. Each partition of a set has an associated equivalence relation.

Poset — A partially ordered set. A set that has a partial order relation defined on its elements.

Power set — The set of all subsets of a given set. $P(S) = \{X \mid X \subseteq S\}$. If S has n elements, $P(S)$ has 2^n elements. For every set S, be it finite or infinite, $|S| < |P(S)|$.

Prime Number — an integer greater than 1 whose only positive factors are 1 and itself.

Principle of mathematical induction — Let $p(n)$ be an open statement. Let c be a fixed integer. If for every integer $n \geq c$, $p(n) \Rightarrow p(n+1)$, and $p(c)$ is also true, then $p(n)$ is true for all integers $n \geq c$. *Stronger Version*: If for every positive integer n, $[p(1) \wedge p(2) \wedge p(3) \wedge \ldots \wedge p(n)] \Rightarrow p(n+1)$, and $p(1)$ is also true, then $p(n)$ is true for all positive integers n.

Proof — A linearly ordered structure of interwoven valid arguments where each sentence is one of the following: 1) an assumption used in a valid argument. 2) an axiom, previous theorem, or definition. 3) a sentence that can be derived from previous sentences by a valid argument. The final stand-alone conclusion is the theorem that has been proved.

Proof by cases — Subdividing a proof into special cases, one of which must be true. The conclusion in a proof by cases is the disjunction of the subconclusions within each case.

Proof by contradiction — A method of proof in which we assume the negation of what we want to derive and then derive a contradiction.

Proper subset relation — A is a proper subset of B if and only if $A \subseteq B$ and $A \neq B$.

Range The set of all second terms of the ordered pairs in a relation R. y is in the range of a function f if and only if there exists an x in the domain such that $y = f(x)$.

Rational number A number that can be represented in the form $\frac{p}{q}$, where p and q are integers with $q \neq 0$. The decimal form of a rational number must have a repeating cycle.

Real number A number that can be represented as a decimal with a finite or infinite number of places. The visual picture of the real numbers is the points on a number line. The real numbers were created in order to provide numerical measurements for all possible lengths of line segments. The set of real numbers is composed of rational numbers and irrational numbers.

Reflexive relation A relation R on a set S that has the following property: for every a in S, $a \mathrel{R} a$. Each element is related to itself.

Relation Embeds a structure between two sets by giving a connection between various elements. A relation R from X into Y can be described as a mapping where some of the elements in X are mapped to some of the elements in Y. It can also be viewed as a set of ordered pairs whose first terms come from X and whose second terms come from Y. Mathematical activity has always focused on relations. In the reasoning process, we are usually trying to figure out how various objects may be related to each other. When we work with a set, we do not individually analyze its elements; instead, we compare the set with other sets by looking for relations between them. Within the grand house of mathematics, there are many diverse areas of study, but within each area, the focus is on relations. Mathematics can be described as the study of relations.

Restriction of a function A function whose domain is restricted to a subset of the original domain, but using the same mapping.

Sentence A string or sequence of words that satisfy the language rules for being a sentence. A well-formed sentence must have both a subject and a verb. Theorems, definitions, and axioms are sentences. A proof is a list of sentences.

Sequence A function whose domain is the set of natural numbers. The notation s_n indicate $s(n)$, the nth term in the sequence.

Set A collection of objects. In formal set theory, a set is undefined since there are no simpler concepts with which to define it. After an axiomatic foundation was constructed for set theory, sets became a major unifying concept in 20th century mathematics. Sets provide the framework for mathematical discourse; they are the building blocks for all quantitative and spatial concepts.

Set subtraction A binary operation on two sets that produces a new set by removing the elements in one set from another set:

$$A - B = \{\, x \mid x \in A \text{ and } x \notin B \}$$

If A and B are finite sets and $B \subseteq A$, then $|A - B| = |A| - |B|$.

Size relation — A has the same size as B if and only if there exists a one-to-one function that maps A onto B. Two sets that have the same size are said to have the same *cardinality*. The natural numbers were created because of the need for a standardized reference set for comparing sizes of finite sets.

Statement — A sentence that is either true or false, but not both. In formal logic, a statement is called a proposition.

Strict order — A transitive and antisymmetric relation that has no element related to itself. Every partial order ($\preceq$) has an associated strict order:

$$a \prec b \text{ if and only if } a \preceq b \text{ and } a \neq b.$$

Subset relation — A is a subset of B if and only if every element in A is also in B. $A \subseteq B$ if and only if for every x, $x \in A \Rightarrow x \in B$. The subset relation is reflexive, transitive and antisymmetric.

Substitution principle — In a sentence with a variable, another letter or legitimate expression may be substituted for a universally quantified variable as long as all occurrences of the variable are replaced by the same substitution. Substitutions can be made for existentially quantified variables if the substituted letters are not used with other variables.

Surjection — An onto function.

Symbol — A letter or figure used to represent something. Phonetic symbols, such as "plus," give pronunciation information. Ideographic symbols like + give a more concise representation that is easier to manipulate.

Symmetric relation — A relation R on a set S that has the following property: for every a and b in S, if $a \mathrm{R} b$, then $b \mathrm{R} a$. The order of the elements does not affect the relation.

Theorem — A statement that has been proved.

Topological sorting — The embedding of a poset in a totally ordered set. Let $\preceq$ be a partial order on S and $\trianglelefteq$ be a total order on S. $\trianglelefteq$ is a *topological sorting* for $\preceq$ if and only if for every x and y in S, if $x \preceq y$, then $x \trianglelefteq y$.

Total order — A partial order in which each pair of elements are comparable. For every a and b in the set S, $a \preceq b$ or $b \preceq a$.

Transitive law — A valid argument whose hypothesis has the form, $p \Rightarrow q$ and $q \Rightarrow r$, and whose conclusion is $p \Rightarrow r$.

Transitive relation — A relation R on a set S that has the following property: for every a, b, and c in S, if $a \mathrm{R} b$ and $b \mathrm{R} c$, then $a \mathrm{R} c$.

Translation — The process of converting words, thoughts or ideas from one form, language, or medium to another. Mathematical reasoning involves a continual translation, back and forth, from everyday language to pictures and symbolic representations.

Truth value — Either true or false. Truth value is only used with sentences.

Uncountable — A set that is not countable. An uncountable set is a larger size of infinity than a countably infinite set. The set of irrational numbers is uncountable, which makes the set of real numbers uncountable.

Undecidable A sentence that is not decidable. It is not possible to derive the sentence or its negation from the axioms.

Undefined terms The basic words from which we construct the vocabulary for an axiomatic system. It is impossible to define every word without being circular. The terms selected to be undefined are chosen to represent the simplest concepts possible, concepts that cannot be explained by simpler concepts.

Union A binary operation on two sets that produces a new set by combining their elements: $A \cup B = \{ x \mid x \in A \text{ or } x \in B \}$. If A and B are finite sets, $|A \cup B| = |A| + |B| - |A \cap B|$.

Universal quantifier Asserts that each substitution of an element from the domain of the variable converts an open statement into a true statement. $\forall x, p(x)$ is true if and only if every element in the domain of x converts $p(x)$ into a true statement.

Universal set A set that serves as the universe for a particular discussion. When defining a set, all members of the set must come from a universal set. Otherwise, contradictions arise in set theory.

Upper bound b is an upper bound for S if and only if for every x in S, $x \preceq b$. Unlike a maximal element, an upper bound does not have to be in S. The *least upper bound* of S is the least of all the upper bounds of S.

Valid argument An argument in which the conclusion follows from the hypotheses. Let $h_1, h_2, \ldots, h_n$ represent the hypotheses and c the conclusion. The argument is valid if and only if the following implication is a law of logic: (h_1 and h_2 and h_3 and . . . and h_n) $\Rightarrow c$.

Variable A letter used to represent an arbitrary element of a given set, which is called the domain of the variable.

Well-defined A definition that is logically acceptable. A "well-defined function" means that the definition produces a function. A "well-defined set" means that the definition produces a legitimate set, one whose members can be determined.

Well-ordering principle Every nonempty subset of the set $\mathbb{N}$ of natural numbers has a least element. The well-ordering principle is equivalent to the principle of mathematical induction. Even though the natural numbers are unbounded at the top, the well-ordering principle gives us a firm grip on its lower side. We may not have a largest element in a set of natural numbers, but we will always have a smallest element. It is surprising how useful that can be.

Well-ordered set A partially ordered set in which every nonempty subset has a least element. A well-ordered set must be totally ordered. In a well-ordered set, every element, except for the greatest element, has an immediate successor. The set of natural numbers is well-ordered, but the set of real numbers is not.

◈ Appendix C

Symbols

Category	Symbol	Meaning	Page
5 Logical Operators	$\sim p$	negation of p	36
	$p \wedge q$	p and q	36
	$p \vee q$	p or q	36
	$p \Rightarrow q$	p implies q	36
	$p \Leftrightarrow q$	p is equivalent to q	36
2 Quantifiers	$\forall x$	for all x	24
	$\exists x$	there exists an x such that	24
Set Notation	$\in$	is an element of	216
	$\notin$	is not an element of	216
	U	universal set	218
	$\emptyset$	empty set	219
	$\{a, b\}$	a set whose elements are a and b	217
	$\{x \mid p(x)\}$	the set of all x such that $p(x)$ is true.	217
	$\mid A \mid$	number of elements in the set A	279
	$P(A)$	power set of A	228
	(x, y)	ordered pair	258
Set Relations	$A = B$	A is equal to B	220
	$A \subseteq B$	A is a subset of B	221
	$A \subset B$	A is a proper set of B	225
	$A \approx B$	A has the same size as B	369
Set Operations	$A \cup B$	union of A and B	233
	$A \cap B$	intersection of A and B	233
	$A - B$	A minus B	233
	A'	complement of A	234
	$A \times B$	A cross B	258
	$\bigcup_{i \in I} A_i$	multiple union	246
	$\bigcup_{A \in F} A$	multiple union of all sets in F	248

◈ Appendix D

Suggested Readings

Abbott, Edwin A. *Flatland: A Romance of Many Dimensions.* (original edition 1880). New York: Dover, 1952.

Barrow, John. *Pi in the Sky: Counting, Thinking, and Being.* Oxford: Clarendon Press, 1992.

Beckmann, Petr. *A History of* π. Boulder: The Golem Press, 1970.

Bell, Eric Temple. *The Development of Mathematics.* New York: McGraw-Hill, 1945.

_____. *Men of Mathematics.* New York: Simon and Schuster, 1937.

Boyer, Carl B. *A History of Mathematics.* New York: John Wiley & Sons, 1989.

Burger, Dionys. *Sphereland.* New York: Thomas Crowell, 1965.

Burton, David M. *A History of Mathematics: An Introduction.* Boston: Allyn & Bacon, 1985.

Davis, Philip J., and Reuben. Hersh. *The Mathematical Experience.* Boston: Birkhäuser, 1981.

Devlin, Keith. *Life by the Numbers.* New York: John Wiley & Sons, 1998.

_____. *Mathematics: The New Golden Age.* London: Penguin Books, 1988.

Dunham, William. *The Mathematical Universe: An Alphabetical Journey through the Great Proofs, Problems, and Personalities.* New York: John Wiley & Sons, 1994.

_____. *Journey through Genius: The Great Theorems of Mathematics.* New York: John Wiley & Sons, 1990.

Euclid. *Elements.* T. L. Heath, ed. New York: Dover, 1956.

Gamow, George. *One, Two, Three . . . Infinity.* New York: Bantam Books, 1961.

Gardner, Martin. *The Night is Large: Collected Essays 1938–1995.* New York: St. Martin's Press, 1996.

Halmos, P. R. *Naive Set Theory.* New York: Springer-Verlag, 1974.

Henderson, David. *Experiencing Geometry.* Upper Saddle River, NJ: Prentice Hall, 1996.

King, Jerry. *The Art of Mathematics.* New York: Plenum, 1992.

Kline, Morris. *Mathematics in the Modern World.* San Francisco: W. H. Freeman and Company, 1968.

_____. *Mathematical Thought from Ancient to Modern Times.* Oxford: Oxford University Press, 1972.

Newman, James R., ed. *The World of Mathematics.* Vols. 1–4. New York: Simon & Schuster, 1956.

Osserman, Robert. *Poetry of the Universe: A Mathematical Exploration of the Cosmos.* New York: Doubleday, 1995.

Quine, Willard Van Orman. *Methods of Logic.* Cambridge, MA: Harvard University Press, 1982.

Rosen, Kenneth H. *Discrete Mathematics and its Applications.* New York: McGraw-Hill, 1995.

Rozsa, Peter. *Playing with Infinity.* New York: Dover, 1961.

Rucker, Rudolf. *Geometry, Relativity and the Fourth Dimension.* New York: Dover, 1977.

_____. *The Fourth Dimension.* Boston: Houghton Mifflin, 1984.

Schattschneider, Doris. *M. C. Escher: Visions of Symmetry.* San Francisco: W. H. Freeman and Company, 1990.

Steen, Lynn Arthur, ed. *Mathematics Today: Twelve Informal Essays.* New York: Springer-Verlag, 1978.

Stoll, Robert R. *Set Theory and Logic.* San Francisco: W. H. Freeman and Company, 1963.

Suppes, Patrick. *Axiomatic Set Theory.* New York: Dover, 1972.

Valens, Evans G. *The Number of Things: Pythagoras, Geometry and Humming Strings.* New York: E. P. Dutton and Company, 1964.

Weeks, Jeffrey R. *The Shape of Space.* New York: Marcel Dekker, 1985.

Index